Impact Studies

Series Editor: Christian Koeberl

AF786725

Springer-Verlag Berlin Heidelberg GmbH

Henning Dypvik
Mark Burchell
Philippe Claeys (Eds.)

Cratering in Marine Environments and on Ice

With 119 Figures and 23 Tables

ESF IMPACT

Springer

Dr. Henning Dypvik
Department of Geology
University of Oslo
P.O. Box, Blindern
0316 Oslo, Norway
Email:
henning.dypvik@geo.uio.no

Prof. Dr. Philippe Claeys
Department of Geology
Vrije University Brussel
Peinlaan 2
1050 Brussels, Belgium
Email: *phclaeys@vub.ac.be*

Dr. Mark Burchell
Centre for Astrophysics
and Planetary Sciences
University of Kent, Canterbury
Kent CT2 7NR
United Kingdom
Email:
M.J.Burchell @kent.ac.uk

ISBN 978-3-642-07376-2 ISBN 978-3-662-06423-8 (eBook)
DOI 10.1007/978-3-662-06423-8

Cataloging-in-Publication Data applied for

Bibliographic information published by die Deutsche Bibliothek
Die Deutsche Bibliothek lists this publication in the Deutsche Nationalbibliographie;
detailed bibliographic data is available in the Internet at <http://dnb.ddb.de>.

springeronline.com

Originally published by Springer-Verlag Berlin Heidelberg New York in 2004

Cover Design: *Kirchner*, Heidelberg

Printed on acid free paper 32/2132 AO - 5 4 3 2 1 0

Preface

The workshop "Submarine craters and ejecta-crater correlation with a special session on ”Icy Impacts and Icy targets” was held on Svalbard, Norway from August 29 to September 3, 2001, with a pre-meeting excursion to Gardnos, Norway (August 28). Forty-five scientists from 17 nations participated in the workshop. This meeting was the seventh in a series of workshops organized by the European Science Foundation programme "Response of the Earth System to Impact Processes" (http://pssri.open.ac.uk/esf). With such topics as submarine craters, ejecta-crater correlation, and icy targets, the high Arctic locality of Svalbard was an obvious choice: One of the rare marine impacts (Mjølnir) took place nearby in the Barents sea, and on Svalbard, after a steep climb, its ejecta can be examined in the field; furthermore permafrost, snow, ice, and glaciers are an integral part of the Svalbard surroundings.

The conference was arranged at the facilities of the University Centre on Svalbard (UNIS) in Longyearbyen, Svalbard. The abstracts were published in a special abstract volume of the Norwegian Geological Society (ISBN 823-519-1701-8). Twenty-three talks and twenty posters were presented. The peer-reviewed conference papers are published in this volume.

Through the presentations of the workshop the formation of marine and icy impacts was examined and the importance of understanding ejecta distribution and correlation with the source crater was emphasized. The idea behind this meeting was to clarify the following questions: What is the role and what are the influences of the presence of water liquid or solid in the target? How does water and ice contribute to the ejecta formation and distribution on Earth and the other planets? In addition, biological processes and changes of the depositional conditions resulting from a marine impact were also discussed, and the post-impact situation was characterized. Modeling is an important and integral part of marine crater research. To foster a better understanding of this field, often seen as complex if not Byzantine by some geologists, the editors have decided to include an article by Elizabetta Pierazzo and Gareth Collins explaining the basic principles of hydrocode modeling.

It was a great pleasure for the Norwegian geological community to host this international meeting and to have advanced, engaged, and challenging discussions of this hot topic in such a cool place.

The 9th meeting in the ESF IMPACT programme was arranged in October 12 to October 16, 2002 and two related papers from that meeting (Vajda et al. and Pesonen et al.) have been included in this book.

Henning Dypvik
Department of Geology
University of Oslo
Norway
henning.dypvik@geologi.uio.no

Philippe Claeys
Department of Geology
Vrije Universiteit Brussel
Belgium
phclaeys@vub.ac.be

Mark Burchell
Centre for Astrophysics
and Planetary Sciences
University of Kent
United Kingdom
M.J.Burchell@ukc.ac.ukr

Acknowledgements

The editors would like to thank the ESF IMPACT programme, the steering committee, and the University Centre on Svalbard (UNIS) for making this workshop possible. We would also like to thank all who contributed to this volume by submitting manuscripts and, in particular, all the referees. The co-convenors Filippos Tsikalas, Morten Smelror, Jan Inge Faleide, Sverre Ola Johnsen, Atle Mørk and Jenö Nagy should also be thanked for all their efforts. We would in particular mention Sverre Ola Johnsen, Atle Mørk, Jenö Nagy and Johan Naterstad for their engaged excursion guiding. This meeting and the proceedings benefited very much from the invaluable help of the series editor Christian Koeberl. The meeting was financially supported by the ESF IMPACT programme, University of Oslo (Industrial Liasion), and the following local Longyearbyen-based companies: Ing. Paulsen, Svalbardbutikken, Svalbard Polar Travel, SpareBank 1- Nord Norge and Rebekka Gave & Suvenir. The abstract volume was kindly sponsored by Geological Survey of Norway, University of Oslo, Norsk Agip, RWE-DEA, DNO, Norwegian Polar Institute, SINTEF Petroleum Research, Esso Norge AS, Volcanic Basin Petroleum Research, Norske Conoco AS, Shell and Statoil.

Finally, we would like to thank the Svalbard weather for a true arctic experience.

Contents

List of Contributors

Pedro Brandão
BioSciences Laboratory
University of Kent at Canterbury
Canterbury, Kent, CT2 7NJ,United Kingdom
(pfd1@kent.ac.uk)

Franz Brandstätter
Department of Mineralogy
Natural History Museum
P.O. Box 417, A-1014 Vienna, Austria
(franz.brandstaetter@nhm-wien.ac.at)

Gerd Merethe A.Bremer
Department of Geology
University of Oslo
P.O. Box 1047, Blindern N-0316 Oslo, Norway
(g.m.a.bremer@geologi.uio.no)

Alan W. Bunch
BioSciences Laboratory
University of Kent at Canterbury
Canterbury, Kent, CT2 7NJ, United Kingdom.
(A.W.Bunch@kent.ac.uk)

Mark J. Burchell
Centre for Astrophysics and Planetary Sciences
University of Kent
Canterbury, Kent CT2 7NR, United Kingdom.
(M.J.Burchell@kent.ac.uk)

Gareth Collins
Lunar and Planetary Laboratory
University of Arizona
Tucson, AZ 85721, USA
(gareth@lpl.arizona.edu)

Philippe Claeys
Department of Geology
Vrije Universiteit Brussel
Pleinlaan 2, B-1050 Brussels, Belgium
(phclaeys@vub.ac.be)

Fabio Donadini
Division of Geophysics
University of Helsinki
P.O. Box 64, FIN-00014 Helsinki, Finland
(fabio.donadoni@helsinki.fi)

Henning Dypvik
Department of Geology
University of Oslo
P.O. Box 1047, Blindern, N-0316 Oslo, Norway
(henning.dypvik@geologi.uio.no)

Jan Inge Faleide
Department of Geology
University of Oslo
P.O. Box 1047 Blindern, N-0316 Oslo, Norway
(j.i.faleide@geologi.uio.no)

Ivan Grey
Centre for Astrophysics and Planetary Sciences
University of Kent
Canterbury, Kent CT2 7NR, United Kingdom
(I.D.Grey@kent.ac.uk)

Arnold Gucsik
Department of Applied Physics
Okayama University of Science
1-1 Ridai-cho, Okayama 700-0005, Japan
(arnold@physics.dap.ous.ac.jp)
and
Institute of Geological Sciences
University of Vienna
Althanstrasse 14, A-1090 Vienna, Austria
(Argul986@hotmail.com)

Eugene P. Gurov
Institute of Geological Sciences
Ukrainian Academy of Sciences
55-b Oles Gonchar Street, Kiev 01054, Ukraine
(ep_gurov@ukr.net)

Robert Handler
Institute of Geology and Paleontology
University of Salzburg
Hellbrunnerstrasse 34, A-5020 Salzburg, Austria
(Robert.handler@sbg.ac.at)

Christopher J. Hollis
Institute of Geological and Nuclear Sciences
P.O. Box 30368
Lower Hutt, New Zealand
(c.hollis@gns.cri.nz)

Ellen Johnson
Centre for Astrophysics and Planetary Sciences
University of Kent
Canterbury, Kent CT2 7NR, United Kingdom.
(emj1@kent.ac.uk)

Christian Koeberl
Department of Geological Sciences
University of Vienna
Althanstrasse 14, A-1090 Vienna, Austria
(christian.koeberl@univie.ac.at)

David T. Jr. King
Department of Geology
Auburn University
Auburn, AL 36849-5305, USA
(kingdat@auburn.edu)

Kari A. Kinnunen
Geological Survey of Finland
P.O. Box 96, FIN-02151 Espoo, Finland
(kari.kinnunen@gsf.fi)

Jacek Leliwa-Kopystynski
Institute of Geophysics
University of Warsaw
ul. Pasteura 7, 02-093 Warszawa, Poland
and
Space Research Center of the Polish Academy of Sciences
ul. Bartycka 18A, 00-716 Warszawa, Poland
(jkopyst@mimuw.edu.pl)

Eugen Libowitzky
Institute of Mineralogy and Crystallography
University of Vienna
Althanstrasse 14, A-1090 Vienna, Austria
(eugen.libowitzky@univie.ac.at)

Ralph D. Lorenz
Lunar and Planetary Laboratory
1629 E. University Blvd.
University of Arizona
Tucson, AZ 85721-0092, USA
(rlorenz@lpl.arizona.edu)

Dieter Mader
Department of Geological Sciences
University of Vienna
Althanstrasse 14, A-1090 Vienna, Austria
(a9010937@unet.univie.ac.at)

Jo R. Mann
Centre for Astrophysics and Planetary Science
School of Physical Sciences
University of Kent at Canterbury
Canterbury, Kent CT2 7NR, United Kingdom
(jrm10@kent.ac.uk)

Ljerka Marjanac
Institute of Quaternary Paleontology and Geology
Croatian Academy of Sciences and Arts
Ante Kovačića 5, 10000 Zagreb, Croatia
(ljerka@hazu.hr)

Tihomir Marjanac
Department of Geology
Faculty of Science
University of Zagreb
Kralja Zvonimira 8, 10000 Zagreb, Croatia
(tmarjan@public.srce.hr)

Alessandro Montanari
Osservatorio Geologico di Coldigioco
I-62020 Frontale di Apiro, Italy
(sandro.ogc@fastnet.it)

Jenö Nagy
Department of Geology
University of Oslo
P.O. Box 1047, Blindern, N-0316 Oslo, Norway
(jeno.nagy@geologi.uio.no)

Thornton L. Neathery
Neathery and Associates
1212-H Veteran's Memorial Parkway, Tuscaloosa, AL 35404, USA
(TLNEATHERY@prodigy.net)

Jósef Pálfy
Hungarian Natural History Museum
Department of Geology and Paleontology
P.O.Box 137, Budapest H-1431, Hungary
(palfy@paleo.nhmus.hu)

Lauri J. Pesonen
Division of Geophysics
University of Helsinki
P.O. Box 64 , FIN-00014 Helsinki, Finland
(Lauri.Pesonen@helsinki.fi)

Lucille W. Petruny
Department of Curriculum and Teaching
Auburn University
Auburn, AL 36849, USA
and
Astra-Terra Research
Auburn, AL 36831-3323, USA
(lpetruny@att.net)

Elisabetta Pierazzo
Planetary Science Institute
620 N. 6th Avenue, Tucson, AZ 85705, USA
(betty@psi.edu)

Ian Raine
Institute of Geological and Nuclear Sciences
P.O. Box 30368, Lower Hutt, New Zealand.
(i.raine@gns.cri.nz)

Wolf Uwe Reimold
Impact Cratering Research Group
School of Geosciences
University of Witwatersrand
Private Bag 3, P.O. 2050, Johannesburg, South Africa
(reimoldw@geosciences.wits.ac.za)

Morten Smelror
Geological Survey of Norway
N-7491 Trondheim, Norway
(Morten.smelror@ngu.no)

C. Percy Strong
Institute of Geological and Nuclear Sciences
P.O. Box 30368, Lower Hutt, New Zealand
(p.strong@gns.cri.nz)

Kalle Suuroja
Geological Survey of Estonia
Kadaka tee 82, Tallinn 12168, Estonia.
(k.suuroja@egk.ee)

Sten Suuroja
Department of Mining
Tallinn Technical University
Kopli 82, Tallinn, Estonia.
(s.suuroja@egk.ee)

Ana Marija Tomša
Department of Geology
Faculty of Science
University of Zagreb
Kralja Zvonimira 8, 10000 Zagreb, Croatia
(mariana3_hr@yahoo.com)

Filippos Tsikalas
Department of Geology
University of Oslo
P.O. Box 1047, Blindern, N-0316 Oslo, Norway
(filippos.tsikalas@geologi.uio.no)

Vivi Vajda
GeoBiosphere Science Centre
Department of Geology
University of Lund
Sölvegatan 12, 223 62 Lund, Sweden
(vivi.vajda@geol.lu.se)

Jorunn O. Vigran
Hans Hagerupsgt. 10, N-7012 Trondheim, Norway
(vigran@online.no)

Impacts into Marine and Icy Environments – A Short Review

Henning Dypvik[1], Mark J. Burchell[2], and Philippe Claeys[3]

[1]Department of Geology, University of Oslo, P.O.Box 1047, Blindern, N-0316 Oslo Norway (henning.dypvik@geologi.uio.no)
[2]Centre for Astrophysics and Planetary Sciences, University of Kent, Canterbury, Kent CT2 7NR, United Kingdom
[3]Department of Geology, Vrije Universiteit Brussel, Pleinlaan 2, B-1050 Brussels, Belgium

Abstract. In this review we discuss the current knowledge of impact events into marine and icy targets. This includes the major consequences of impact in marine depositional basins and on icy targets. We also discuss some of the future fields of research that could be of interest, in particular questions regarding triggering of volcanic activity, tsunami generation, and impact-associated petroleum reservoirs. In the discussion of the icy impact craters a summary of the exploration history is presented, both discussing comets and icy targets. An updated schedule for related missions is given, expressing the importance of knowing these processes also for understanding the development of the solar system.

1 Marine Impacts

1.1 Introduction

Because only few examples are known, impacts into marine environments and icy targets are amongst the least understood and studied parts of impact crater geology. This is, however, in contradiction to their global importance. More than 70 % of the earth is covered by water and 14% (of the earth) by ice (10.5% sea-ice included). These proportions have varied significantly over geological time, when numbers of asteroids and comets have struck both rock, water, or icy targets. Impacts on icy targets are also of great importance in understanding the developments of the various planets and the satellites of the outer planets, e.g., Mars or Europa. In addition, impact mechanisms, crater formation and collapse, melt production and ejecta distribution are poorly known for impact on targets other than solid silicates; the response of water and ice to impact events clearly deserves more thorough studies. So far the general focus of most research has been limited to the more easily accessible sub-aerial (on-land) craters and impacts into solid basement terrain. Moreover, collisions with icy bolides have also received too little attention. It should be added, however, that the last few years this trend has changed somewhat.

Consequently, it is of great interest to summarize information on this topic and point out some of the main, recent developments to better understand the formation of marine craters and ejecta production. Such an approach will result in improved ejecta - crater correlations. A specific aspect of marine cratering, which is not present in subaerial impacts, is the significant water effect. The generated waves and/or tsunamis can severely modify the morphology of the crater and influence post-impact sedimentation. The consequences of marine impacts on the biological evolution, their potential effects on oceanic circulation and the development of short or long-term hydrothermal systems at the bottom of the ocean also deserve attention. Documenting impact on water and ice targets not only contributes to our understanding of the geological record of impact events and enables us to foresee the environmental and hazardous consequences of these events, but it also helps in reconstructing the evolution of other important planetary bodies.

1.2
The Status of Marine Impacts Events

Currently about 170 terrestrial impact craters are known on Earth (Figure 1) Grieve et al. 1995; Gersonde et al. 2002). Twenty five have been recognized as original marine impacts (Dypvik and Jansa 2003, Table 1), the majority of them into continental crust. So far the Eltanin event (Kyte et al.1981; Gersonde et al. 2002) is the only known impact that has occurred in the deep ocean (5000 m water depth in the south Pacific). In the case of Eltanin, the about 1-km-diameter bolide did not reach the sea-floor; only ejecta and no crater have been found. Today, six of the 25 known marine impacts sites are still located in the oceanic environment, whereas, as result of subsequent tectonic processes, the remaining nineteen sites are, presently found on land (Dypvik and Jansa 2003, Table 1). The submarine craters represent consequently about 15% of the crater record: much too little for a planet which is two thirds ocean. The limited marine-crater representation is due to several reasons, e.g., the fact that because of plate tectonics no "old" ocean floor (> 200 million years old) is preserved, the limited knowledge of fine scale topography and structural characteristics of many deep ocean basins, and the lack of constrains on the morphology expected for impact structure formed on the thin oceanic crust. It is difficult, at this point, to estimate if the geophysical characteristics established for craters on land fully apply to marine craters, especially for larger events where the excavation cavity extends beyond the oceanic crust. Ivanov and Melosh (2003) claim that such a huge event is possible, but is highly improbable to have occurred in the past 3.3 Gyr.

Marine impact craters are expected to be buried soon after formation. The immediate infill of the crater should limit erosion, but also quickly hide the structure beneath a veneer of marine sediments. Consequently marine craters are expected to be well preserved in comparison to sub-aerial structures. Subtracting

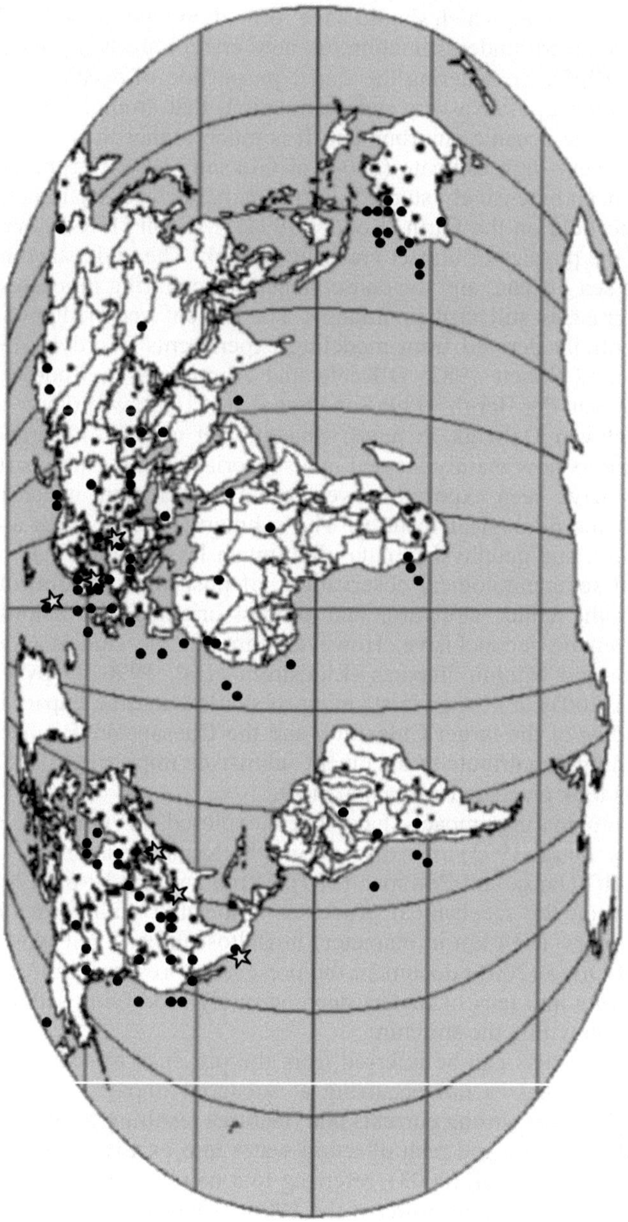

Fig.1 The distribution of impact craters on the Earth. Modified from French (2000). The Chicxulub, Chesapeake Bay, Montagnais, Mjølnir, Kärdla and Neugrund, and Kaluga are marked with star symbols.

the known marine impacts (25) from the number of identified terrestrial craters (170) gives 145 craters, which should have formed on land in the last 3.5 Ga. This is of course a severe underestimation, as new craters are being identified every year (Claeys 1995). Considering the 2 to 1 proportion of ocean on this planet, it would then indicate, after a very rude estimation, that around 300 craters should be expected in the oceanic environment. It is much higher than the 25 found, but far below the very high estimate (8104) of Glikson (1999). In the last 40 years, craters have been extensively studied on planetary surfaces by remote sensing or directly in the field on the Earth (Rondot 1994). Melosh (1989) gave a thorough treatment of the physics of impact cratering in rock. The understanding of impact events in the sea / ocean, and the consequences of cratering processes excavating the oceanic crust, is still in their infancy. The current knowledge of these processes is essentially derived from modeling experiments (Nordyke 1977; Strelitz 1979; Gault and Sonett 1982; O'Keefe and Ahrens 1982; Roddy et al. 1987; Melosh 1989; van der Bergh 1989; Sonett et al. 1991; Crawford and Mader 1998; Shuvalov and von Dalwigk in press; Shuvalov et al. 2002) extrapolated from submarine craters now mainly located in a subaerial setting. These structures may consequently have been exposed to weathering and erosion, which have altered their original morphological features. The submarine crater cores are few, field studies are rare, and geophysical data often much less detailed than for land craters. Detailed sedimentological observations of post-impact sedimentary successions within the crater structure, and process-oriented discussions of ejecta-production, remain inconclusive. However, through the studies of the medium sized Lockne and Mjølnir impacts (Lindstrøm et al. 1996; Dypvik et al.1996; Smelror et al. 2001), a new understanding of shallow marine impact is emerging. Ongoing studies of the larger Chicxulub and the Chesapeake Bay structures will in the close future contribute to document submarine impacting. Unfortunately all these impacts took place on continental crust.

Cometary submarine impacts have been explored by Ormø and Lindstrøm (2000) and by Jansa (1993) after discovering the Montagnais impact crater on the Canadian shelf (Jansa and Pe-Piper 1987). Ormø and Lindstrøm (2000) concentrated mainly on the mechanical processes associated with formation of small submarine craters (<14 km in diameter) in Baltoscandia. Their work and that of Dalwigk and Ormø (2001) document the presence of resurge gullies, about 1 km wide, 3 km long and tens of meters deep, strongly modifying sediment distribution and erosion within the structure.

Submarine impacts can be inferred from the presence of sedimentary features resulting from processes not occurring at subaerial impacts; e.g., formation of tsunamis, high waves, strong currents and features resulting from collapse of central high and crater rim and rush of return water into excavating crater ("resurge" activity). Recently Melosh (2003), referring to a newly released report (van Dorn et al. 1968), claimed that asteroids of the 100 to 1000 m diameter range will produce waves with periods between storm–waves and earthquake-produced tsunamis. According to Melosh, their hazard has been over-rated and they do not pose as great a threat as previously believed.

Impact of a large bolide into marine environments will also generate tremor-like earthquakes, which could result in fluidization of sediments, slope instability, slides, slumping, generation of turbidites, mass flows and debris flows and avalanches. During the past 10 years an increasing number of studies have focused on biological and sedimentological consequence at the periphery of marine impacts (e.g., Alvarez et al. 1992; Smit et al. 1992; Smit et al. 1996; Bohor 1996; Sharpton et al. 1996; Norris et al. 1999; Monteiro et al. 2000; Smelror et al. 2002; Stewart and Allen 2002). In particular, the deposition of massive sand, and debris flows associated with the Chicxulub crater at the K/T boundary studies should be mentioned. The formation of these coarse units, often in a deep sea setting were attributed to the erosion, transport and deposition of sediment generated by tsunamis caused by the collapse of the platform margin or, at more distal sites, to failure of the slope and generation of massive debris flows. These major magnitude sedimentological processes recognized in the field or on seismic lines, were all attributed to the nearby impact and the propagation of shock waves. Discussions of impact structures, such as the Montagnais (Jansa 1993; Jansa and PePiper 1987), Mjølnir (Dypvik et al. 1996; Smelror et al. 2001; Tsikalas et al. 1998), Lockne (Lindstrøm et al. 1996; Sturkell 1998; Sturkell and Ormø 1997), and the Chesapeake Bay craters (Poag 1997) have revealed similar associated sedimentary processes. In addition, ejecta layers identified in the Precambrian Hamersley Group of Australia (Simonson and Hassler 1997; Simonson et al. 1998) are associated with resurge, suspension currents and debris flow deposition. It should also be mentioned that only limited attention has been given to related effects of waves and tsunamis generated by impacts (Bourgeois et al. 1988; Oberbeck et al. 1993; Smit et al.1996; Warme and Sandberg 1996; Warme and Kuehner 1998; Poag et al. 1999; Ward and Asphaug 2000; Shuvalov et al. 2002), and even less to the study of margin collapse, which can be associated with meteorite impacts on continental margins and shelves (Norris et al. 1999). It is likely that the sedimentary record still contains undiscovered traces of impact-induced sedimentary disturbances masquerading as breccias or coarse clastic sequences or diamictites. The Devonian Alamo breccia, which for years was mapped as a regular breccia unit, is a clear illustration of this fact (Warme and Sandberg 1996; Warme and Kuehner 1998).

A wide range of geological and geophysical data have recently been presented from three small to medium sized craters from the East European platform: the 380 Ma old Kaluga (Russia), which is 15 km in diameter (Masaitis 2002), the 20 km in diameter Neugrund (Estonia) of early Cambrian age (535 Ma) (Suuroja and Suuroja 2002) and the 4 km in diameter, 455 Ma old Kärdla crater (Estonia) (Suuroja et al. 2001). These craters are all well preserved, marine, complex craters formed on continental crust, rapidly filled and hidden beneath younger sediments. They show severe resurge effects and probable tsunami / wave effects resulting in locally high rates of sedimentation. This often occurred after the major post-impact modification stage, as was also the case in e.g. the Mjølnir Crater, indicating that local sediment disturbance extended for long after the impact itself. The slumps and avalanches following the resurge and tsunamis were important mechanisms in the post-impact filling of the crater structures. These processes considerably modified the shape and morphology of the structure. It is likely that they also affected the distribution/succession of the impactites, in particular the suevite. They may have hampered the formation or led to a rapid cooling of the melt. The Kaluga, Kärdla and Neugrund craters were formed in shallow (50 – 500 m) waters in epicontinental sea settings, with target areas characterized by rather thin (0-200 m) sedimentary covers on Precambrian basement. The amounts of melt material is minor and only in the Kaluga case has a lense of melt-rock (20 m tagamite) been detected. The consequences of such processes on the development of hydrothermal cells in the crater still remain to be examined. In the three cases mentioned above widespread ejecta have been found, in the Kaluga event the so-called Nava Breccia has been recognized as an ejecta unit up to 500 km away from the crater. In the Nava Breccia the redistribution can be explained by tsunami / wave reworking of ejecta.

The presence of water in the target area also has an influence on the ejecta formation. According to (Melosh 1989) water vapor formed will both accelerate and increase the ejecta formation and distribution, and result in wider distribution of material from the marine impacts than from comparable subaerial impacts.

Simonson and Harnik (2000) suggested that there could be significant and systematic differences in the composition of ejecta from a continental crust target versus pure oceanic crust impacts. Because the vast majority of the mass ejecta comes from the upper part of the target, major differences are expected between a mafic basaltic and a felsic granitic source. It is unlikely for example that vast amounts of the characteristic shocked quartz will be produced by an impact on a pure oceanic target. The produced shock minerals, e.g., olivine, will be much less stable than quartz and much more difficult to characterize as ejecta. As these criteria might be applicable on Archean impacts, such differentiation is not fully applicable in the Phanerozoic cases. In the Phanerozoic, all known marine impact craters so far discovered were formed in shallow seas on continental crust, and the ejecta composition is comparable for terrestrial and shallow marine impacts. The fact that we are missing the pure oceanic impacts and/or fail to identify them in sedimentary sequences, may be due to our tools being aimed at characterizing ejecta from continental crust.

As mentioned earlier, it must be clearly emphasized that so far no impact crater has been identified in oceanic crust. A good example of an impact into the deep ocean is the Eltanin event, 2.5 Myr ago in the Southern Ocean (Eltanin Impact, Kyte et al. 1981; Kyte 2002a; 2002b; Gersonde et al. 1997; 2002). No crater was, however, formed on the ocean floor, but evidences of sediment disturbance in the area where the impact took place, are wide-spread and visible in piston cores and on seismic profiles (Gersonde et al. 1997; 2002). Impact ejecta is encountered in 23 cores drilled over the whole area (80,000 km^2). It is essentially composed of material, such as various solid chuncks of the stony–iron or iron meteorites (howardites or mesosiderites), vesicular impact melt, and glass spherules with Ni-rich spinels most likely condensed from the cloud of vaporized projectile (Kyte 2002a; 2000b). There is no evidence that the Eltanin impact melt contains a significant terrestrial silicate component that might have been incorporated by mixing of the projectile with oceanic crust (Kyte 2000a). Sea water contamination is attested by the presence of high Na content and detectable Cl in the glass (Margolis et al. 1991).

1.3 Marine Impacts - Future Research

Can an oceanic impact, and the formation of a crater in the deep sea, affect the oceanic circulation or can it modify the thermal distribution in the oceans? So far no study has addressed this type of problem. Another fundamental aspect of oceanic crater research is how to identify a crater in the deep ocean; what will it look like? There has been some discussion and speculation (Jones 2000), on the production of huge volcanism as the thin oceanic crust is ruptured during an impact. Jones (2000) advocates that decompression melting would generate massive mafic volcanism. The crater would then lie undetected because hidden by a huge volcanic province. However the debate is active and there is little agreement as to the effects of impact on oceanic crust. According to impact models (Ivanov and Melosh 2003) the production of huge melt volume would be effective only for large (50 km in diameter bolide) sized impact; an very unlikely event in the Phanerozoic. Bolides around 5 km in diameter are well expected to have impacted the ocean in the Phanerozoic. Based on sub-aerial cratering models, it would seem likely that the thin oceanic crust could be perforated by such an event. Even without the production of massive volcanism, the shape, structure and internal morphology of the expected crater is unclear at present time.

Marine impacts and their related effects (submarine slides, avalanches etc.) may result in the formation of tsunamis, e.g., in the case of the Mjølnir impact into the paleo-Barents Sea (Shuvalov et al. 2002). The formation of tsunamis and possible generation of tsunami-related currents occur both along the sea floor in open waters and in the coastal regions (Ward and Asphaug 2000; Ward and Asphaug 2002). Tsunami deposits in deep water as well as run-up situations, have been discussed and recognized around the world in relation with studies of tsunamis (Chague-Goff et al. 2002; Cita et al. 1996; Cita and Aloisi 2000; Bondevik et al. 1997; Goff et al. 1998). Smit et al. (1996) claimed that tsunami deposits would be more easily found in deeper water, below storm wave base. They can be recognized by special sequences of sedimentary structures and textures indicative of high energy deposition and alternating current directions. So far the tsunami deposits described in the literature have very different appearances dependant on a whole range of parameters and characteristics of the different locations and basins (see, e.g., Cita et al. 1996; Cita and Aloisi 2000; Takyama et al. 2000). No doubt this is an extremely complex and poorly known subject.

The search for new craters on the sea bed is an important, but expensive and time-consuming task, which in the future will be a part of the marine science programs. It is clear that the influence of marine impacts for man and biota has had an important influence and naturally its influence for the development of life. At the present only the Chicxulub impact has been shown to have global influence on the biological distribution, but several of the smaller marine impact craters are shown to locally have had serious, but only short time, influence on the local biota (Smelror et al. 2002).

The search for the economic potential of impact craters should also be mentioned. In the marine environment petroleum reservoir rocks are often associated with organic, rich claystone or shales. Donofrio (1998) described several American petroleum carrying impact craters, his example craters, however, are mainly of subaerial origin. The craters are highly fractured, and possess increased porosities in the target area. They may even to some extent carry increased maturation of the possible source rocks. It is possible that the marine impact craters posses a greater potential as traps for petroleum than the subaerial ones, the high exploration expenses and demand for large submarine reservoirs to be economical, may be, however, serious obstacles. Grajales et al. (2000) have elegantly demonstrated that one of the major oil field in Mexico is most likely linked to the Chicxulub impact. The breccia formed by the impact-induced collapse of the Yucatan platform led to the deeper water formation of a porosity-rich reservoir, sealed by the ejecta layer. It is likely that this reservoir extends to the whole Yucatan margin and that similar high porosity brecciated lithologies occur associated with other marine craters.

2 Impact Craters on Ice

Ice is widespread in the Solar System. It can be found on planetary surfaces (Earth, Mars), on the surfaces of planetary satellites (various moons of Jupiter and Saturn for example), and if we take an extreme example the whole surface of Pluto may be ice (Brown 2002). In the case of Pluto the mean mass of the body implies a substantial rocky component to the body, but the surface may be entirely covered by ice.

Ice in this context does not necessarily mean water ice. In the case of Pluto, N_2 and methane ices are more plausible based on observations of reflectance spectra. Minor Solar System bodies such as comets are also commonly described as icy. For comets however, the ice may again not just be pure water ice but will include substantial amounts of silicate materials, other volatiles etc mixed into it. Whether this mixing is uniform, layered or in clumps (i.e., many rocky boulders, each covered in ice all bound together) is one of the great mysteries of Solar System science. Questions such as these will be investigated in detail during the next decade by various space missions (principally, Stardust, Rosetta and Deep Impact, see below).

Like any exposed Solar System surfaces, icy bodies undergo impacts and thus exhibit impact craters. Some of the earliest work concerning impact cratering in ices was that of Croft et al. (1979), who fired rifle bullets into sand-ice mixtures to see what craters would be like in an analogue of an icy Martian regolith. Although the impact speeds were less than 1 km s^{-1}, this represented a new dimension to impact cratering studies with planetological implications. Subsequently Croft (1981a) made studies of impacts at 2 to 6 km s^{-1} into ice and ice-saturated sands, thus moving ice impacts into the hypervelocity regime[1]. The motivation was not just to study Martian impact cratering processes, but also to consider impact records on the icy satellites of Jupiter and Saturn. Since 1981, there have been several laboratory studies of hypervelocity impacts on ices.

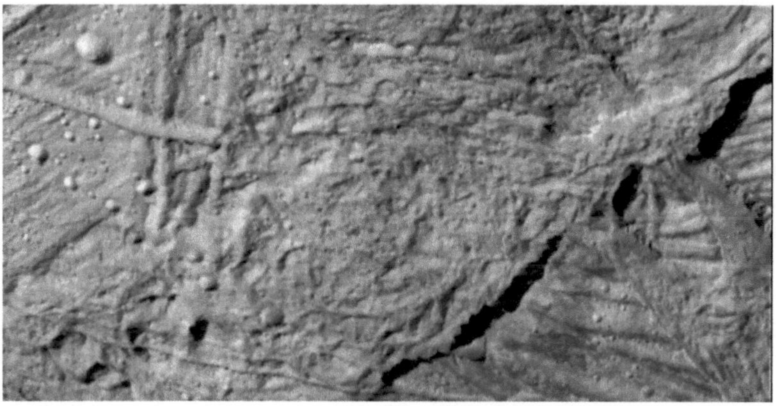

Fig. 2. Small icy craters on Europa. Image size is 8.4 km wide and the largest crater visible is approximately 300 m across (Source NASA/JPL).

The Voyager missions to the outer Solar System sent back many pictures of icy bodies, some heavily cratered, some less so. Immediately upon receipt of these images, relative dating of surfaces could be attempted. For example, after allowing for any local perturbation/enhancement of the impact flux at each planet and indeed at different orbital radii from the parent planet, the degree of resurfacing on the icy satellites could be estimated. Craters were thus playing a role in helping understand the nature of these icy bodies. Classification of craters by their morphology was an early activity by researchers. For example, Croft (1981b) reported on size distributions of craters on Ganymede and Callisto. Questions then arose as to whether thermal and/or viscous relaxation were altering crater shape. This was an area where it was thought that impact experiments in the laboratory could provide insights and accordingly Greeley et al. (1982) reported on a series of shots in the laboratory into layered targets with a surface of ice and a clay subsurface. However, the influence of relaxation of ice on crater morphology is still not fully understood. For a recent review see Durham and Stern (2001).

More recently, the Galileo mission to Jupiter has provided a wealth of data concerning the Jovian satellites (and the Cassini mission to Saturn will similarly

[1] The hypervelocity regime is crudely where the impact speed is similar to or exceeds the speed of the resultant compression waves in both target and projectile. For most materials such wave speeds are typically between 1 and 3 km s^{-1}, so impacts at speeds of more than 1 km s^{-1} are usually, and somewhat loosely, referred to as hypervelocity.

revolutionize our knowledge of the Saturnian system). Understanding the craters on these satellites (Figures 2 and 3) will help reveal details of their structure that could otherwise remain obscure. For example, several of these satellites (e.g., Europa) are not considered to be icy rocks, but rather to have a rocky core, a liquid ocean and finally a solid icy surface. Impact craters on such bodies present a fascinating means to probe this structure. Modeling can help understand how deep an ice shell must be to support the observed craters (e.g., see Turtle and Pierazzo 2001). This can present constraints not only on the thickness of the ice, but by implication on the methods and degree of internal heating which are required to sustain the liquid sub-surface ocean and hence the thickness of the covering ice layer. Equally, impacts generate ejecta, and it should be possible for impact produced ejecta to escape from one Jovian satellite and travel to another. So just as we find Martian meteorites on Earth, there may be Europan "meteorites" on other satellites
of Jupiter. And, in return, other bodies may contribute materials to Europa. Understanding these possibilities requires a detailed knowledge of impact mechanics involving ice targets. This has added interest given that Europa is held to be a candidate for searches for life (i.e., it is held to possess sub-surface water oceans and water is a key ingredient of life), see, e.g., Chyba and Phillips (2002) for a recent review. If there is life on Europa one can ask if it has become frozen in the

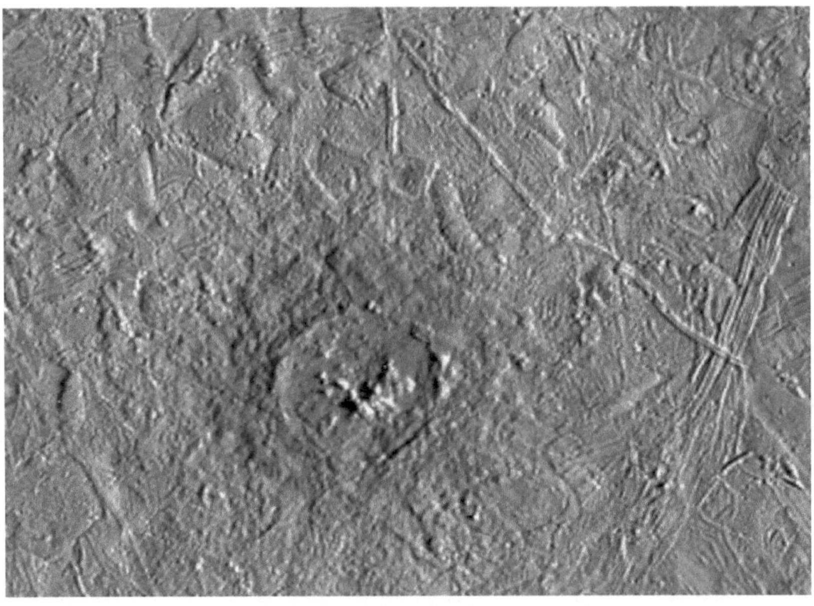

Fig. 3. Crater Pwyll on Europa. (Source NASA/JPL) Crater is approximately 26 km across.

ice, and then if it can be knocked off trapped in icy ejecta as a result of an impact of another body onto the surface of Europa. This intriguing speculation is discussed in a recent paper by Burchell et al. (2003). To consider it in detail one must know if bacteria can survive hypervelocity impacts (see Burchell et al. 2001) and then generalize this to impacts on ice.

When the Cassini mission arrives at Saturn, new questions will arise concerning impact cratering on icy satellites. Titan is a particular satellite of interest for the Cassini mission. Cassini will map it from space and the Huygens probe will be dropped through its atmosphere onto its surface. It has already been hypothesized as to what craters on Titan might look like (e.g., see Lorenz, this volume and references therein).

Comets also present an interesting subject for impact studies. The best images of comet nuclei are shown in Figure 4 (Halley's comet) and Figure 5 (comet Borelly). High resolution images of asteroids (e.g., Eros, see Figure 6) reveal that they are subject to bombardment just like any other Solar System body. Comets are no different in this respect. They have been bombarded after formation. Comets are however different to other icy bodies in at least one important respect, their low density. Estimates of comet density vary, but typically range from 200 to 1500 kg m^{-3}, with a value of 500 kg m^{-3} being typical. Such a value implies porosity. Therefore an impact on a comet is not just an impact on an icy body,

Fig. 4. Halley's comet observed from the Giotto spacecraft. Nucleus size is 8 x 7 x 1.5 km (Source ESA/Giotto).

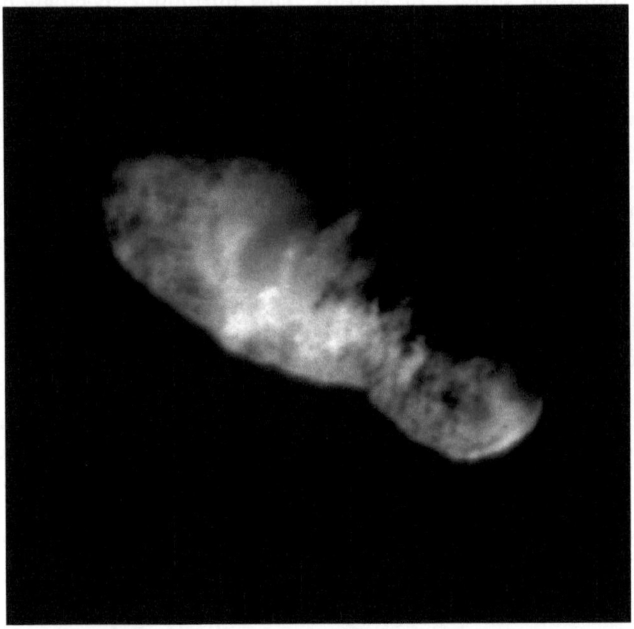

Fig. 5. Comet Borelly observed from the DS-1 spacecraft. Nucleus is approximately 8 x 4 km (source NASA/JPL).

but also on one that is not well consolidated. This considerably complicates the impact mechanics. The picture is made even more obscure by the lack of detailed knowledge of cometary composition and structure. Worse, these may change quite significantly with time. Here one thinks of comets which pass through the inner Solar System. The material that is ejected to form the beautiful cometary tails that we observe from Earth, is not just depleting the comet nucleus, but in all probability is doing so in a biased fashion, building up volatile depleted crusts over less processed interiors. The dark surfaces of comets in the inner Solar system (given by their low albedo) (Figures 4 and 5) immediately cautions against considering comets with surfaces of pure water ice. Further, the presence of other volatile materials suggests that during their long stay in the outer Solar System, significant amounts of astrochemistry may have occurred on cometary surfaces driven by Solar and interstellar radiation (e.g., galactic cosmic rays etc.). The widely accepted Whipple model of comets as dirty snowballs (based on observation and speculation about comet Encke in the 1950s, see Whipple 1950,1951) probably no longer serves today to adequately describe comets in that our questions have become much more detailed and sophisticated. But what replaces it?

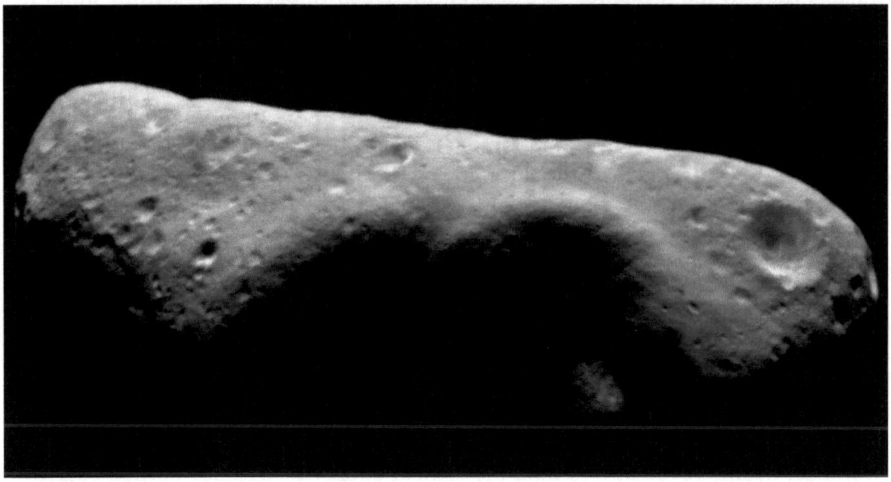

Fig. 6. Asteroid Eros observed from the NEAR spacecraft. Largest dimension is approximately 33 km (Source NASA/JPL).

Several space missions to comets are now underway or soon to be launched. The Stardust mission was launched in February 1999. It will fly past comet P-Wild 2 in January 2004, collect some grains of freshly emitted cometary dust and return them to Earth in January 2006. In addition, it is hoped that the spacecraft's cameras will collect pictures of the comet nucleus with ten times the resolution of those made in previous missions (i.e., the nucleus of Halley's comet, Figure 4, by the Giotto spacecraft in 1986, and the nucleus of comet Borelly, Figure 5, by the Deep Space 1 spacecraft). At that resolution it should be possible to see craters. Indeed even the absence of any traces of cratering will in itself be a statement about the evolution of the surface of the comet.

The next mission to be launched was due to be Rosetta, in January 2003. Due to problems with the launch vehicle the launch has been delayed to February 2004. This mission will rendezvous with comet Churyumov-Gerasimenko in November 2014, orbit it and accompany it on its journey into the inner Solar System. In addition, a small lander will be deployed onto the surface. Comet science will be revolutionized by the Rosetta mission.

But before the Rosetta results will come those from Deep Impact. This mission (see A'Hearn 1999) will also launch in 2004 and in 2005 will drop a 370 kg mass (composed of 46% copper) at 10.1 km s^{-1} into the nucleus of comet Temple 1. The resulting crater (assumed diameter $\approx$ 100 m, see Belton and A'Hearn 1999) will be imaged by the main spacecraft as it approaches the comet and flies past. Spectroscopy of the expanding vapour plume from the impact site will yield information on comet composition. The growth, final size and shape of the resultant impact crater will yield details of the structure of the comet. Although the mechanisms for crater growth on planetary scales differ from that in the laboratory for impacts (due to the influence of gravity), the Deep Impact event will be at an intermediate scale. The relative roles in crater growth of target material properties and gravity will be interesting (Burchell et al. this volume).

One intriguing possibility of how impacts may affect icy bodies is raised by a consideration of catastrophic disruption of a target body in an impact. Traditionally, this is held to occur when the impact energy density per unit target mass exceeds a threshold related to the strength of the body. A basic rule of thumb for rocky bodies is that if the diameter of the impact crater that would form on a semi-infinite body exceeds 50% of the target body diameter, catastrophic disruption of the target occurs. Is this rule the same for icy bodies? And what if the body is porous, or of low internal strength? For example, see Benz and Asphaug (1999) for a discussion. There is much here that requires investigation. In particular, it has recently been suggested that as comets approach the sun, the release of volatiles from their heated interior may be impeded by the presence of a volatile depleted mantle. An internal pressure may build up (this is advanced for the appearance of sudden jets of activity in discrete locations on a comet nucleus). In which case, if an impact occurs (even at less than the critical energy density for normal catastrophic disruption) will the whole comet explode? This idea has been advanced from recent laboratory impact experiments involving foam targets with an internal pressure and low strength. It would be fascinating to see similar experiments with heated ice targets.

It is reasonably clear that impact cratering on icy bodies will figure substantially in the analysis of data from a variety of space missions in the near future. This analysis will be guided by what is learnt from impact studies at laboratory scales and by detailed hydrocode simulations at Solar System scales. Indeed, further such studies will be triggered by new questions arising from the data from space missions. Some aspects of current such studies are presented in the papers in these proceedings. In the future more experiments and modelling will be required over a wide range of conditions.

3 Conclusions

It may seem odd that the study of marine impacts has been neglected in terrestrial cratering investigations for so long. However, impact craters on land are so much more accessible and obvious and, thus, studied in most detail. This situation is no doubt reinforced by the observation of craters on other solar system bodies, e.g., the Moon, Mars, Venus, asteroids etc., where the exposed surface is rocky. Thus, most studies implicitly assume that impacts occur on rock, not into water, ice or soft sediment successions. Also, some aspects of the impact process are particular to marine impacts (e.g., re-surge currents, tsunamis, etc.), and others directly affect the crater development (i.e., fast refill by water cooling craters in shallow oceans much more rapidly than craters formed on land). In addition, an impact in a marine environment has been shown to a have a direct effect on, e.g., ejecta formation and distribution. Their influence on the geology of the target areas for example formation of porosity rich reservoirs, which can potentially be oil bearing.

Given that, as pointed out in this paper, only 15% of the 170 craters found on the Earth today are believed to have occurred in marine environments, and that water covers such a large part of the Earth, oceanic impacts should not be ignored. In a wider Solar System context ocean impacts may appear to be limited to the Earth. However, it should be remembered that there is evidence for possible oceans or large expanses of water on Mars in earlier times. Thus, some of the discussions about terrestrial oceanic impact may be applicable to Mars as well.

Impacts into ice represent a special field from a terrestrial viewpoint. However, at times in its history the Earth had more extensive coverage of ice than today. Looking further afield in the Solar System, all the outer planets have ice-covered satellites of various sizes. A more complete picture of solar system impact processes requires an understanding of impacts into ice (and water), as well as into rock. A variety of ongoing and planned space missions will yield many observations of craters in icy bodies. These will await detailed interpretation until impacts in ice are better understood. Indeed, there is speculation that one body to be studied (Titan, the largest moon of Saturn) may have some liquid on its surface, permitting the possibility (albeit a small one) of impact craters in a marine environment with an icy sub-surface (rather than rocky subsurface as on Earth).

In conclusion, the old view that impact processes are solely a geologic process involving rock targets, can be seen to be too restrictive. There are many significant effects of the impact process due to the presence of a marine environment. They alter the picture of impacts and result in somewhat different ejecta distributions and compositions than from purely rocky targets. This field is not a new one, but has increasingly matured such that it has now moved into the limelight as regards research into impacts and can no longer remain neglected. Possible areas for future research (e.g., tsunami generation etc.) have been outlined. What is clear is that many new and interesting discoveries in the fields of oceanic and ice impacts await us.

Acknowledgements

Christian Koeberl is gratefully acknowledged for a detailed and constructive review.

References

Alvarez W, Smit J, Lowrie W, Asaro F, Margolis SV, Claeys P, Kastner M, Hildebrand AR (1992) Proximal impact deposits at the Cretaceous – Tertiary boundary in the Gulf of Mexico: A study of DSDP Leg 77, Sites 536 and 540. Geology 20: 697-700

Belton MJS, A'Hearn MF (1999) Deep sub-surface exploration of cometary nucleii. Advances in Space Research 24: 1167-1173

Benz W, Asphaug E (1999) Catastrophic disruptions revisited. Icarus 142: 5-20

Bohor BF (1996) A sediment gravity flow hypothesis for siliciclastic units at the K/T boundary, Northeastern Mexico. In: Ryder G, Fastovsky D, Gartner S (eds) The Cretaceous –Tertiary Event and other Catastrophes in Earth History. Geological Society of America, Special Paper 307: 183 – 196

Bondevik S, Svendsen JI, Mangerud J (1997) Tsunami sedimentary facies deposited by the Storegga tsunami in shallow marine basins and coastal lakes, western Norway. Sedimentology 44: 1115 - 1131

Bourgeois J, Hansen TA, Wiberg PL, Kauffman EG (1988) A tsunami deposit at the Cretaceous-Tertiary boundary in Texas. Science 241: 567- 570

Brown ME (2002) Pluto and Charon: Formation, seasons, comparison. Annual Review of Earth and Planetary Sciences 30: 307 – 345

Burchell MJ, Mann J, Bunch AW, Brandão P (2001) Survivability of bacteria in hypervelocity impact. Icarus 154: 545-547

Burchell MJ, Galloway JA, Bunch AW, Brandão P (2003) Survivability of bacteria ejected from icy surfaces after hypervelocity impact. Origins of Life and Evolution of the Biosphere 30: 53 - 74

Burchell MJ, Johnson E, Grey I (this volume) Estimating crater size for hypervelocity impacts on small icy bodies (e.g., comet nucleus). In: Dypvik H, Burchell M, Claeys P (eds) Cratering in Marine Environments and on Ice. Impact Studies vol. 5, Springer, Heidelberg, pp 197 - 210

Chaqué–Goff C, Dawson S, Goff JR, Zachariasen J, Berryman KR, Garnett DL, Waldron HM, Mildenhall DC (2002) A tsunami (ca. 6300 years BP) and other Holocene environmental changes, northern Hawke's Bay, New Zealand. Sedimentary Geology 150: 89 - 102

Chyba CF, Phillips CB (2002) Europa as an abode of life. Origins of Life and Evolution of the Biosphere 32: 47-68

Cita MB, Aloisi G (2000) Deep-sea tsunami deposits triggered by the explosion of Santorini (3500y BP), eastern Mediterranean. Sedimentary Geology 135: 181 – 203

Cita MB, Camerlenghi A, Rimoldi B (1996) Deep-sea tsunami deposits in the eastern Mediterranean: new evidence and depositional models. Sedimentary Geology 104: 155-173

Claeys P (1995) When the sky fell on our heads: Identification and interpretation of impact products in the sedimentary record. Reviews of Geophysics vol. 33. Supplement to US National report to International Union of Geodesy and Geophysics, 1991 – 1994, pp 95-100.

Crawford DA, Mader C (1998) Modeling asteroid impact and tsunami. Science of Tsunami Hazards 6: 21-30

Croft SK (1981a) Hypervelocity impact craters in icy media [abs.]. Lunar and Planetary Science 12: 190-191

Croft SK (1981b) Cratering on Ganymede and Callisto: Comparisons with the terrestrial planets [abs.]. Lunar and Planetary Science 12: 187-188

Croft SK, Kieffer SW, Ahrens TJ (1979) Low velocity impact craters in ice and ice-saturated sands with implications for Martian crater count ages. Journal of Geophysical Research 84 (B14): 8023-8032

Dypvik H, Jansa LF (2003) Sedimentary signatures and processes during marine bolide impacts: a review. Sedimentary Geology 161: 309-337

Dypvik H, Gudlaugsson ST, Tsikalas F, Attrep MJr, Ferrell REJr, Krinsley DH, Mørk A, Faleide JI, Nagy J (1996) The Mjølnir structure – An impact crater in the Barents Sea. Geology 24: 779 – 782

Donofrio RR (1998) North American impact structures hold giant field potential. Oil and Gas Journal 96: 69 – 83

Durham WB, Stern LA (2001) Rheological properties of water ice – Applications to satellites of the outer planets. Annual Review of Earth and Planetary Sciences 29: 295-330

Gault DE, Sonett CP (1982) Laboratory simulation of pelagic asteroidal impact: Atmospheric injection, benthic topography and the surface wave radiation field. In: Silver LT, Schultz PH (eds) Geological Implications of Impacts of Large Asteroids and Comets on the Earth. Geological Society of America, Special Paper 190: 69 - 92

Gersonde R, Kyte FT, Bleil U, Diekman B, Flores JA, Gohl K, Grahl G, Hagen R, Kuhn G, Sierro FJ, Volker D, Abelmann A, Bostwick JA (1997) Geological record and reconstruction of the late Pliocene impact of the Eltanin asteroid in the Southern Ocean. Nature 390: 357 – 363

Gersonde R, Deutsch A, Ivanov BA, Kyte FT (2002) Oceanic impacts – a growing field of fundamental geoscience. Deep-Sea Research II 49: 951 – 957

Glikson AY (1999) Oceanic mega impacts and crustal evolution. Geology 27: 387 – 390

Goff JR, Crozier M, Sutherland V, Cochran U, Shane P (1998) Possible tsunami deposits of the 1855 earthquake, North Island, New Zealand. In: Stewart IS, Vita-Finzi C (eds) Coastal Tectonics. Geological Society (London) Special Publication 133: 353 – 374

Grajales-Nishimura JM, Cedillo PE, Rosales DC, Moran ZD, Alvarez W, Claeys P, Ruiz-Morales J, Garcia JH, Padilla AP, Sanchez RA (2000) Chixculub impact; the origin of reservoir and seal facies in the southeastern Mexico oil fields. Geology 28: 307 – 310

Greeley R, Fink JH, Gault D, Guest JE (1982) Experimental simulation of impact cratering on icy satellites. In: Morrison DM (ed) Satellites of Jupiter. University of Arizona Press, Tuscon, Arizona: pp 340 – 378

Grieve RAF, James R, Smit J, Therriault A (1995) The record of terrestrial impact cratering. Geological Society of America Today 5(10): 193 - 196

Ivanov BA, Melosh HJ (2003) Impacts do not initiate volcanic eruptions [abs.]. Lunar and Planetary Science 34: abs #. 1338 (CD-ROM)

Jansa LF (1993) Cometary impacts into ocean: their recognition and threshold constraint for biological extinctions. Paleogeography Paleoclimatology Palaeoecology 104: 271 – 286

Jansa LF, Pe-Piper G (1987) Identification of an underwater extraterrestrial impact crater. Nature 327: 612 – 614.

Jones A (2000) Impact induced volcanism on Earth; searching for the evidence; crustal magma chambers [abs.]. In: Catastrophic events and mass excinctions; impacts and beyond. LPI Contribution # 1053, Lunar and Planetary Institute, Houston, pp 87 - 88

Kyte FT (2002a) Composition of impact melt debris from the Eltanin impact strewn field, Bellinghausen Sea. Deep-Sea Research II 49: 1029 – 1047

Kyte FT (2002b) Unmelted meteoritic debris collected from Eltanin ejecta in Polarstern cores from expeditions ANT XII/4. Deep-Sea Research II 49: 1063 – 1071

Kyte F, Zhou Z, Wasson JT (1981) High noble metal concentrations in a late Pliocene sediment. Nature 292: 417 - 420

Lindstrøm M, Sturkell EFF, Tørnberg R, Ormø J (1996) The marine impact crater at Lockne, central Sweden. Geologiske Föreningen Förhandlingar 118: 193 – 206

Lorenz RD (1997) Impact and cratering on Titan: a pre-Cassini view. Planetary and Space Science 45: 1009-1019

Lorenz RD (this volume) Titan: A new world covered in submarine craters In: Dypvik H, Burchell M, Claeys P (eds) Cratering in Marine Environments and on Ice. Impact Studies vol. 5, Springer, Heidelberg, pp 185 - 195

Masaitis VL (2002) The Middle Devonian Kaluga impact event (Russia): New interpretation of marine setting. Deep Sea Research Part 2, 6: 1157 - 1169

Margolis SV, Claeys PF, Kyte FT (1991) Microtektites, microkrystites, and spinels from a late Pliocene asteroid impact in the Southern Ocean. Science 251: 1594 - 1597

Monteiro JF, Rampino MR, Ribeiro A, Munha J (2000) Evidence from Iberia and the central Atlantic Ocean for an oceanic impact near the Cenomanian – Turonian boundary [abs.]. In: Catastrophic events and mass excinctions; impacts and beyond. LPI Contribution # 1053, Lunar and Planetary Institute, Houston, p 143

Melosh HJ (1989) Impact Cratering. A Geologic Process. Oxford University Press, Oxford, 245 pp

Melosh HJ (2003) Impact-generated tsunamis: An over-rated hazard [abs.]. Lunar and Planetary Science 34: abs. # 2013 (CD-ROM)

Nordyke MD (1977) Nuclear cratering experiments; United States and Soviet Union. In: Roddy DJ, Pepin RO, Merrill RB (eds) Impact and Explosion Cratering; Planetary and Terrestrial Implications. Pergamon Press, New York, pp 103 – 124

Norris RD, Huber BT, Self-Trail JM (1999) Synchroneity of the K-T oceanic mass extincttion and meteorite impact; Blake Nose, western North Atlantic. Geology 27: 419 – 422

Oberbeck VR, Marshall JR, Aggarwal H (1993) Impacts, tillites, and the breakup of Gondwanaland. Journal of Geology 101: 1-19

O'Keefe JD, Ahrens TJ (1982) The interaction of the Cretaceous/Tertiary extinction bolide with the atmosphere, ocean and solid Earth. In: Silver LT, Schultz PH (eds) Geological Implications of Impacts of Large Asteroids and Comets on the Earth. Geological Society of America, Special Paper 190: 103 – 120

Ormø J, Lindstrøm M (2000) When a cosmic impact strikes the sea bed. Geological Magazine 137: 67 – 80

Poag CW (1997) The Chesapeake Bay bolide impact: a convulsive event in Atlantic coastal plain evolutions. Sedimentary Geology 108: 45 - 90

Poag CW, Hutchinson DR, Colman SM, Lee MY(1999) Seismic expression of the Chesapeake Bay impact crater, structural and morphologic refinements based on new seismic data. In: Dressler BO, Sharpton VL (eds) Large Meteorite Impacts and Planetary Evolution II. Geological Society of America, Special Paper 339: 149 – 164

Roddy DJ, Schuster SH, Rosenblatt M, Grant LB, Hassing PJ, Kreyenhagen KN (1987) Computer simulations of large asteroid impacts into oceanic continental sites; Preliminary results on atmospheric cratering and ejecta dynamics. International Journal of Impact Engeneering 5: 525 – 541

Rondot J (1994) Recognition of eroded astroblems. Earth - Science Reviews 35: 331 – 365

Sharpton VL, Marin LE, Carney JL, Lee S, Ryder G, Schuraytz BC, Sikora P, Spudis PD (1996) A model of the Chicxulub impact basin based on evaluation of geophysical data, well logs and drill core samples. In: Ryder G, Fastovsky D, Gartner S (eds) The Cretaceous-Tertiary Event and other Catastophes in Earth History. Colorado, Geological Society of America Special Paper 307: 55 – 74

Shuvalov VV, Dypvik H, Tsikalas F (2002) Numerical simulations of the Mjølnir marine impact crater. Journal of Geophysical Research 107: doi: 10.1029/2001JE001698, 1-1-1-12

Simonson BM, Harnik P (2000) Have distal impact ejecta changed through geologic time? Geology 28: 975-978

Simonson BM, Hassler SW (1997) Revised correlations in the Early Precambrian Hamersley Basin based on a horizon of resedimented impact spherules. Australian Journal of Earth Sciences 44: 37 – 48

Simonson BM, Davies D, Wallace M, Reeves S, Hassler SW (1998) Iridium anomaly but no shocked quartz from Late Archean microkrystite layer: Oceanic impact ejecta? Geology 26: 195 – 198

Smelror M, Kelley S, Dypvik H, Mørk A, Nagy J, Tsikalas F (2001) Mjølnir (Barents Sea) meteorite impact ejecta offers a Boreal Jurassic-Cretaceous boundary marker. Newsletters in Stratigraphy 38: 129 - 140

Smelror M, Dypvik H, Mørk A (2002) Phytoplankton blooms in the Jurassic-Cretaceous boundary beds of the barents sea possibly induced by the Mjølnir impact. In: Buffetaut E, Koeberl C (eds) Geological and Biological Effects of Impact Events. Impact Studies, vol. 1, Springer, Heidelberg, pp 69 – 81

Smit J, Montanari A, Swinburne NHM, Hildebrand AR, Margolis SV, Claeys P, Lowrie W, Asaro F (1992) Tektite-bearing, deep-water clastic unit at the Cretaceous-Tertiary boundary in northeastern Mexico. Geology 20: 99 - 103

Smit J, Roep TB, Alvarez W, Montanari A, Claeys P, Grajales-Nishimura JM, Bermudez J (1996) Coarse – grained, clastic sandstone complex at the K/T boundary around the Gulf of Mexico: Deposition by tsunami waves induced by the Chicxulub impact? In: Ryder G, Fastovsky D, Gartner S (eds) The Cretaceous –Tertiary Event and other Catastrophes in Earth History. Geological Society of America, Special Paper 307: 151 – 182

Sonett CP, Pearce SJ, Gault DE (1991) The oceanic impact of large objects. Advances in Space Research 11: 77 - 86

Stewart SA, Allen PJ (2002) A 20-km-diameter multi-ringed impact structure in the North Sea. Nature 418: 520-523

Strelitz R (1979) Meteorite impact in the ocean. Proceedings of the 10th Lunar and Planetary Science Conference, pp 2799 - 2813

Sturkell EFF (1998) Resurge morphology of the marine Lockne impact crater, Jämtland, central Sweden. Geological Magazine 135: 121 - 127

Sturkell EFF, Ormø J (1997) Impact-related clastic injections in the marine Ordovician Lockne impact structure, Central Sweden. Sedimentology 44: 793 – 804

Suuroja K, Suuroja S (this volume) Neugrund structure – the newly discovered early Cambrian impact crater at the entrance of the Gulf of Finland, Estonia. In: Dypvik H, Burchell M, Claeys P (eds) Cratering in Marine Environments and on Ice. Impact Studies vol. 5, Springer, Heidelberg, pp 75 - 95

Suuroja K, Suuroja S, All T, Flodèn T (2001) Kärdla (Hiiuma Island, Estonia) - the buried and well-preserved Ordovician marine impact structure. Deep Sea Reseach Part II 46: 1121 – 1144

Takayama H, Tada R, Matsui T, Iturralde-Vinent MA, Oji T, Tajika E, Kiyokawa S, Garcia D, Okada H, Hasegawa T, Toyda K (2000) Origin of the Peñalvar Formation in northwestern Cuba and its relation to K/T boundary impact event. Sedimentary Geology 135: 295 – 320

Tsikalas F, Gudlaugsson ST, Faleide JI (1998) The anatomy of a buried complex impact structure: the Mjølnir Structure, Barents Sea. Journal of Geophysical Research 103: 30,469 – 30,483

Turtle EP, Pierazzo E (2001) Thickness of a Europan ice shell from impact crater simulation. Science 294: 1326-1328

van der Bergh S (1989) Life and death in the inner solar system. Publications of the Astronomical Society of the Pacific 101: 500 - 509

van Dorn WG, LeMéhauté B, Hwang LD (1968) Handbook of Explosion - generated Water Waves, Volume 1 - State of the Art. Tetra Tech. Pasadena, California, 174 pp

von Dalwigk I, Ormø J (2001) Formation of resurge gullies at impacts at sea: The Lockne crater, Sweden. Meteoritics and Planetary Science 36: 359 – 369

Ward SN, Asphaug E (2000) Asteroid impact tsunami: a probabilistic hazard assessment. Icarus 145: 64-78

Ward SN, Asphaug E (2002) Impact tsunami - Eltanin. Deep - Sea Research II 49: 1073-1080

Warme JE, Kuehner HC (1998) Anatomy of an anomaly; the Devonian catastrophic Alamo impact breccia of southern Nevada. International Geological Review 40: 189 - 216

Warme JE, Sandberg CA (1996) Alamo megabreccia; a record of a Late Devonian impact in southern Nevada. Geological Society of America Today 6(1): 1 – 7

Whipple FL (1950) A comet model. I. The acceleration of Comet Encke. Astrophysical Journal 111: 375 – 394

Whipple FL (1951) A Comet Model. II. Physical relations for comets and meteors. Astrophysical Journal 113: 464 – 474

Biotic Responses to the Mjølnir Meteorite Impact, Barents Sea: Evidence from a Core Drilled within the Crater

Gerd Merethe A. Bremer[1], Morten Smelror[2], Jenö Nagy[1] and Jorunn O. Vigran[3]

[1]Department of Geology, University of Oslo, P.O. Box 1047, Blindern, N-0316 Oslo, Norway. (g.m.a.bremer@geologi.uio.no)
[2]Geological Survey of Norway, N-7491 Trondheim, Norway.
[3]Hans Hagerupsgt. 10, N-7012 Trondheim, Norway.

Abstract. The Mjølnir meteorite crater in the Barents Sea was formed at the Juassic/Cretaceous (Volgian-Ryazanian) boundary. The meteorite impacted the organic-rich clays assigned to the Hekkingen Formation (Volgian-Ryazanian age) and penetrated into underlying Middle Jurassic to Triassic rocks. Studies of macro- and microfossil assemblages from core 7329/03-U-01 drilled at the edge of the crater's central high have revealed the following: 1) a lower unit consisting of strongly disturbed crater fill deposits with a microfossil content similar to that found in the Wilhelmøya Subgroup (Svalbard) and microfloral assemblage similar to those from the Sassendalen Group (Botneheia Formation) and the Storfjorden and Wilhelmøya subgroups on Svalbard; 2) a series of gravity flow deposits containing a mixture of reworked microfloras and -faunas of Middle Triassic/Middle Jurassic origin, and dinoflagellates of Volgian-Ryazanian age; 3) post-impact sediments of the Hekkingen Formation showing anomalous biotal features. In the lowermost post-impact sediments a conspicuous acme of *Leiosphaeridia* combined with an influx of abundant juvenile freshwater algae (*Botryococcus*) occurs. This points to brackish surface water conditions. In the same interval, only a few foraminiferids are found while the bivalve *Buchia* are frequent. High freshwater supply, stratified water-masses and high influx of released nutrients are considered as main factors acting on the post-impact depositional environment. Open marine conditions were restored in the Early Ryazanian.

1 Introduction

The 40 km diameter Mjølnir meteorite crater, located in the central Barents Sea (Fig. 1), was first described by Gudlaugsson (1993). The crater was formed close to the Volgian-Ryazanian boundary (142 ± 2.6 Ma) by the impact of a 1.5 to 2 km diameter iron meteorite in the palaeo-Barents Sea (Dypvik and Attrep 1999, Dypvik et al. 1996, Gudlaugsson 1993, Smelror et al. 2001b). At that time sedimentation on this 300-400 m deep shelf target area was dominated by dark organic-rich clays deposited under hypoxic to anoxic conditions. These deposits have been assigned to the Hekkingen Formation (Worsley et al. 1988).

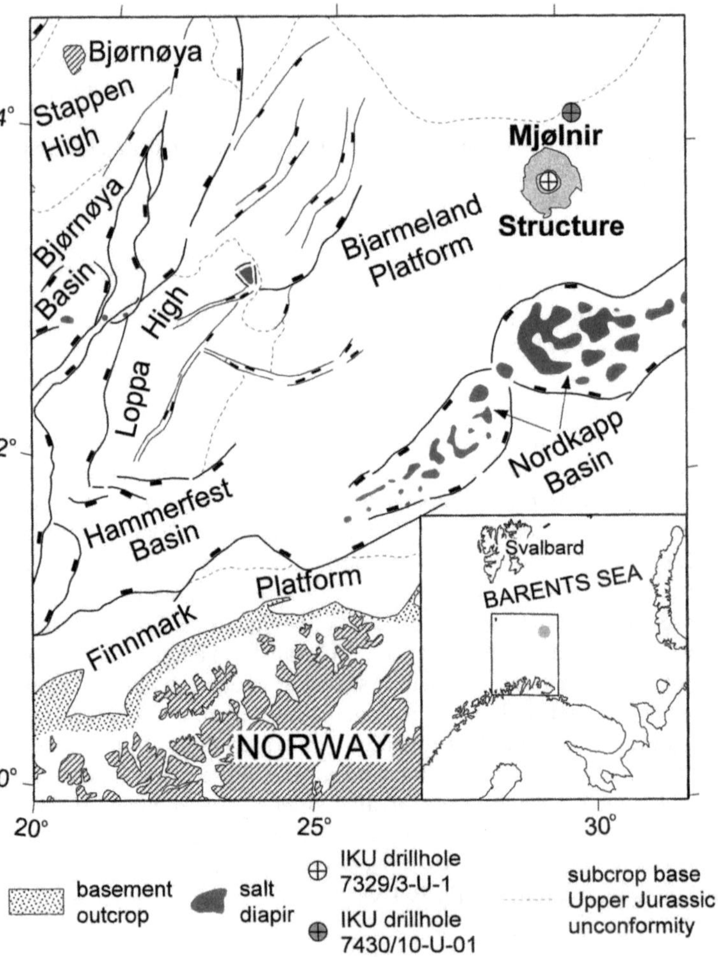

Fig. 1. Location map of the south-western Barents Sea showing the Mjølnir structure and the two drilled wells 7329/03-U-01 and 7430/10-U-01 (Tsikalas et al. 1998)

The marine Mjølnir crater is one of the 25 largest impact structures discovered on Earth, and the drilled cores provide an opportunity for detailed studies of both pre- and post-impact sediments and marine biotas. A 121 m long core (7329/03-U-01), located at the edge of the crater's central high (Fig. 2), is divided into 3 main units (Fig. 3), based on the lithological features: 1) strongly deformed crater fill sediments of the *"Ragnarok formation"*; 2) gravity flow deposits; 3) post-impact dark organic-rich shales of the Hekkingen Formation overlain by condensed carbonates and marls of the Klippfisk Formation.

The Hekkingen Formation on the Barents Shelf is correlative with the Agardhfjellet Formation (late Jurassic age) on Svalbard (Mørk et al. 1999, Nøttvedt et al. 1993) and its marine fossil assemblages closely resemble those described from the Agardhfjellet Formation (Nagy and Basov 1998, Nagy et al. 1988, Smelror and Below 1993, Smelror et al. 2001b, Wierzbowski and Århus 1990). There are no indications of any significant biotic extinctions associated with the Mjølnir impact (Bremer et al. 2001). This observation is in accordance with Jansa (1993) who proposed a minimum bolide size necessary for extinctions of >4 km in nucleus diameter. Low diversity invertebrate faunas of ammonites and bivalves (mainly *Buchia*) and low diversity foraminiferal assemblages (mainly agglutinated species) occur both below and above the impact influenced strata (Bremer et al. 2001).

In the ejecta-bearing beds in a core drilled 30 km northeast of the Mjølnir Crater rim (borehole 7430/10-U-01, Fig. 1), Smelror et al. (2001b) found a distinctly high abundance of prasinophycean algae. They suggested that the algal bloom possibly was induced by the large amounts of nutrients released into the water column by the impact. In the present study we have further explored the biotal response to the Mjølnir event and document the results of faunal and microfloral analyses of core 7329/03-U-01 drilled within the crater.

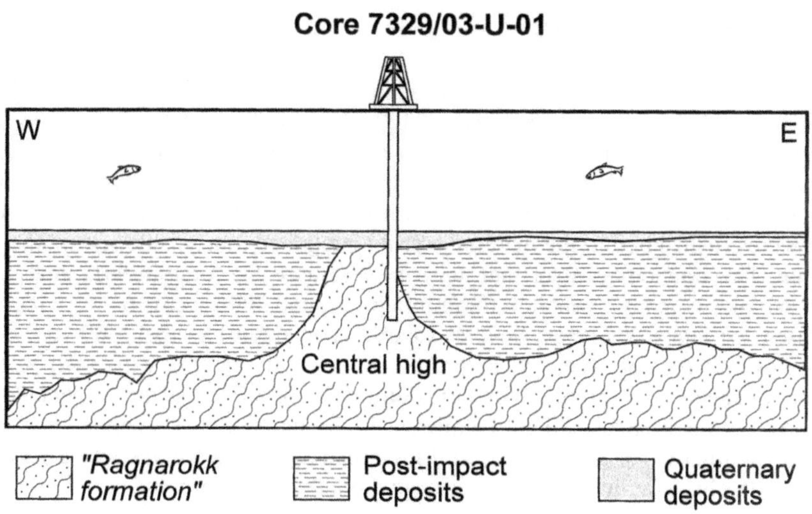

Fig. 2. Diagramatic cross-section of the central part of the Mjølnir crater showing the position of core 7329/03-U-01 within the crater.

2
Lithology

Lithological descriptions of core 7329/03-U-01 have been published by Smelror et al. (2001b) and Sandbakken & Dypvik (2001). Based on lithological features, the core can be subdivided into several depositional intervals (Fig 3), as mentioned earlier. The interval from the base of the core at 171.08 m up to 74.02 m is defined as the *"Ragnarok formation"* (Sandbakken and Dypvik 2001). This interval is divided into two depositional units: unit I (171.08-88.31 m) consists of a mixture of Middle and Upper Triassic to Lower Jurassic target rocks impacted by the meteorite and redeposited as fallout into the crater. The succeeding, and much thinner unit II (88.31-74.02 m) is interpreted as gravity flow deposits with three main subunits recognised: IIA (88.31-87.42 m) is a conglomeratic debris flow of sand and small pebbles, IIB (87.42-75.69 m) represents a mudflow deposit, while IIC (75.69-74.02 m) consists of at least three separate gravity flows of a sand, silt and clay mixture. The sediments composing unit II most likely originated from the uplifted central high of the crater. The upper, post-impact part of the core (74.02-57.30 m) is unit III and consists of Lower Ryazanian dark brown to black organic-rich shales of the Hekkingen Formation, capped by Valanginian condensed carbonates and marls of the Klippfisk Formation from 57.30 m to 50.00 m.

3
Material and Methods

A total of 41 samples for both micropalaeontological and palynological analyses were taken at 1-4 m intervals from core 7329/03-U-01. The palynological samples were processed using standard preparation methods, including HCl and HF treatment. A total of around 300 palynomorph specimens where counted in each residual sample. For the micropalaeontological investigation, 15-25 grams of dry sediment were crushed before the material was disintegrated in a tenside and methanol mixture. The fractions coarser than 63 μm were used for foraminiferal analyses. From samples containing enough foraminifera, around 300 specimens were picked and used for faunal analyses. Several samples contained less than 300 tests and from those all the specimens were picked. Species diversity is expressed as the number of species per sample, while the dominance is defined as the percentage of the most common species. The abundance is the number of specimens per 100 gram of sediment. The radiolarians and the green algae *Tasmanites* >90 μm were picked from the foraminiferal preparations where they were present.

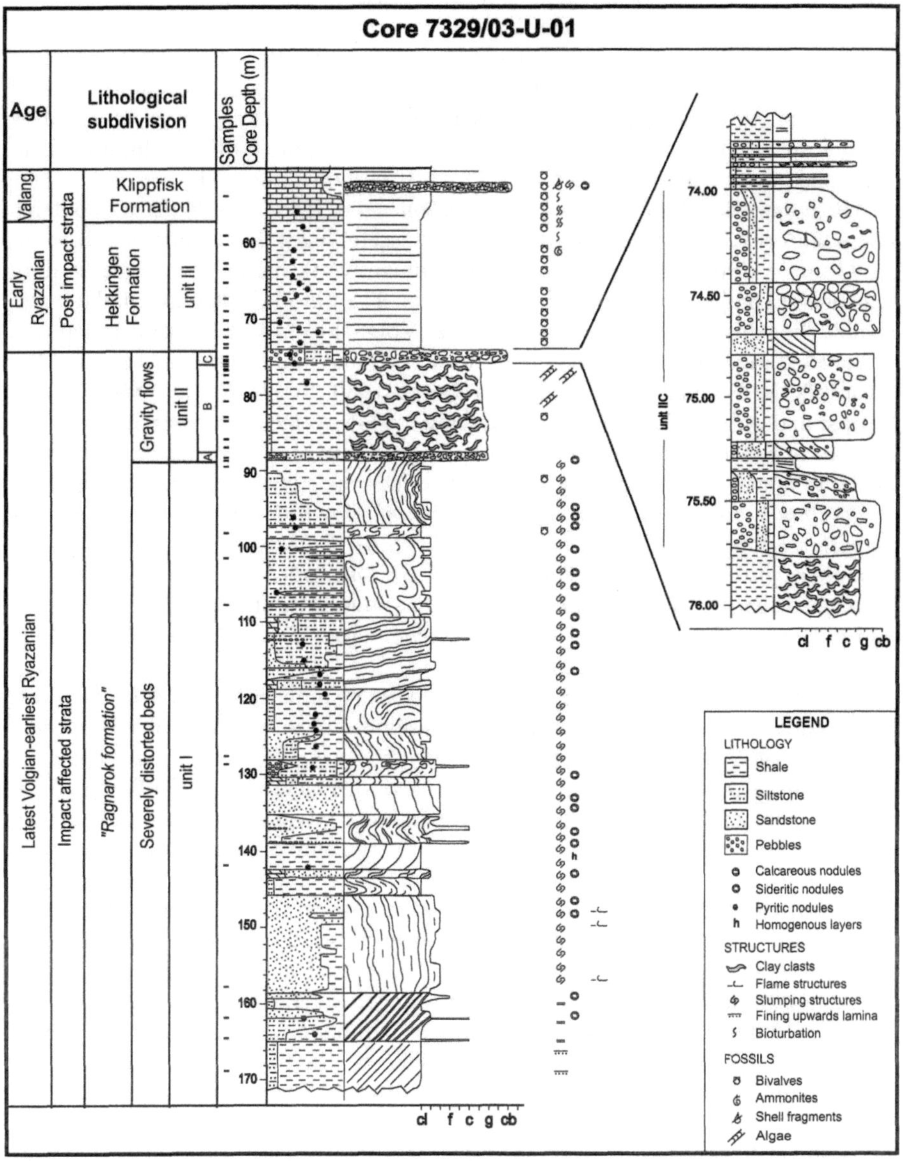

Fig. 3. Lithostratigraphic log of the Mjølnir crater core 7329/03-U-01 (modified after Sandbakken 2002)

3.1 Marine Microfloras

The microfloral assemblages recovered in the impact-affected *"Ragnarok formation"* (171.08-74.02 m) are comparable to those found in the Wilhelmøya and Storfjorden subgroups on Svalbard and the correlative Realgrunnen Subgroup on the Barents Shelf (Bjærke 1977, Bjærke and Dypvik 1977, Smelror and Below 1993, Smelror et al. 1998). The investigated samples contain low abundance and low diversity assemblages of acritarchs and prasinophycean algae (i.e. tasmanitids and leiospheres). The relative abundance of the marine microflora (dinoflagellate cysts and prasinophytes) in core 7329/03-U-01 is shown in Fig. 4.

The upper part of the *"Ragnarok formation"*, unit II (88.31 –74.02 m), contains a less uniform development of palynomorph assemblages. Four samples (88.11 m, 87.93 m, 74.80 m and 74.10 m) contain dominantly prasinophycean algae (mainly leiospheres) and common dinoflagellate cysts. The dinoflagellate assemblages resemble those found elsewhere in the Upper Volgian to Lower Ryazanian deposits of the Hekkingen Formation on the Barents Shelf and in the uppermost parts of the Agardhfjellet Formation on Svalbard (Bjærke 1980, Smelror et al. 1998, Wierzbowski and Århus 1990). They are typically of low to medium diversity, with the number of species found in each sample varying between 4-17 species. The abundance of species is generally low. Characteristic and most common taxa include *Atopodinium haromense*, *Cassiculosphaera magna*, *Cribroperidinium spp.*, *Escharisphaeridia spp.*, *Paragonyaulacysta borealis*, *Isthmocystis distincta*, *Senoniasphaaera jurassica*, *Sirmiodinium grossii* og *Tubotuberella apatela*. The common occurrence *Paragonyaulacysta borealis* suggest a stratigraphic correlation and biogeographic relation to the *Paragonyaulacysta*? *borealis* assemblage Zone of Lebedeva and Nikitenko (1999).

The samples from 76.16-75.96 m revealed very high numbers of tasmanitids, all of equal size (about 200 μm). These resemble those found in the Botneheia Formation on Svalbard (Forsberg et al. 1984).

The oldest post-impact deposits of the Hekkingen Formation (74.02-68.55 m) contain more than 60 % leiospheres and relatively low amounts of dinoflagellate cysts. In addition, these samples also contain very common freshwater algae identified as juvenile specimens of *Botryococcus*. This unique combination of very abundant leiospheres and abundant freshwater algae in the Hekkingen Formation is previously only reported from the ejecta-bearing strata recovered from borehole 7430/10-U-01 (Fig. 1) by Smelror et al. (2001b) and in the upper Agardhfjellet Formation at Janusfjellet on Central Spitsbergen by Dypvik et al. (2000). The overlying topmost deposits of the Hekkingen Formation (sampled at 64.93-58.72 m) contain significantly less frequent leiospheres; the marine microflora is here dominated by dinoflagellate cysts. These marine palynomorph assemblages are comparable to those found in the uppermost Hekkingen Formation elsewhere on the Barents Shelf and on Svalbard (Bjærke 1980, Smelror et al. 1998, Wierzbowski and Århus 1990). A single occurence of *Gochteodinia villosa* is recorded at 62 m, suggesting a biostratigraphic correlation to the *Gochteodinia villosa* Interval Biozone (RPJ17) of Riding et al. (1999).

3.2 Pollen and Spores

The microfloral assemblages (Fig. 4) from of the impact affected strata (171.08-88.31 m) of core 7329/03-U-01 are totally dominated by terrestrial palynomorphs. The recovered pollen and spore assemblages include a stratigraphical mixture of species like *Accinctisporites circumdatus*, *Aratrisporites macrocavatus* , *Densoisporites nejburgii*, *Doubingerispora filamentosa*, *Echinitosporites iliacoides*, *Illinites chitonoides*, *Jerseyiaspora punctispinosa*, *Leschikisporis aduncus*, *Proprisporites pococki*, *Protodiploxypinus ornatus*, *Rewanispora foveolata* and *Triadispora obscura*. The species have previously been reported from assemblages of the Botneheia Formation on Svalbard (Vigran unpublished), and from correlative Steinkobbe and Snadd formations on the Barents Shelf (Vigran et al. 1998), and from the Storfjorden (Vigran unpublished) and Wilhelmøya subgroups (Bjærke 1977). The pollen and spores found in the gravity flow deposits of unit II between 88.31-74.02 m contain a similar mixture of mainly Middle to Upper Triassic species. Younger Mesozoic (indeterminate Jurassic) pollen and spores are also present at some levels.

In the oldest post-impact deposits of the Hekkingen Formation, non-diagnostic Jurassic pollen and spores are present, but they are totally outnumbered by the prasinophycean algae in the Volgian-Ryazanian palynological assemblages. In the youngest deposits of the Hekkingen Formation pollen, and to a minor extent spores, again become the most common palynomorphs.

3.3 Freshwater Algae

Freshwater algae (Fig. 4) are a minor constituent of all samples of the *"Ragnarok formation"* (from 171.08 m up to 74.02 m). In the oldest post-impact deposits there is a prolific bloom of juvenile freshwater algae of the genus *Botryococcus*. As mentioned above, a similar acme has previously been recorded from the ejecta bearing strata of the Hekkingen Formation in borehole 7430/10-U-01 (Smelror et al. 2001a) and it is also observed in the Agardhfjellet Formation at Janusfjellet on Central Spitsbergen (Dypvik et al. 2000), about 400 km away from the Mjølnir Crater.

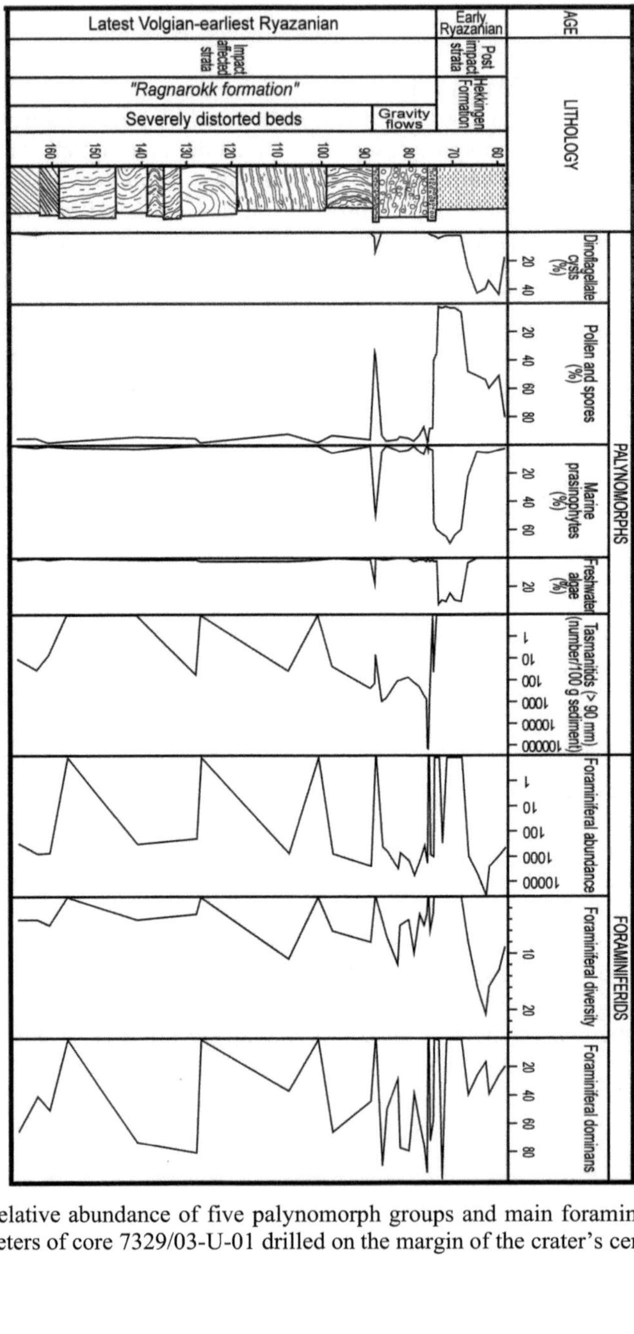

Fig. 4. Relative abundance of five palynomorph groups and main foraminiferal assemblage parameters of core 7329/03-U-01 drilled on the margin of the crater's central high.

3.4 Microfaunas

The analysed samples from the impact-affected sediments between 171.08 and 74.02 m contain well preserved, but strongly impoverished, benthic foraminiferal assemblages, with *Ammodiscus* aff. *yonsnabensis*, *Trochammina* aff. *minutissima* and *Evolutinella sp. 1* as the dominating species. All specimens are generally small in size, less than 100 μm in diameter. This is also typical for the faunas found in the Wilhelmøya Subgroup (Knorringfjellet Formation) on Svalbard, with which the recovered assemblages show close similarities. A few samples from the Stø Formation (equivalent to the Upper Wilhelmøya Subgroup) off Troms were analyzed for foraminifera, and they also contain low diversity faunas with small specimens of *Ammodiscus* and *Trochammina* similar to those found in the impact-affected interval in core 7329/03-U-01.

The post-impact sediments (74.02-57.30 m) overlying the disturbed strata contain rich and unaffected Ryazanian faunas dominated by *Evolutinella schleiferi, Evolutinella vallata, Gaudryina rostellata, Recurvoides obskiensis, Trochammina* cf. *annae* and *Trochammina* aff. *quinquelocularis*. These assemblages closely resemble those found in the Agardhfjellet Formation on Svalbard. A chart showing the occurences of the most important species in the studied strata is shown in Fig. 5.

The main faunal parameters (diversity, dominance and abundance) of the strata penetrated by core 7329/03-U-01 are shown in Fig. 4. The abundance of foraminifera in the impact-affected interval (*"Ragnarok formation"*) is, with a few exceptions, low and the assemblages consist of generally very small (<100 μm) and exclusively agglutinant specimens. The dominance is very high, while the diversity is extremely low in these assemblages. In all except three samples the number of species is 6 or less (Table 1).

The oldest post-impact sediments of the Hekkingen Formation do not contain any foraminiferids except in a single sample at 73.00 m were a few specimens of *Trochammina* aff. *septentrionalis* are recorded. Above 67.00 m, in the Hekkingen Formation, both the diversity and abundance increase while the dominance decreases. In spite of this faunal expansion, the diversity is still relatively low. The assemblages are dominated by agglutinating taxa but there are also a few calcareous forms present in the samples at 66.00 m and 60.00 m.

The uppermost part of the core comprises the marls of the Klippfisk Formation. It contains almost exclusively calcareous assemblages with high diversity, high abundance and low dominance.

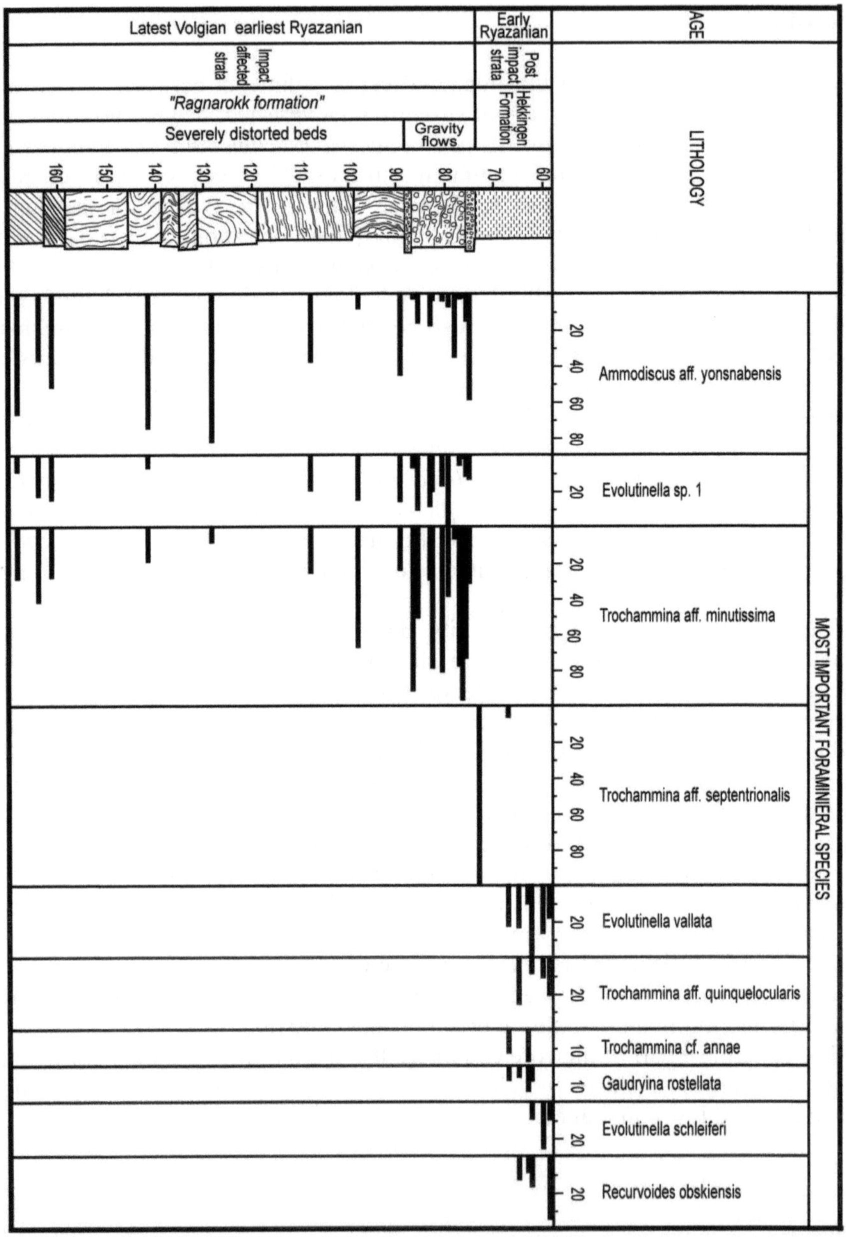

Fig. 5. Distribution of the quantitatively most important foraminiferal species in the Mjølnir crater core (7329/03-U-01).

Table 1. Maximum, minimum and mean values of main parameters of foraminiferal faunas. Barren samples are excluded from these calculations.

MAIN FAUNAL PARAMETERS

				Mean	Min.	Max.
Hekkingen Formation	Unit III	Post impact strata	Abundance	6534	271	35812
			Diversity	12	1	21
			Dominance	38	16	100
"Ragnarok formation"	Unit II	Gravity flow interval	Abundance	1504	342	5489
			Diversity	6	3	12
			Dominance	66	28	95
	Unit I	Impact distorted interval	Abundance	736	178	2297
			Diversity	6	3	11
			Dominance	57	37	81

To elucidate differences in the depositional conditions, the foraminiferids have been divided into 3 groups, based on their supposed environmental preferences: 1) marginal marine, 2) normal marine, 3) no or unknown preferences. Marginal marine taxa are considered to be *Ammodiscus* and *Ammobaculites*. Rich occurrences of these species have been attributed to marginal environments in the Jurassic (Løfaldli and Nagy 1980, Nagy et al. 1990, Nagy and Seidenkrantz submitted). The normal marine taxa includes *Cribrostomoides*, *Evolutinella*, *Gaudryina*, *Glomospirella*, *Haplophragmoides*, *Recurvoides*, *Saccammina*, *Textularia* and all the calcareous genera (Løfaldli and Nagy 1980, Nagy et al. 1990). The normal marine nature of these genera is indicated by their common association with ammonites and belemnites on Svalbard and in the Barents Sea. All other taxa are considered to have unknown or to be without any depositional environmental preferences.

From recent environments Murray (1991) lists *Ammobaculites* as marginal marine, *Trochammina* as both marginal and normal marine, while all the other genera mentioned above (except *Evolutinella*) are considered normal marine.

Figure 6 illustrates the distribution of the marginal marine and the normal marine species groups. The marginal marine species dominate in the lower and middle part of the core (171.08-88.31 m), while the normal marine taxa dominate in the upper part (67.00-58.72 m). Unit II (88.31-74.02 m) reveals marked variations.

In addition to the foraminiferids, radiolarians occur in small amounts throughout the core, except for the sample at 73.00 m where they are prolific. A similar peak was noted in core 7430/10-U-01 (Bremer, personal observation). Such peaks may indicate increased influx of open marine water. Radiolarians are otherwise poorly documented from the Jurassic of the Barents Sea area. To our knowledge there is no published information, but some sporadic records can be found in restricted commercial reports.

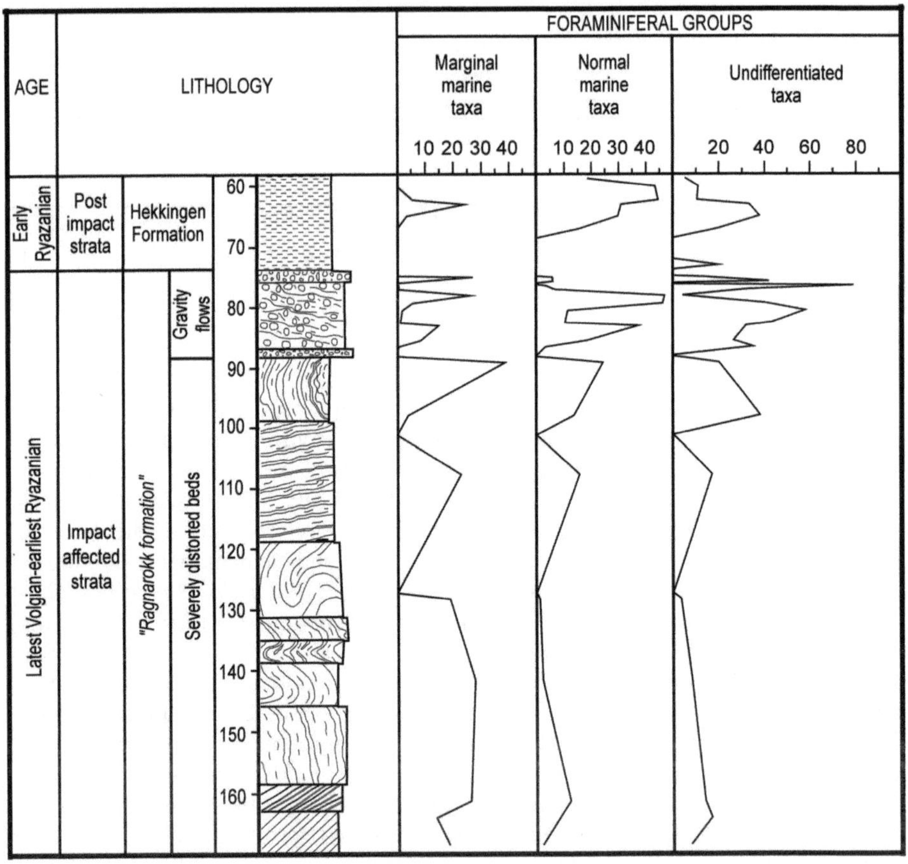

Fig. 6. Distribution of the foraminiferal groups based on assessed environmental preferences in core 7329/03-U-01. (All values are in vol. percent).

3.5 Macrofaunas

In core 7329/03-U-01 macrofossils were only recovered from the post-impact deposits. They include thin-shelled representatives of the bivalve genus *Buchia* and ammonites of the genus *Borealites.* The bivalves occur scattered throughout both the Hekkingen and Klippfisk formations whereas the ammonite findings are restricted to the Hekkingen Formation. These fossils allow a relatively accurate biostatigraphic age-determination of the post-impact strata, and indicate a Volgian-Ryazanian boundary age for the Mjølnir impact (Smelror et al. 2001c). The macrofaunas recorded in 7329/03-U-01 closely resemble faunas typically for the upper Hekkingen Formation on the Barents Shelf (Smelror et al. 2001b, Århus 1991, Århus et al. 1990) and in the Agardhfjellet Formation on Svalbard (Birkenmajer et al. 1982).

4 Discussion

High dominance and low diversity microfloral and microfaunal assemblages are generally associated with extreme or changing environmental conditions. The generally low diversity microfossil assemblages present in unit I of core 7329/03-U-01 (171.08-88.31 m) loosely resemble those found in the marginal marine Wilhelmøya Subgroup and Stø Formation (Løfaldli and Nagy 1981, Mørk et al. 1982, Smelror et al. 2001c). We therefore assume that fragmented deposits equivalent to these strata comprise the disturbed and chaotic *"Ragnarok formation"*. The dominantely Middle to Late Triassic terrestrial spores and pollen suggest that deposits of the Steinkobbe and Snadd formations in particular, have been fragmented.

The deposits composing Unit II between 88.31 m and 74.02 m contain a mixture of reworked microfloras and microfaunas. The recovered assemblages are comprised of Middle Triassic to Early/Middle Jurassic foraminifera and palynomorphs, together with Volgian-Ryazanian dinoflagellates. The wide range in stratigraphic age and environmental preferences represented by the recovered phytoplankton and foraminiferal taxa reinforces the proposal that these microfloras and microfaunas of the gravity flow deposits are to a considerable extent reworked.

The very high numbers of tasmanitids recovered at 76.37 m, 76.16 m and 75.96 m in combination with reworked Triassic pollen, indicate a mud flow origin where the tasmanitids and the pollen have been washed out, sorted and resedimented at the top of the flow. Immediately below and above this mud flow with its high numbers of tasmanitids, the samples contain microfloral assemblages that resemble those typical for the Hekkingen Formation. These findings support the threefold subdivision of unit II proposed by Sandbakken and Dypvik (2001).

The dark, organic-rich, post-impact sediments are assigned to the Hekkingen Formation, which was deposited in open marine to restricted shelf environments with hypoxic to anoxic bottom conditions (Leith et al. 1993, Smelror et al. 2001c, Worsley et al. 1988). The diversities shown by the macrofauna, microfauna and marine microflora are relatively low.

The core interval between 74.02 m and 67.00 m shows biotal features which are unique within the Hekkingen Formation. These strata are almost totally devoid of foraminifera and other marine microfossils, but thin-shelled bivalves belonging to the genus *Buchia* are present throughout. Even more striking, these strata which immediately overlie the impact-affected interval contain a conspicuous acme of *Leiosphaeridia* combined with the influx of very abundant juvenile freshwater algae of the genus *Botryococcus*. A virtually identical combination of marine and freshwater algae was also recognised in the ejecta-bearing deposits penetrated in core hole 7430/10-U-01 (Smelror et al. 2001a) and in the Janusfjellet section on Central Spitsbergen (Dypvik et al. 2000).

Smelror et al. (2001a) proposed that the bloom of *Leiosphaeridia* recorded in the ejecta-bearing sediments of core 7430/10-U-01 was induced by the large amounts of newly released nutrients in the water-column by shock-waves and tsunamis created by Mjølnir impact. This may also be the case for the blooms recovered in the gravity flow interval of core 7329/03-U-01. The bloom involving the juvenile fresh-water algae is, however, less easy to explain. The presence of *Botryococcus* requires a considerable freshwater influx to the depositional region. Increased run-off from a land area, possibly caused by the Mjølnir impact, is suggested, but independent evidence for such an event is not recognised.

The scarcity of foraminiferids in the algal bloom interval (74.02-67.00 m) can be explained either by hypoxic to anoxic conditions or by strongly reduced salinity extending to the bottom waters. The latter assumption is less plausible because the presence of ammonites indicates marine salinities in at least a part of the water column. Oxygen depleted conditions are supported by extremely high TOC-values (10-20 wt.%) measured in the few scattered samples which were analyzed. High organic productivity, a prerequisite of bottom water hypoxia, is suggested by the above mentioned algal blooms in the planktonic habitat. The *Buchia* shells found in these strata can possibly be explained by transportation as pseudoplankton into the area similar to the so-called paper pectens. The pseudoplanktonic nature of such thin-shelled bivalves in black shales is suggested by Wignall (1994). The presence of both ammonites and thin-shelled bivalves as body fossils in this unit further suggest that the lack or very sparse occurrence of foraminifera is not due to diagenetic destruction of calcareous taxa and agglutinated forms with calcareous cement.

Above 66.0 m in core 7329/03-U-01 a foraminiferal and microfloral succession typical for the uppermost Hekkingen Formation elsewhere on the Barents Shelf and in the upper Agardhfjellet Formation on Svalbard is re-established. Measured TOC values in this unit are low (1-2 %), and the species diversity is intermediate in the almost exclusively agglutinated assemblages. This return to “normal conditions” takes place just above the stratigraphic level with the last occurrence of *Buchia unchensis* (67.14 m) and the first appearance of *Buchia okensis* (66.80 m) suggesting that normal oceanographic conditions were restored in the Early Ryazanian, prior to the time corresponding to the *Hectoceras kochi* ammonite zone.

5 Conclusions

The marine Mjølnir impact crater was formed in a marine shelf environment preserving relatively much information on both pre- and post-impact sediments and marine biotas. The crater succession recovered by core 7329/03-U-01 consists of three main units: 1) chaotic and strongly disturbed crater fill deposits, mainly derived from Middle Triassic to Lower Jurassic target rocks; 2) a series of gravity flow deposits (mainly debris and mud flows), most likely derived from the elevated central high of the crater; 3) post-impact dark organic-rich shales of the Hekkingen Formation, capped by condensed carbonates and marls of the Klippfisk Formation.

The microfossil assemblages found in unit I, between the base (171.08 m) and 88.31 m of core 7329/03-U-01, closely resemble those found in the Wilhelmøya and Realgrunnen subgroups and we therefore assume that the generally low diversities found in the disturbed and chaotic impact target rocks of the *"Ragnarok formation"* are due to the originally marginal marine nature of these sedimentary target rocks.

The deposits in unit II between 88.31 m and 74.02 m contain a mixture of reworked microfloras and microfaunas and the recovered assemblages consist of Middle Triassic to Early/Middle Jurassic foraminifera and palynomorphs, and Volgian-Ryazanian dinoflagellates.

The oldest post-impact sediments of the Hekkingen Formation contain algal blooms of leiospheres (marine prasinophytes) and juvenile *Botryococcus* (freshwater algae). Blooms of *Botryococcus* points to an environment with surface water conditions that, if not fresh, were at least brackish although an independent evidence for this development is lacking. Deposition of organic-rich shales in oxygen-depleted conditions is in accordance with high productivity of phytoplankton. The interval contain only a few foraminiferids (i.e., between 74.02-67.00 m), but frequent, possibly pseudoplanktonic, bivalves of the genus *Buchia.* High freshwater supply possibly caused as a result of the Mjølnir impact, leading to stratified water-masses with a brackish surface layer and high influx of released nutrients may have resulted in the anomalous post-impact depositional environment recognized in the Mjølnir Crater, the nearby shelf area at core hole 7430/10-U-01, and much farther afield at the Janusfjellet section on Svalbard. Open marine conditions, typical for the upper Hekkingen Formation elsewhere on the Barents Shelf, were restored in the Early Ryazanian.

Acknowledgements

Samples from the three investigated cores were made available for analyses by IKU-Sintef Group. The study was funded by the Statoil-VISTA Programme and the Norwegian Research Council. The authors wish to thank Henning Dypvik, Atle Mørk and Filippos Tsikalas for inspiring collaboration in the Mjølnir Project. Acknowledgement is given to Pål Sandbakken for valuable discussions concerning the sedimentology of the cores. Appreciation is extended to Adrian Read for corrections and comments on the manuscript at various stages. We also thank the reviewers, Eric Buffetaut and B. Mohr for helpful suggestions that improved the paper.

References

Birkenmajer K, Pugaczewska H, Wierzbowski A (1982) The Janusfjellet Formation (Jurassic-Lower Cretaceous) at Myklegardfjellet, East Spitsbergen. Palaeontologica Polonica 43: 107-140

Bjærke T (1977) Mesozoic palynology of Svalbard; II, Palynomorphs from the Mesozoic sequence of Kong Karls Land. Norsk Polarinstitutt Årbok 1976: 83-99

Bjærke T (1980) Mesozoic palynology of Svalbard; V, Dinoflagellates from the Agardhfjellet Member (Middle and Upper Jurassic) in Spitsbergen. In: Geological and Geophysical Research in Svalbard and on Jan Mayen Norsk Polarinstitutt, Oslo, Norway, pp 145-167

Bjærke T, Dypvik H (1977) Sedimentological and palynological studies of Upper Triassic-Lower Jurassic sediments in Sassenfjorden, Spitsbergen. Norsk Polarinstitutt Årbok 1976: 131-150

Bremer GMA, Smelror M, Nagy J (2001) Biotic responses to the marine Mjølnir meteorite impact (Volgian-Ryazanian boundary, Barents Sea) [abs.]. In: Smelror M, Dypvik H, Tsikalas F (eds) 7th Workshop of the ESF Impact Program. Norwegian Geological Society, Trondheim, Norway, pp 11-12

Dypvik H, Attrep M Jr. (1999) Geochemical signals of the late Jurassic, marine Mjølnir impact. Meteoritics and Planetary Science 34: 393-406

Dypvik H, Kyte FT, Smelror M (2000) Iridium peaks and algal blooms - The Mjølnir impact. [abs.] Lunar and Planetary Science 31, Abstract #1538, CD - ROM

Dypvik H, Gudlaugsson ST, Tsikalas F, Attrep M Jr, Ferrell RE Jr, Krinsley DH, Mørk A, Faleide JI, Nagy J (1996) Mjølnir structure; an impact crater in the Barents Sea. Geology 24: 779-782

Forsberg AW, Mørk A, Vigran JO (1984) Triassic source rock potential influenced by the green alga *Tasmanites*. Continental Shelf Institute (IKU), Trondheim, Norway, Report, 36 pp

Gudlaugsson ST (1993) Large impact crater in the Barents Sea. Geology 21: 291-294

Jansa LF (1993) Cometary impacts into ocean: their recognition and the threshold constraint for biological extinctions. Palaeogeography, Palaeoclimatology, Palaeoecology 104: 271-286

Lebedeva NK, Nikitenko BL (1999) Dinoflagellate cysts and microforaminifera of the Lower Cretaceous Yatria River Section, Subarctic Ural, NW Siberia, Russia. Grana 38: 134-143

Leith TL, Weiss HM, Mørk A, Århus N, Elvebakk G, Embry AF, Brooks PW, Stewart KR, Pchelina TM, Bro EG, Verba ML, Danyushevskaya A, Borisov AV (1993) Mesozoic hydrocarbon source-rocks of the Arctic region. In: Vorren TO, Bergsager E, Dahl-Stamnes OA, Holter E, Johansen B, Lie E, Lund TB (eds) Arctic geology and petroleum potential: Proceedings of the Norwegian Petroleum Society Conference, Amsterdam, pp 1-25

Løfaldli M, Nagy J (1980) Foraminiferal stratigraphy of Jurassic deposits on Kongsøya, Svalbard. In: Geological and geophysical research in Svalbard and on Jan Mayen. Norsk Polarinstitutt, Oslo, Norway, pp 63-95

Løfaldli M, Nagy J (1981) Agglutinating foraminifera in Jurassic and Cretaceous dark shales in southern Spitsbergen. In: Verdenius JG, Van-Hinte JE, Fortuin AR (eds) Proceedings of the First Workshop on Arenaceous Foraminifera 7.-9. September, 1981. Continental Shelf Institute (IKU), Trondheim, Norway, pp 91-107

Murray JW (1991) Ecology and Palaeoecology of Benthic Foraminifera. Longman Scientific & Technical, Essex, England, 397 pp

Mørk A, Dallmann WK, Dypvik H, Johannessen EP, Larssen GB, Nagy J, Nøttvedt A, Olaussen S, Pcelina TM, Worsley D (1999) Mesozoic lithostratigraphy. In: Dallmann, WK (ed) Lithostratigraphic lexicon of Svalbard: review and recommendations for nomenclature use: upper Palaeozoic to Quaternary bedrock. Norsk Polarinstitutt, Tromsø, Norway, pp 127-214

Mørk A, Knarud R, Worsley D (1982) Depositional and diagenetic environments of the Triassic and Lower Jurassic succession of Svalbard. In: Embry AF, Balkwill HR (eds) Arctic Geology and Geophysics: Proceedings of the Third International Symposium on Arctic Geology. Canadian Society of Petroleum Geologists, Calgary, Canada, pp 371-398

Nagy J, Basov VA (1998) Revised foraminiferal taxa and biostratigraphy of Bathonian to Ryazanian deposits in Spitsbergen. Micropaleontology 44: 217-255

Nagy J, Løfaldli M, Bäckström SA (1988) Aspects of foraminiferal distribution and depositional conditions in Middle Jurassic to Early Cretaceous shales in eastern Spitsbergen. In: Roegl F, Gradstein FM (eds) Second workshop on Agglutinated foraminifera: proceedings of the second Workshop on Agglutinated Foraminifera Vienna 1986. Abhandlungen der Geologischen Bundesanstalt, Vienna, Austria, pp 287-300

Nagy J, Pilskog B, Wilhelmsen RM (1990) Facies controlled distribution of foraminifera in the Jurassic North Sea basin. In: Hemleben C, Kaminski MA, Kuhnt W, Scott DB (eds) Proceedings of the NATO Advanced Study Institute on Paleoecology, Biostratigraphy, Paleoceanography and Taxonomy of Agglutinated Foraminifera. Reidel Publishing Company, Dordrecht-Boston, pp 621-657

Nagy J, Seidenkrantz M-S (submitted) New foraminiferal taxa and revised biostratigraphy of Jurassic marginal marine deposits on Anholt, Denmark. Micropaleontology

Nøttvedt A, Cecchi M, Gjelberg JG, Kristensen SE, Lonoy A, Rasmussen A, Rasmussen E, Skott PH, van Veen PM (1993) Svalbard-Barents Sea correlation; a short review. In: Vorren TO, Bergsager E, Dahl-Stamnes OA, Holter E, Johansen B, Lie E, Lund TB (eds) Arctic geology and petroleum potential: Proceedings of the Norwegian Petroleum Society Conference. Elsevier, New York, pp 363-375

Riding JB, Fedorova VA, Ilynia VI (1999) Jurassic and lowermost Cretaceous dinoflagellate cyst biostratigraphy of the Russian Platform and Northern Siberia, Russia. American Association of Stratigraphic Palynologists Contribution Series 36: 1-179

Sandbakken PT (2002) A geological investigation of the Mjølnir Crater core (7329/03-U-01): with emphasis on shock metamorphosed quartz. Cand. Scient. (Master) thesis, Department of Geology, University of Oslo, 141 pp

Sandbakken P, Dypvik H (2001) The Mjølnir crater - A core description.[abs.] In: Smelror M, Dypvik H, Tsikalas F (eds) 7th Workshop of the ESF Impact Programme. Submarine Craters and Ejecta-Crater Correlations and Icy Impacts and Icy Targets. Norwegian Geological Society, Trondheim, Norway, pp 69-70

Smelror M, Below R (1993) Dinoflagellate biostratigraphy of the Toarcian to Lower Oxfordian (Jurassic) of the Barents Sea region. In: Vorren TO, Bergsager E, Dahl-Stamnes OA, Holter E, Johansen B, Lie E, Lund TB (eds) Arctic geology and petroleum potential: Proceedings of the Norwegian Petroleum Society Conference. Elsevier, New York, pp 495-513

Smelror M, Mørk A, Montail E, Rutledge D, Leereveld H (1998) The Klippfisk Formation: a new lithostratigraphic unit of Lower Cretaceous platform carbonates on the western Barents Shelf. Polar Research 17: 181-202

Smelror M, Dypvik H, Mørk A (2001a). Phytoplankton blooms in the Jurassic-Cretaceous boundary beds of the Barents Sea possibly induced by the Mjølnir meteorite impact. In: Buffetaut E, Koeberl C (eds) Geological and Biological Effects of Impact Events. Springer, Berlin, pp 66-81

Smelror M, Kelly SRA, Dypvik H, Mørk A, Nagy J, Tsikalas F (2001b) Mjølnir (Barents Sea) meteorite impact offers a Volgian-Ryazanian boundary marker. Newsletter on Stratigraphy 38: 129-140

Smelror M, Mørk A, Mørk MBE, Weiss HM, Løseth H (2001c) Middle Jurassic-Lower Cretaceous transgressive-regressive sequences and facies distribution off northern Nordland and Troms, Norway. In: Martinsen OJ, Dreyer T (eds) Sedimentary environments offshore Norway - palaeozoic to recent: proceedings of the Norwegian Petroleum Society Conference, 3-5 May 1999, Bergen, Norway. Elsevier, Amsterdam, pp 211-232

Tsikalas F, Gudlaugsson ST, Faleide JI (1998) Collapse, infilling and post-impact deformation at the Mjølnir Impact Structure, Barents Sea. Geological Society of America Bulletin 110: 537-552

Vigran JO, Mangerud G, Mørk A, Bugge T, Weitschat W (1998) Biostratigraphy and sequence stratigraphy of the Lower and Middle Triassic deposits from the Svalis Dome, Central Barents Sea, Norway. Palynology 22: 89-141

Wierzbowski A, Århus N (1990) Ammonite and dinoflagellate cyst succession of an Upper Oxfordian-Kimmeridgian black shale core from the Nordkapp Basin, southern Barents Sea. Newsletter on Stratigraphy 22: 7-19

Wignall PB (1994) Black Shales. Clarendon Press, Oxford, 127 pp

Worsley D, Johansen R, Kristensen SE (1988) The Mesozoic and Cenozoic succession of Tromsøflaket. In: Dalland A, Worsley D, Ofstad K (eds) A Lithostratigraphic scheme for the Mesozoic and Cenozoic succession offshore mid- and northern Norway. Oljedirektoratet, Stavanger, pp 42-65

Århus N (1991) The transition from deposition of condensed carbonates to dark claystones in the Lower Cretaceous succession of the southwestern Barents Sea. Norsk Geologisk Tidsskrift 71: 259-263

Århus N, Kelly SRA, Collins JSH, Sandy MR (1990) Systematic palaeontology and biostratigraphy of two Early Cretaceous condensed sections from the Barents Sea. Polar Research 8: 165-194

Near-field Erosional Features at the Mjølnir Impact Crater: the Role of Marine Sedimentary Target

Filippos Tsikalas and Jan Inge Faleide

Department of Geology, University of Oslo, P.O. Box 1047 Blindern, N-0316 Oslo, Norway. (filippos.tsikalas@geologi.uio.no)

Abstract. Detailed seismic interpretation and structural mapping reveal evidence of near-field erosional features and their outcomes resulting from the Mjølnir impact at ~142 Ma. In particular, in the vicinity of the 40-km-diameter crater three erosive/depositional gullies are identified cutting through the rim wall faults. The gullies have meandering patterns and reach 70 m depth, 5 km width, and 25 km length. Mjølnir lacks a prominent, raised rim wall and exhibits, instead, a moderate crater rim locally tilted by 7-10° towards the crater. The Mjølnir impact into a shallow-water sedimentary basin in the Barents Sea generated a water cavity, which often collapsed, caused resurge water and material to flow back into the crater site. The back-rushing flow gave rise to gullies that are shallower and fewer in number, compared to other craters, swept out the rim wall without leaving considerable erosional scours and brought an extensive, ~50 km^3, volume of surrounding sediments mixed with excavated/ejected target material back to the crater. The highly unconsolidated and largely water-saturated sediments at the shallowest target levels had very low strength and great healing capacity resulting in preservation of modulated near-field erosional features. Extensive post-impact burial of the area may have further reduced the impact-related relief at the crater vicinity.

1 Introduction

During the past years extensive studies of the 40-km-diameter Mjølnir structure in the Barents Sea (Fig. 1) conclusive geological and geophysical evidence of its impact crater origin revealed (Tsikalas et al. 1999). In particular, it has been documented that the structure was formed by a meteoritic bolide impact into a shallow-water sedimentary basin, giving rise to extensively brecciated strata and a near-field ejecta layer with impact signatures of shocked quartz grains and iridium enrichments (Dypvik et al. 1996; Tsikalas et al. 1998a; Dypvik and Attrep 1999).

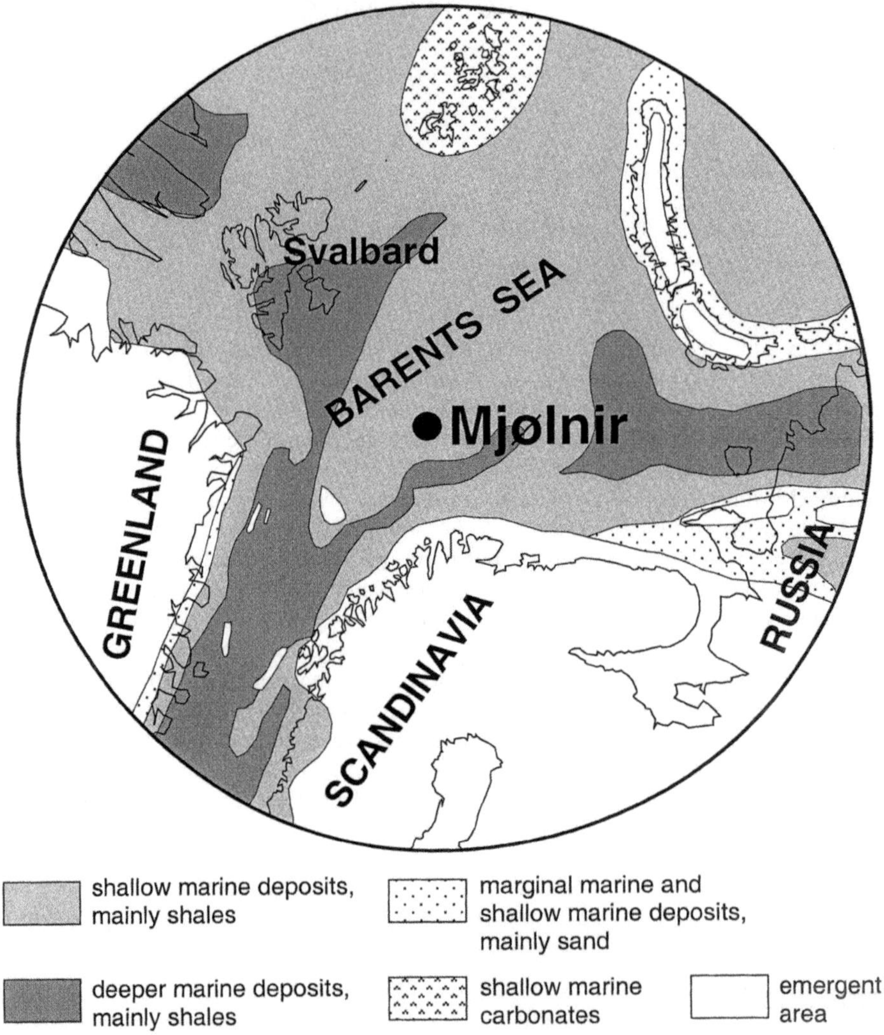

Fig. 1. Mjølnir impact location shown at a ~142 Ma plate reconstruction by Lawver et al. (1999) overlain on a simplified paleogeographic synthesis by Brekke et al. (2001) approximately at the time of impact. Emergent area means dry land.

The Mjølnir crater is excellently imaged in seismic reflection profiles as a large volume of extensive seismic disturbance bounded on top by a prominent structural relief (Figs. 2 and 3): a 8-km-wide central high surrounded by a 4-km-wide annular depression, and a 12-km-wide complex outer zone that includes a small peak ring and a marginal fault zone terminating at a ~150-m-high near-circular rim wall (Tsikalas et al. 1998a). Two shallow boreholes, one near the center and another ~30 km from the crater periphery (Fig. 3), have provided a detailed seismic stratigraphy for the impact-related and post-impact strata (Tsikalas et al. 2002a). In particular, the dating of the impact event has been placed between a prominent lower Barremian limestone bed (UB, upper boundary) and a top Callovian unconformity (LB, lower boundary) (Fig. 3). The impact horizon (TD, top disturbance) within the LB-UB unit corresponds to the top of the impact-induced seismic disturbance within the crater and to the proximal ejecta layer outside the periphery. Reflector TD is overlain at the central crater by a unit, TD-R3 (Fig. 3), which contains current-reworked, fine grained, fall-out debris related to the final stages of the impact process, whereas the biostratigraphic Volgian-Ryazanian age (142±2.6 Ma) for the impact (Smelror et al. 2001) was determined just at the bottom of the R3-R2 unit (Fig. 3).

A greatly expanded geophysical and geological database presently exists and classifies the well-preserved Mjølnir crater as a natural laboratory for marine impact studies. Based on updated detailed structural mapping in this study we search for impact-related near-field erosional features, we re-assess and estimate key volumetric parameters associated with these, and we investigate their relation to the marine sedimentary target.

2 Near-field Erosional Features and Outcomes

Recent studies of marine target impacts and systematic compilation of their structural and morphological features have shown that there are several marked differences among impact craters formed on land and those at sea (e.g., Gersonde and Deutsch 2000; Ormö and Lindström 2000). The primary cause for several of these features is that at a marine impact the growing crater rim and ejecta curtain push the water outside and form a water cavity. Collapse of this cavity starts at its base and causes a flow direction towards the crater (Shuvalov 2002). When the water depth is sufficient enough to overflow and cut through any uplifted rim, then characteristic erosional/depositional resurge gullies are formed acting as inlets of water and material flow back to the crater site. This flow greatly affects the crater rim and leads to extensive infilling.

The seismic investigation of Mjølnir's impact-related and post-impact structure and stratigraphy is based on a seismic database that consists of three types of reflection profiles, namely, 872-km-long high-resolution single-channel, 174-km-long shallow multi-channel, and 1081-km-long conventional multi-channel profiles (Tsikalas et al. 1998a, b). In this study, we probe into the same dataset focusing on structural and morphological features associated with processes operating at the threshold between the last impact-related and the immediately-after-impact stages. The single-channel profiles retain frequencies from 70 to 500

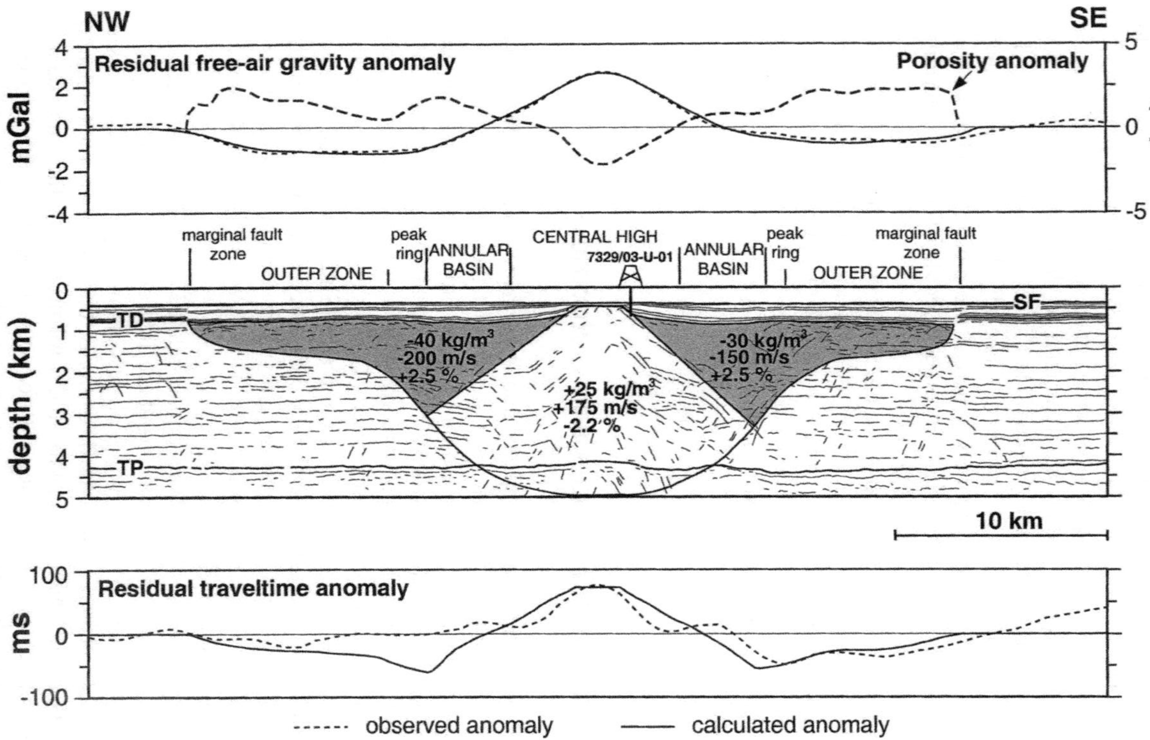

Fig. 2. Impact crater model and observed and calculated free-air gravity and seismic traveltime anomalies, and calculated porosity anomaly. The model geometry is superimposed on the seismic interpretation corrected for regional tilt. Density contrasts are given in kg/m^3, seismic velocity anomalies in m/s, and porosity anomalies in vol. %. SF, sea floor; TD, impact horizon; TP, Top Permian (modified extensively from Tsikalas et al. 1998c and Tsikalas et al. 2002b).

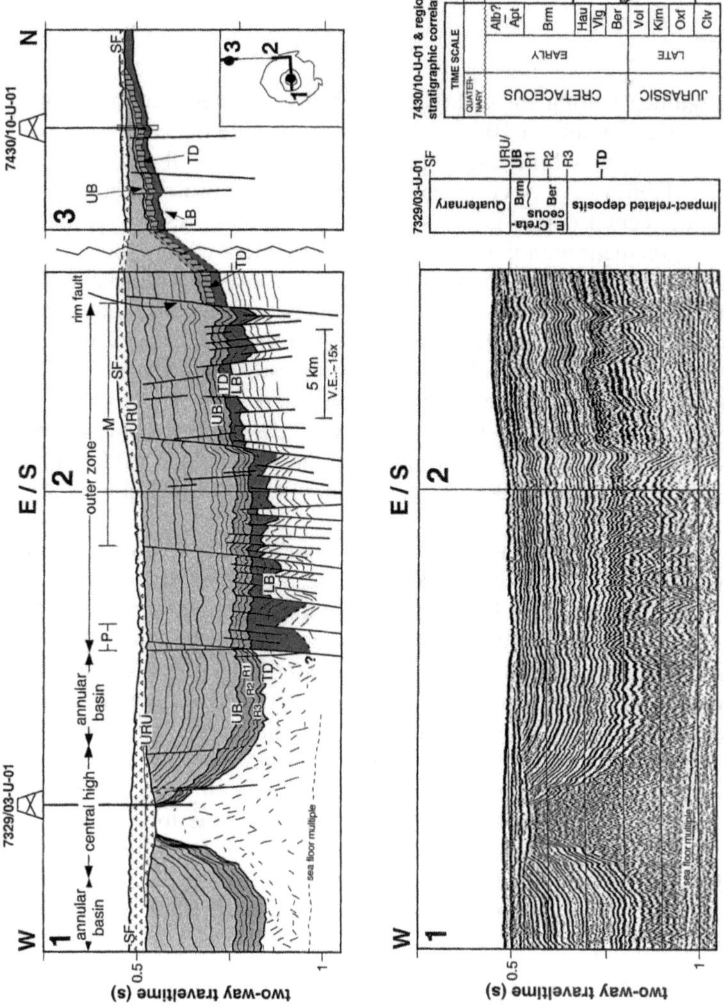

Fig. 3. Interpreted (top) and uninterpreted (bottom) examples of shallow multi-channel profiles crossing the structure and providing the seismic correlation between the two shallow boreholes. Stratigraphic relationships between the main seismic units at Mjølnir and the boreholes are also shown. Reflectors UB and LB bound the time of impact. SF, sea floor; URU; late Cenozoic upper regional unconformity; UB, lower Barremian; TD (impact horizon), the first continuous reflector above the seismic disturbance; LB, base Upper Jurassic. Vertical-bars raster denotes the uncertainty in the seismic tie of the impact horizon. Qrz/Ir, position of shocked quartz grains and iridium peak, respectively. P, peak ring; M, marginal fault zone.

Hz with a vertical resolution of 4-8 m, while the shallow multi-channel profiles have a frequency range of 8-180 Hz with a vertical resolution of 5-10 m. The deep-penetrating multi-channel profiles exhibit a vertical resolution of 20-50 m and have been used together with shallow refracted arrivals recorded from sonobuoys and shallow multi-channel profile stacking velocities to establish a velocity function for the Mjølnir area (Tsikalas et al. 1998a, b). We use this velocity function to convert traveltimes to depths and layer thicknesses.

2.1 Resurge Gullies

Three prominent gullies cutting through the rim faults are identified at Mjølnir (Fig. 4). They are located within a 10 to 15-km-radius outside the crater rim wall and exhibit meandering and bifurcating patterns. In addition, the location of gullies at Mjølnir exhibit a general concentric form, i.e., their prolongation points out towards the crater center (Fig. 4). The spatial distribution of resurge gullies at Mjølnir is in good agreement with that of the few other craters with gullies (4-km-diameter Kärdla crater, Puura and Suuroja 1992; 20-km-diameter Kamensk crater, Movshovich and Milyavsky 1990; and 14-km-diameter Lockne crater, von Dalwigk and Ormö 2001) extending from less than half to a maximum one crater diameter outside the crater rim. The gully on the northwest side of Mjølnir (Fig. 4) reaches 60-70 m in depth just outside the crater rim where it is ~5 km wide, and has a length of ~25 km at its deepest parts. The one on the southeast side reaches 30-40 m depth and ~5 km width in the vicinity of the rim, being ~10 km long. Finally, the one in the south is 30-40 m deep, ~1 km wide and 5-10 km long (Fig. 4).

Figure 5 shows a seismic profile across part of the gully at the southeast side of the crater periphery. The gully shown here has a concave form at its top at the level of impact horizon (reflector TD) where several small to minor erosional undulations locally cut through reflections of the underlying ejecta layer (Fig. 5). The undulations may indicate final reworking of material by the resurge water flow. The sediments within the gully exhibit dispersed and scattered seismic reflectivity patterns (Fig. 5). They are confined within undisturbed sub-horizontal reflections of pre-impact platform sediments below, and immediately-after-impact layers above that imprint the impact-generated relief. Although there is no direct geological information, the gullies at Mjølnir are expected, due to the observed dispersed seismic reflectivity character and similarities to other craters with gullies, to be filled with resurge sediments from the surrounding platform mixed with near-field excavated/ejected target material and the denser/heavier portions of the fall-back ejecta.

2.2 Extensive Infilling

Utilizing the well-established empirical relationships of Schmidt and Holsapple (1982) and Melosh (1989) supplemented by seismic observations, it was estimated that the volume of the material displaced from the crater, which is equivalent to

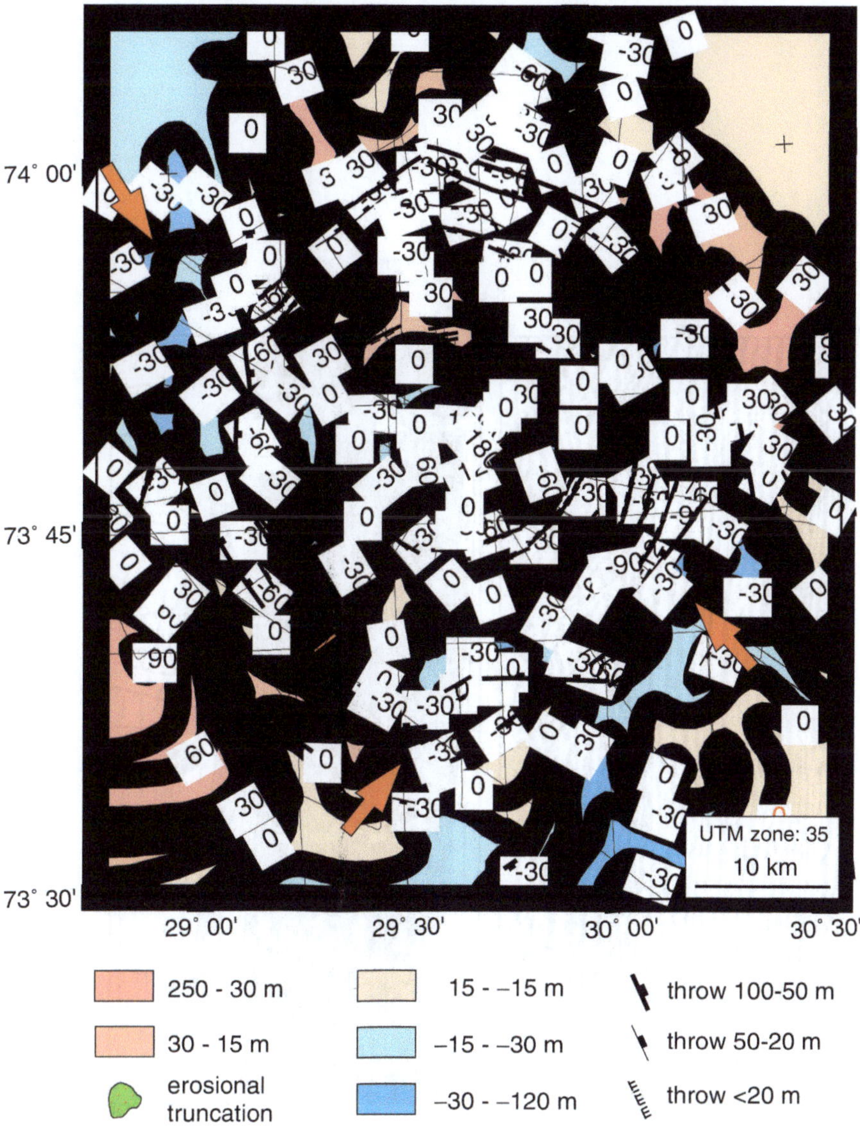

Fig. 4. Morphology and structure at the level of reflector UB, expressed as depth residuals with reference to a regional reflector surface dipping to the south. The regionally prominent reflector UB has been used as a marker horizon in the entire seismic reflection dataset because it is located just above and excellently images the top of seismic disturbance (TD, impact horizon), and in contrast to reflector TD can be confidently traced throughout the dataset. Therefore, the depth residuals image the impact-generated relief at the sea bottom after the end of the impact-related processes. Arrows point at the resurge gullies. Map is based on the entire seismic reflection database available (thin solid lines). Velocity used for depth conversion at the level of reflector UB is 3.0 km/s (range 2.7-3.4 km/s, Tsikalas et al 1998b). Contour interval 15 m.

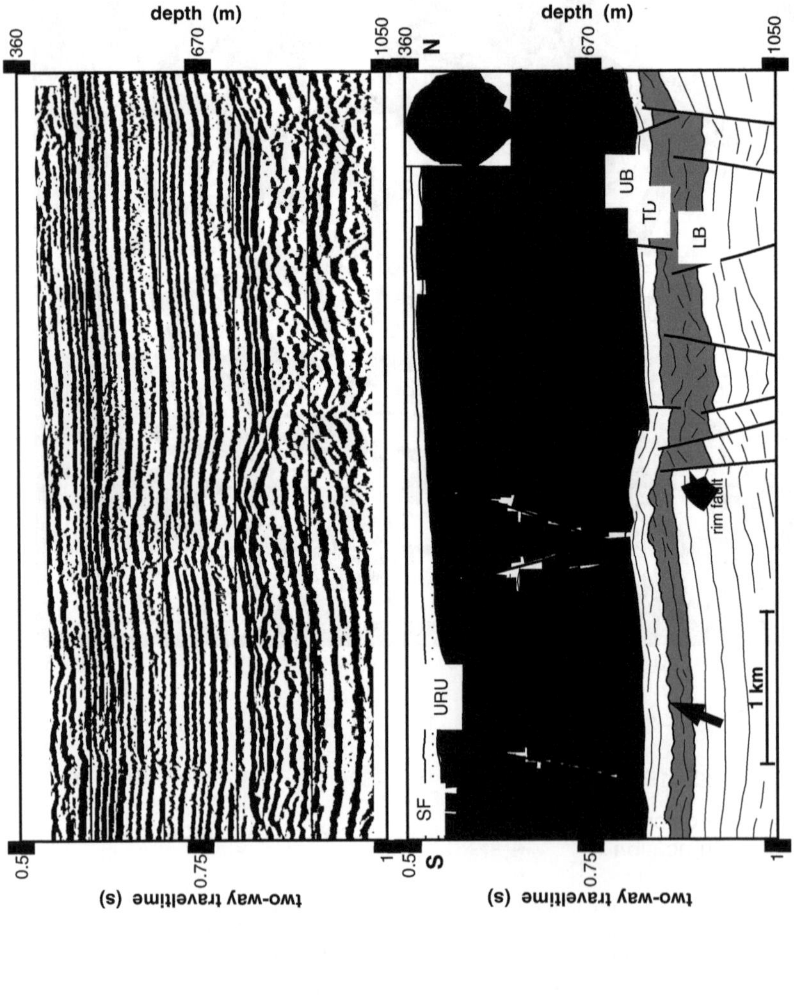

Fig. 5. Shallow multi-channel seismic profile, and interpretation, across the crater rim. Arrow points at part of the gully on the southeast side of Mjølnir (cfr. Fig. 4 and its caption).

the excavated crater volume, is ~180 km^3 (Tsikalas et al. 1998b). Recently, numerical simulations increased the displaced volume to ~230 km^3 and showed that the thickness of the primary ejecta layer is expected to decrease rapidly with distance from the crater center; more than 60% of the ejecta volume will be deposited within the crater's final diameter (Shuvalov et al. 2002). Moreover, the first effort to compare the theoretically predicted and the reconstructed crater reliefs across Mjølnir at the time of impact was conducted by Tsikalas et al. (1998a). In the light of the numerical simulation results (Shuvalov et al. 2002), we now provide updated and better constrained estimates for the predicted Mjølnir crater depth. Here, crater depth refers to the apparent crater depth, i.e., the depth relative to the pre-impact surface to reflector TD, and not to the true crater depth corresponding to its base. We also estimate the volume of excess infilling experienced by Mjølnir as a result of the impact-generated water-cavity collapse and formation of resurge gullies that channellized material flow back to the crater site.

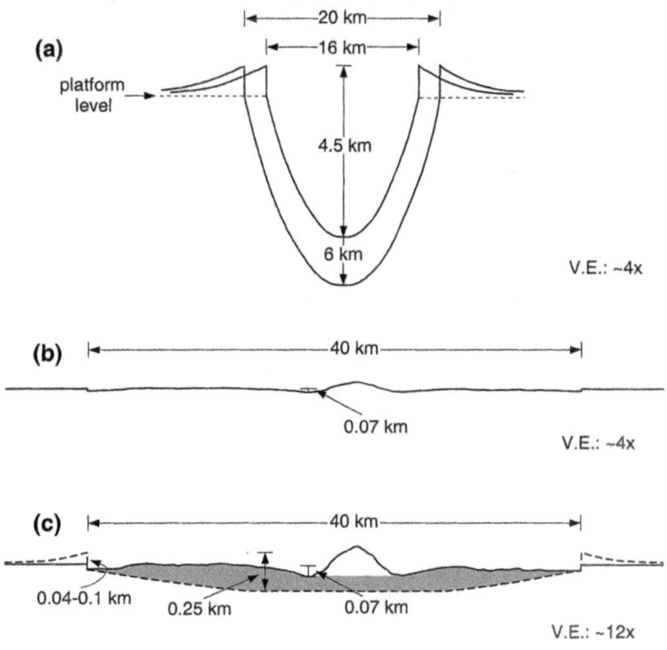

Fig. 6. Schematic diagram of crater collapse and infilling at the Mjølnir crater. (a) Estimated range of transient cavity dimensions. (b) Reconstructed relief from Tsikalas et al. (1988a). (c) Theoretically predicted relief utilizing the volume balance method of Croft (1979, 1985) and Melosh (1989) for a collapse factor of 2-2.5 (dashed line). Reconstructed relief from (b) (solid line) is shown for comparison. We neglected the volume of the central high because it is considerably less than the final crater volume, and thus does not affect the calculations. Shading denotes the additional infilling affecting Mjølnir (modified extensively from Tsikalas et al. 1998a). v.e. = vertical exaggeration.

We estimate the theoretically predicted relief across the Mjølnir crater (Fig. 6) using the volume balance method of Croft (1979, 1985) and Melosh (1989). The method is based on a geometrical model that equates the volume of the parabolic-shaped transient cavity to the volume of the flat-floored final crater, assuming mass conservation under collapse of the material surrounding the transient crater. On the basis of numerical simulations integrated with seismic observations a transient cavity of 16-20 km in diameter was determined, translating to a collapse factor of 2-2.5 for the 40-km-final-diameter Mjølnir crater (Gudlaugsson 1993; Tsikalas et al. 1998b; Shuvalov et al. 2002). Using these values, an average empirical 0.28 ratio between the transient crater depth and diameter (Melosh 1989) implies depths for the transient cavity in the range of 4.5-6 km (Fig. 6). On the basis of this range of transient crater dimensions and assuming balanced volumes, we calculate a theoretically expected crater depth for Mjølnir of 250±100 (Fig. 6). This value differs considerably from the average depth of ~30-40 m and a maximum depth of ~70 m in the annular basin obtained through reconstruction of the original crater relief (Tsikalas et al. 1998a). We also calculate the volume of the excess resurge infilling affecting Mjølnir, i.e., the volume difference between the predicted and the reconstructed crater surfaces to be ~50 km^3 which approximates one-third to one-fifth of the total excavated/ ejected volume (Fig. 6).

2.3 Crater Rim

The presence of a raised crater rim is inferred from the study of terrestrial and planetary impact craters (Melosh, 1989; Spudis, 1993; Grieve and Pesonen, 1996). The volume-balance model indicates that a crater with the structural/ morphological parameters and extensive collapse as Mjølnir should have a rim height of 40-100 m (Fig. 6). However, Mjølnir not only lacks a raised crater rim but it locally exhibits a small, but distinct, inward bending of strata on the hanging wall crest of the prominent rim faults (Figs. 5 and 7). The tilting is in the order of 7-10°, 15-20 m relative to the surrounding platform level and it is present both at the ejecta layer (LB-TD) and the immediately-after-impact deposited layer TD-UB. Although the Mjølnir crater experienced structural reactivation and differential subsidence as a result of extensive post-impact burial (Tsikalas et al. 1998a, 2002b), Figures 5 and 7 clearly show that post-impact reactivation of the rim fault is not related to the observed rim tilting. This is because the pre- and post-impact strata outside the crater appear horizontal, and because the post-impact structural reactivation and sediment thickness variations took place solely within the crater boundaries. Thus, the observed tilting of the ejecta layer is primarily an impact-related feature. Figure 7 also shows several small-scale undulations of possible erosional origin on top of the ejecta layer outside the crater, indicating final reworking of material by the resurge water flow.

Fig. 7. (opposite) Shallow multi-channel seismic profile, and interpretation, across the crater rim.

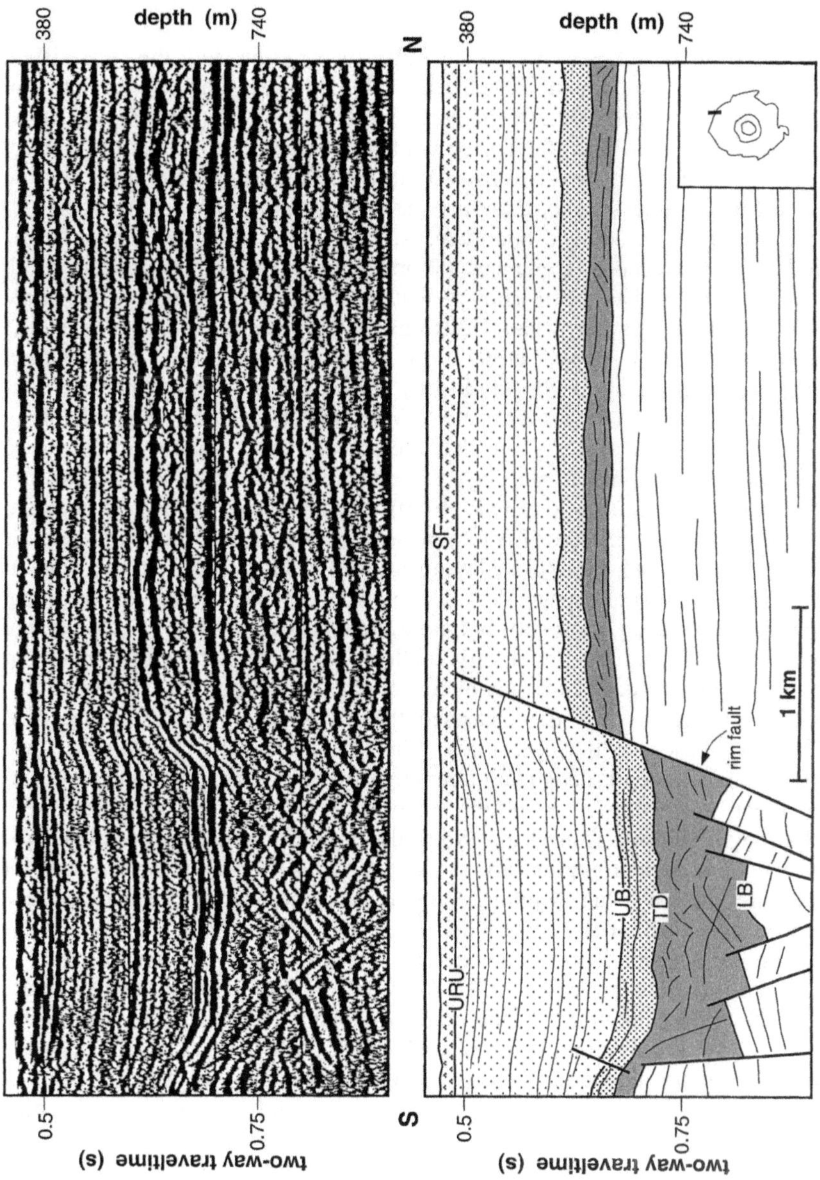

depth (m)
380
740
0.5
0.75
two-way traveltime (s)
N
S
SF
URU
UB
TD
LB
rim fault
1 km

3 Impact into a Shallow Marine Sedimentary Basin

The two shallow boreholes in the vicinity of Mjølnir and the established lithostratigraphy of the area reveal that the impact took place into an epicontinental basin with 300-500 m paleo-water depth (Fig. 1) (Dypvik et al. 1996). At that time, ~142 Ma, the region contained upper Paleozoic strata, mainly carbonates and evaporites, overlain by 4-5 km thick Mesozoic siliciclastic marine sediments (Fig. 2); Triassic cyclic marine shales and fluviodeltaic sandstones, and Lower to Middle Jurassic sand-dominated strata passing above to shaley Upper Jurassic-Lower Cretaceous sequences (Worsley et al. 1988; Gabrielsen et al. 1990). Moreover, it has been shown that at the time of impact the Late Kimmeridgian to Early Berriasian Hekkingen Formation was deposited at the shallowest target levels (Dypvik et al. 1996; Smelror et al. 2001). The formation consists of brownish-grey to very dark grey shales and claystones that are fine-grained, organic-rich and well-laminated (Worsley et al. 1988; Dypvik et al. 1991). Mineralogical analyses show that the shales consist mainly of illite and kaolinite, exhibiting some mixed layered clay minerals of smectite/illite and probably some vermiculite (Dypvik 1980). Furthermore, from the regional correlation we interpret reflector LB to represent the top of the sandy sequences of the Middle Jurassic Stø Formation laying at ~100-150 m depth at the time of impact. This value represents the decompacted depth based on regional stratigraphic correlation close to borehole 7430/10-U-01 (Fig. 3). Both the Hekking and Stø formations are at very shallow depths at the time of impact and are thus expected to have been highly unconsolidated and largely water-saturated.

Numerical modelling has demonstrated that in order to match the observed crater structure based on the lithostratigraphy of the area, the target rocks must exhibit a composite depth-dependent strength structure; very low strength values for the upper 3 km depth resembling unconsolidated and soft sediments, and gradual strength increase from 3 to 6.5 km depth before obtaining basement-like strength values at greater depths (Shuvalov et al. 2002). A major consequence of this strength structure is that Mjølnir, compared to other land and marine craters, experienced an exceptionally strong slumping and gravitational collapse of the transient crater. Mjølnir is now considered in terms of gravitational collapse to represent an intermediate-case crater between simple and complex, where intensive slumping counteracts to the crater floor uplift and leads to suppression and burial of the central high (Tsikalas et al. 1998b; Shuvalov et al. 2002). In the context of crater forming processes that last only few minutes, the initiation of collapse preceeds the infilling, however at some point the processes become coeval. We infer that resurge gullies at Mjølnir may have also formed close to the initial peak-ring, which approximates the transient crater diameter. However, Mjølnir collapsed 2-2.5 times its transient crater diameter and, thus, such gullies were destroyed. The remaining and most important ones are those operating at the end-threshold of the impact processes and cutting through the final crater rim, similar to the other craters with gullies. Through these gullies a great portion of material was brought back to the crater site and together with the extensive collapse led to an exceptionally shallow crater relatively to its size.

The gullies at Mjølnir are very narrow close to the rim and widen farther outward (Fig. 4). This form is diagnostic for resurge gullies and results from the turbulent and highly erosive resurge flow proceeding as headward erosion disintegrated with distance (Ormö and Lindström 2000). It has been proposed that as the flow approximates the initially prominent crater rim at Mjølnir hydraulic excavation becomes dominant and sweeps out the water-saturated rim wall (Tsikalas et al. 1998a). This may result in local tilting of the rim fault crest (Figs. 5 and 7). A plausible scenario to additionally weaken the surrounding platform strata at the vicinity of rim wall and thus facilitate hydraulic excavation describes that fragmented material was ejected within sedimentary strata at weaker layers opened due to shock-wave decompression (Sturkell and Ormö 1997). Similar processes may have operated at Mjølnir and may explain the lack of a raised crater rim at Chesapeake Bay Crater (Poag et al. 1994; Poag 1996) and the bevelled crater rim form at Montagnais Crater (Jansa et al. 1989; Jansa 1993).

Detailed seismic interpretation and structural mapping have contributed considerably to bring out evidence of the near-field erosional features and their outcomes resulting from the Mjølnir impact (Figs. 4-7). Nonetheless, the observed gullies for a crater of Mjølnir's size are shallower and fewer in number compared to other craters with gullies. In addition, the crater rim is inclined but does not exhibit prominent erosional undulations or deep scours. It seems that all erosional features observed are subtle and superficial in their nature. We attribute this to the relatively unconsolidated and soft marine sedimentary target composed of shales/claystones at the top one-to-two hundred meters. In this context, laboratory experiments have shown that meteorite impacts in unconsolidated, water-rich sedimentary targets may result in a more modulated and greatly dampened crater topography (Gault and Sonnett 1982; McKinnon 1982). This applies particularly to the shallowest target levels at Mjølnir, where the sediments had very low strength, implying low cohesion and a low friction coefficient. Unconsolidated clays with a high percentage of illite and kaolinite behave, thus, as a material with tremendous healing capacity. Such material largely resembles incoherent clay slurry and is therefore not capable of retaining deep erosional features; the scours within the gullies and at the rim wall created by the erosive force of back-rushing flow may have healed very quickly. Furthermore, when we decompact the sediments within the gullies, using the well-established porosity-depth functions of Tsikalas (1992) and Tsikalas et al. (2002b), their maximum relief immediately-after-impact reaches ~50-100 m. This means that the top of the Stø Formation sandy sequences, overlain by the shaley Hekkingen Formation, at ~100-150 m depth may have been only slightly penetrated by the resurge erosion outside the final crater diameter. Even in this case, however, the sandy sequences at such shallow depths are not consolidated enough to sustain deep and permanent erosional features.

Figure 4 shows that by the end of the impact processes the places within the crater boundaries that consumed most of the resurge infill and still preserve a remaining relief are the annular basin and the footwall of the rim faults. Together with the remaining relief at resurge gullies outside the crater, they provide the depression space to be filled at first by subsequently deposited post-impact sediments. The crater and its vicinity were progressively buried by a ~2-2.5 km thick overburden of Cretaceous and Tertiary sediments that were later significantly eroded due to the Cenozoic uplift and erosion (Vågnes et al. 1992; Richardsen et al. 1993; Faleide et al. 1996). It has been shown that the extensive post-impact burial triggered differential compaction of a substratum overlying a 800-1400 km^3 volume of impact-induced disturbance that exhibits lateral changes in physical properties (Fig. 2) (Tsikalas et al. 1998c). In particular, the more porous and less denser brecciated crater periphery compacted more relative to the central crater core and the surrounding platform sediments (Tsikalas et al. 2002b). A similar compaction trend is expected to have taken place within the gullies at the initial stages of post-impact burial as the brecciated sediments there compacted more than the surrounding platform sediments, amplifying probably the original gully relief. However, the amount of brecciated sediments within the gullies is not great enough, as within Mjølnir, to sustain any impact-induced porosity anomaly and thus compact differentially. The extensive post-impact overburden was capable of leveling out such anomaly and original relief, and thus the sediments within the gullies compacted, at the most, similar as the platform sediments. Therefore, post-impact compaction may lead to additional suppression of the initially subtle erosional features at the crater vicinity.

4 Conclusions

Impact of the Mjølnir projectile at ~142 Ma into a 300-500 m paleo-water depth sedimentary basin in the Barents Sea initiated crater forming processes, eventually leading to the 40-km-diameter Mjølnir crater, and gave rise to a water cavity. Collapse of the water cavity caused a resurge water and material flow back to the crater site. The flow resulted in erosive/depositional gullies in the crater vicinity, greatly affected the crater rim, and led to extensive infilling.

Three prominent gullies are identified at Mjølnir located within a 10 to 15-km radius outside the crater rim wall. They vary from 30-70 m, 1-5 km, and 5-25 km in depth, width, and length, respectively. They exhibit meandering and bifurcating patterns and cut the rim faults where they have channelized resurge sediments from the surrounding platform mixed with near-field excavated/ejected target material and the denser/heavier portions of the fall-back ejecta. Mjølnir lacks a raised, prominent crater rim as inferred from statistics of known terrestrial impact craters, and locally exhibits a small, 7-10° tilting. Moreover, we calculate through volume balance analysis a theoretically expected crater depth for Mjølnir on the order of 250±100 m, which differs considerably from the reconstructed average and maximum depths of 30-40 m and 70 m, respectively. The same analysis provides also the excess resurge infilling volume at Mjølnir to be on the order of 50 km^3.

Due to the low strength siliciclastic sedimentary target and the presence of water, the final crater was significantly enlarged due to greater gravitational collapse, and at the same time its morphological expression was considerably reduced due to increased resurge infilling. Moreover, during the erosive resurge flow the unconsolidated shaley sediments at the shallowest target levels resembled slurry material with a tremendous healing capacity. Therefore, only few and subtle near-field erosional features were retained at the end of the impact-related processes that may have been later additionally subdued due to the considerable post-impact burial of the crater area. The study reveals the importance and the dynamic role of water and target lithology and strength to the resulting structure and morphology of a marine impact crater and its surroundings.

Acknowledgments

We gratefully acknowledge the Norwegian Defense Research Establishment, the Norwegian Petroleum Directorate, and SINTEF Petroleum Research for providing the data this study is based on. The study is part of the "Mjølnir Project" financed by the Norwegian Research Council. Statoil is also acknowledged for providing financial support. We thank J. Ormö and an anonymous reviewer for their helpful comments and suggestions.

References

Brekke H, Sjulstad HI, Magnus C, Williams RW (2001) Sedimentary environments offshore Norway - an overview. In: Martinsen OJ, Dreyer T (eds) Sedimentary Environments Offshore Norway - Paleozoic to Recent. Norwegian Petroleum Society Special Publication 10: 7-37

Croft SK (1979) Impact craters from centimeters to megameters. Ph.D. dissertation, Los Angeles, University of California, 264 pp

Croft SK (1985) The scaling of complex craters. Proceedings 15th Lunar and Planetary Science Conference, part 2: Journal of Geophysical Research 90 (supplement): C828-C842

Dypvik H (1980) The sedimentology of the Janusfjellet Formation, Central Spitsbergen (Sassenfjorden and Agatdhfjellet areas). Norsk Polarinstitutt Skriftserie 172: 97-134

Dypvik H, Attrep M Jr (1999) Geochemical signals of the late Jurassic, marine Mjølnir impact. Meteoritics and Planetary Science 34: 393-406

Dypvik H, Nagy J, Eikeland TA, Backer-Owe K, Johansen H (1991) Depositional conditions of the Bathonian to Hauterivian Janusfjellet Subgroup, Spitsbergen. Sedimentary Geology 72: 55-78

Dypvik H, Gudlaugsson ST, Tsikalas F, Attrep M Jr, Ferrell RE Jr, Krinsley DH, Mørk A, Faleide JI, Nagy J (1996) Mjølnir structure: An impact crater in the Barents Sea. Geology 24: 779-782

Faleide JI, Solheim A, Fiedler A, Hjelstuen BO, Andersen ES, Vanneste K (1996) Late Cenozoic evolution of the western Barents Sea-Svalbard continental margin. Global and Planetary Change 12: 53-74

Gabrielsen RH, Færseth RB, Jensen LN, Kalheim JE, Riis F (1990) Structural elements of the Norwegian continental shelf. Part I: The Barents Sea region. Norwegian Petroleum Directorate Bulletin No. 6, 33 pp

Gault DE, Sonett CP (1982) Laboratory simulation of pelagic asteroidal impact: atmospheric injection, benthic topography, and surface wave radiation field. In: Silver LT, Schultz PH (eds) Geological Implications of Impacts of Large Asteroids and Comets on Earth. Geological Society of America Special Paper 190: 69-92

Gersonde R, Deutsch A (2000) New field of impact research looks to the oceans. [abs.] EOS, Transactions, Americal Geophysical Union 81, 20: 221-223

Grieve RAF, Pesonen LJ (1996) Terrestrial impact craters: their spatial and temporal distribution and impacting bodies. Earth, Moon and Planets 72: 357-376

Gudlaugsson ST (1993) Large impact crater in the Barents Sea. Geology 21: 291-294

Jansa LF (1993) Cometary impacts into ocean: their recognition and the threshold constraint for biological extinctions. Palaeogeography, Palaeoclimatology, Palaeoecology 104: 271-286

Jansa LF, Pe-Piper G, Robertson BP, Friedenreich O (1989) Montagnais: a submarine impact structure on the Scotian shelf, eastern Canada. Geological Society of America Bulletin 101: 450-463

Lawver LA, Gahagan LM, Campbell DA, Brozena JM, Childers V (1999) Mid-Jurassic to Recent tectonic evolution of the Arctic region (*Powerpoint* animation, using the *PLATES* animation software) [abs.] In: Lawver LA, Brozena JM, Kovacs LC, Childers V, Compilations in the Canada Basin, Aerogeophysical Anomalies. Eos, Transactions, American Geophysical Union, Fall Meeting 1999, San Francisco, 80 (46): 1000

McKinnon WB (1982) Impact into the Earth´s ocean floor: preliminary experiments, a planetary model, and possibilities for detection. In: Silver LT, Schultz PH (eds) Geological Implications of Impacts of Large Asteroids and Comets on Earth. Geological Society of America Special Paper 190: 129-142

Melosh HJ (1989) Impact cratering - A geologic process. Oxford University Press, New York, 245 pp

Movshovich EV, Milyavsky AE (1990) Morphology and inner structure of Kamensk and Gusev astroblemes. In: Masaitis VL (ed) Impact Craters of the Mesozoic-Cenozoic Boundary. Leningrad, Nauka: pp 110-146 (in Russian)

Ormö J, Lindström M (2000) When a cosmic impact strikes the sea bed. Geological Magazine 137: 67-80

Poag CW (1996) Structural outer rim of Chesapeake Bay impact crater: seismic and borehole evidence. Meteoritics and Planetary Science 31: 218-226

Poag CW, Powars DS, Poppe LJ, Mixon RB (1994) Meteoroid mayhem in Ole Virginny: source of the North American tektite strewn field. Geology 22: 691-694

Puura V, Suuroja K (1992) Ordovician impact crater at Kärdla, Hiiumaa Island, Estonia. Tectonophysics 216: 143-156

Richardsen G, Vorren TO, Tørudbakken BO (1993) Post-Early Cretaceous uplift and erosion in the southern Barents Sea: a discussion based on analysis of seismic interval velocities. Norsk Geologisk Tidsskrift 73: 3-20

Schmidt RM, Holsapple KA (1982) Estimates of crater size for large-body impact: gravity-scaling results. In: Silver LT, Schultz PH (eds) Geological Implications of Impacts of Large Asteroids and Comets on Earth. Geological Society of America Special Paper 190: 93-102

Shuvalov V (2002) Numerical modeling of the impacts into shallow sea. In: Plado J, Pesonen L (eds) Impacts in Precambrian Shields, Impact Studies, vol. 2, Springer Verlag, Berlin-Heidelberg, pp 323-336

Shuvalov V, Dypvik H, Tsikalas F (2002) Numerical simulations of the Mjølnir marine impact crater. Journal of Geophysical Research (Planets) 107 (E7)

Smelror M, Kelly SRA, Dypvik H, Mørk A, Nagy J, Tsikalas F (2001) Mjølnir (Barents Sea) meteorite impact ejecta offers a Volgian-Ryazanian boundary marker. Newsletter on Stratigraphy 38: 129-140

Spudis PD (1993) The geology of multi-ring impact basins. Cambridge, United Kingdom, Cambridge University Press, 263 pp

Sturkell EFF, Ormö J (1997) Impact-related injections in the marine Ordovician Lockne impact structure, central Sweden. Sedimentology 44: 793-804

Tsikalas F (1992) A study of seimic velocity, density and porosity in Barents Sea wells (N Norway). Master's Thesis, University of Oslo, Norway, 169 pp

Tsikalas F, Gudlaugsson ST, Faleide JI (1998a) Collapse, infilling, and post-impact deformation at the Mjølnir impact structure, Barents Sea. Geological Society of America Bulletin 110: 537-552

Tsikalas F, Gudlaugsson ST, Faleide JI (1998b) The anatomy of a buried complex impact structure: the Mjølnir Structure, Barents Sea. Journal of Geophysical Research 103: 30469-30484

Tsikalas F, Gudlaugsson ST, Eldholm O, Faleide, JI (1998c) Integrated geophysical analysis supporting the impact origin of the Mjølnir Structure, Barents Sea. Tectonophysics 289: 257-280

Tsikalas F, Gudlaugsson ST, Faleide JI, Eldholm O (1999) Mjølnir Structure, Barents Sea: a marine impact crater laboratory. In: Dressler BO, Sharpton VL (eds) Large Meteorite Impacts and Planetary Evolution II. Geological Society of America Special Paper 339: 193-204

Tsikalas F, Faleide JI, Eldholm O, Dypvik, H (2002a) Seismic correlation of the Mjølnir marine impact crater to shallow boreholes. In: Plado J, Pesonen L (eds) Impacts in Precambrian Shields. Impact Studies, vol. 2, Springer Verlag, Berlin-Heidelberg, pp 307-321

Tsikalas F, Gudlaugsson ST, Faleide JI, Eldholm O (2002b) The Mjølnir marine impact crater porosity anomaly. Deep-Sea Research Part II 49: 1103-1120

von Dalwigk I, Ormö, J (2001) Formation of resurge gullies at impacts at sea: The Lockne crater, Sweden. Meteoritics and Planetary Science 36: 359-369

Vågnes E, Faleide JI, Gudlaugsson ST (1992) Glacial erosion and tectonic uplift in the Barents Sea. Norsk Geologisk Tidsskrift 72: 333-338

Worsley D, Johansen R, Kristensen SE (1988) The Mesozoic and Cenozoic succession of Tromsøflaket. In: Dalland A, Worsley D, Ofstad K (eds) A lithostratigraphic scheme for the Mesozoic and Cenozoic succession offshore mid- and northern Norway. Norwegian Petroleum Directorate Bulletin No. 4, pp 42-65

Global Effects of the Chicxulub Impact on Terrestrial Vegetation - Review of the Palynological Record from New Zealand Cretaceous/Tertiary Boundary

Vivi Vajda[1], J. Ian Raine[2], Christopher J. Hollis[2] and C. Percy Strong[2]

[1]GeoBiosphere Science Centre, Department of Geology, University of Lund, Sölvegatan 12, 223 62 Lund, Sweden. (vivi.vajda@geol.lu.se)
[2]Institute of Geological & Nuclear Sciences, P.O. Box 30368, Lower Hutt, New Zealand.

Abstract. Analysis of pollen and spore assemblages from both terrestrial and near-shore marine sediments in New Zealand reveal an instant and dramatic mass-kill of the land plants in close association with the Cretaceous-Tertiary boundary (KTB) event. The turnover in the palynoflora is followed by a recovery succession, the most prominent feature of which is an interval dominated by fern spores (fern spike) starting at the boundary in the three sections studied. The duration of the period of fern-dominance is here calculated to have lasted for ca. 8000-20,000 years, based on sedimentation rates calculated from foramineral data. This time span is orders of magnitude greater than seen in normal seral successions following deforestation. We suggest that, whereas low ambient light levels initially favoured communities dominated by pioneering free-sporing plants (ferns and bryophytes), other environmental variables relating to the suppression of seed stocks or seedling growth were probably responsible for the long interval required for re-establishment of gymnosperm- and angiosperm- dominated vegetation. In New Zealand, the devastated vegetation recovered slowly, many of the latest Cretaceous taxa reappearing higher in the succession following their absence in the lowermost Paleocene. One hundred and five miospore taxa were identified in the three sections. Species-level extinction rate at the KTB is below 15%, although more extensive biostratigraphic studies are needed to confirm this. Based on the miospore assemblages and leaf fossil physiognomic studies, the Early Paleocene climate was cooler than prior to the K/T event, perhaps reflecting long-term consequences of the impact.

1 Introduction

The Cretaceous-Tertiary transition records the history of one of the largest mass extinctions on Earth. Palynology has served as an important tool in pinpointing the Cretaceous-Tertiary boundary (KTB) in North American terrestrial sedimentary sections since the 1980s. It is well known that a severe disruption of the flora and some extinctions among flowering plants are characteristic features of the boundary layer in these sections. The effects of the Chixculub impact on the marine biota are now relatively well-resolved in the Northern and Southern Hemispheres through many and detailed studies of marine sections, results summarised in Norris (2001), Kiessling and Claeys (2002), Hollis (2003) and Hollis and Strong (2003). While there have been general studies of the effects on terrestrial plant communities in Antarctica and New Zealand (Couper 1960; Askin 1988; Askin and Jacobson 1996; Raine 1984; Johnson and Raine 1991; Ward 1997), it is only recently that detailed records of terrestrial palynomorphs from well dated KTB sections have been reported from New Zealand (Vajda et al. 2001, 2002, 2003; Raine and Vajda 2002; Vajda and Raine 2003). This paper presents an overview of floristic turnover and recovery as interpreted from the palynoflora from three New Zealand KTB sections, together with a tentative time frame for the vegetation recovery.

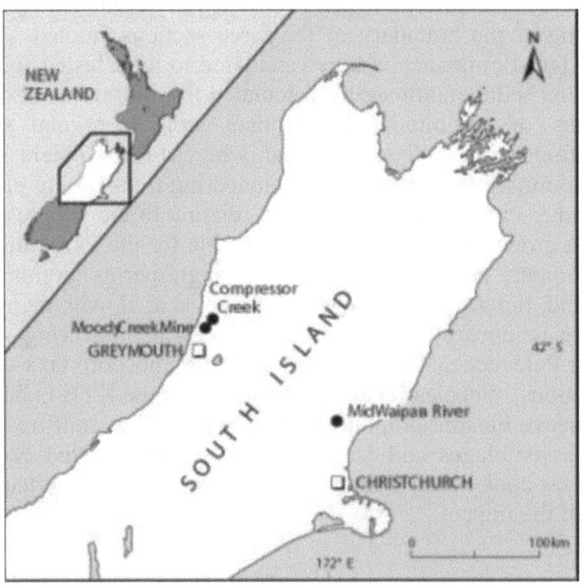

Fig. 1. Location of the Compressor Creek and Moody Creek Mine KTB sections in Greymouth Coalfield, and mid-Waipara River section in north Canterbury.

2
Geological Setting

The Cretaceous-Tertiary nonmarine exposures studied at Moody Creek Mine and Compressor Creek are situated within Greymouth Coalfield, north of Greymouth (Figs. 1 and 2). Here the Cretaceous-Paleocene transition lies within the upper part of the Rewanui Coal Measure Member of the Paparoa Coal Measures (Nathan 1978). The coal measures represent a subsiding floodplain paleoenvironment which was well-vegetated and incorporated peat-forming communities in permanently ponded settings (Bal and Lewis 1994; Ward 1997). However, mires were regularly inundated by silt and mud resulting in intercalation of coal seams and carbonaceous mudstone.

At *Moody Creek Mine* KTB sediments are exposed along Seven Mile Creek (171° 16' 40" E, 42° 23' 18" S, Fig. 2). The sedimentary units incorporate sandstone, carbonaceous mudstone, and coal seams. An iridium anomaly defines the boundary within a coal seam (Vajda et al. 2001).

Compressor Creek is an outcrop section in the upper valley of Seven Mile Creek (171° 18' 35" E, 42° 22' 31" S, Fig. 2), and consists of coal, carbonaceous mudstone and siltstones.

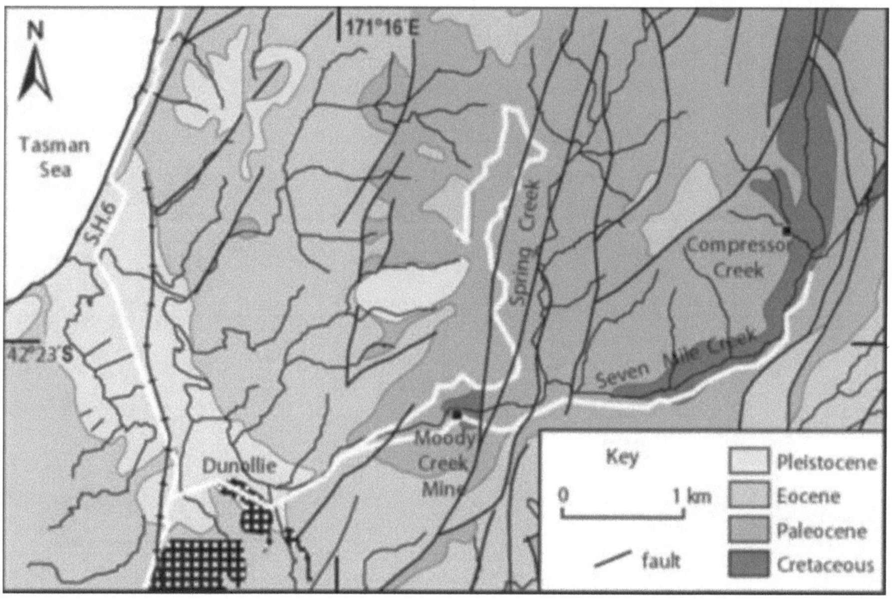

Fig. 2. Simplified geological map of the southern part of Greymouth Coalfield (after Nathan 1978), showing location of the Compressor Creek and Moody Creek Mine sections.

Geochemical results are as yet unavailable for the section, and the position of the KTB is entirely recognised on the basis of palynology (i.e., the fern spike and extinction of key pollen taxa).

At *mid-Waipara* the KTB exposure occurs within glauconitic sandstone of the upper Conway Formation (Browne and Field 1985). The section is located between Doctors Gorge and the Canterbury Plains (172° 34' 56" E, 43° 3' 44" S, Fig. 1), along the middle course of the Waipara River (Fig. 3). The KTB at mid-Waipara is well constrained by biostratigraphy, both from foraminifera and radiolaria (Strong 1984; Hollis and Strong 2003), and miospores and dinoflagellates (Couper 1960; Wilson 1984, 1987; Vajda and Raine 2003) and is further marked by a 5 cm thick, iron-stained zone. The sediments consist of glauconitic sandstones, calcareous below the boundary, but non-calcareous immediately above it (Browne and Field 1985). These were deposited in neritic inner-to-mid-shelf paleoenvironments. Geochemical analyses reveal anomalously elevated abundances of nickel, zinc and chromium, and an iridium anomaly of 0.49 ppb (Brooks et al. 1986) in the iron-stained layer marking the KTB.

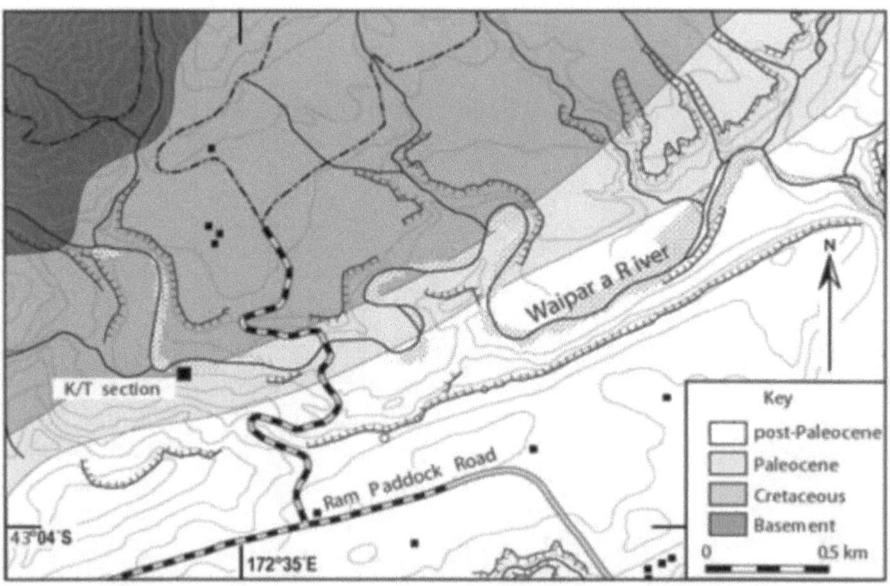

Fig. 3. Simplified geological map of mid-Waipara River (after Wilson 1963), showing location of sampled K/T boundary section.

3 Material and Methods

3.1 Identification of the Cretaceous-Tertiary Boundary (KTB)

The position of the KTB in Greymouth Coalfield sections was initially resolved by coarse palynological sampling, using the miospore zonation of Raine (1984) to provide biostratigraphic control. As there is not any clear visual boundary layer (Fig. 4), the precise position of the K-T boundary was then determined by fine-scale biostratigraphic and geochemical sampling and analyses. Iridium and other trace element assay were carried out by neutron activation analysis. The KTB position at mid-Waipara was already known from previous foraminiferal, palynological and geochemical studies.

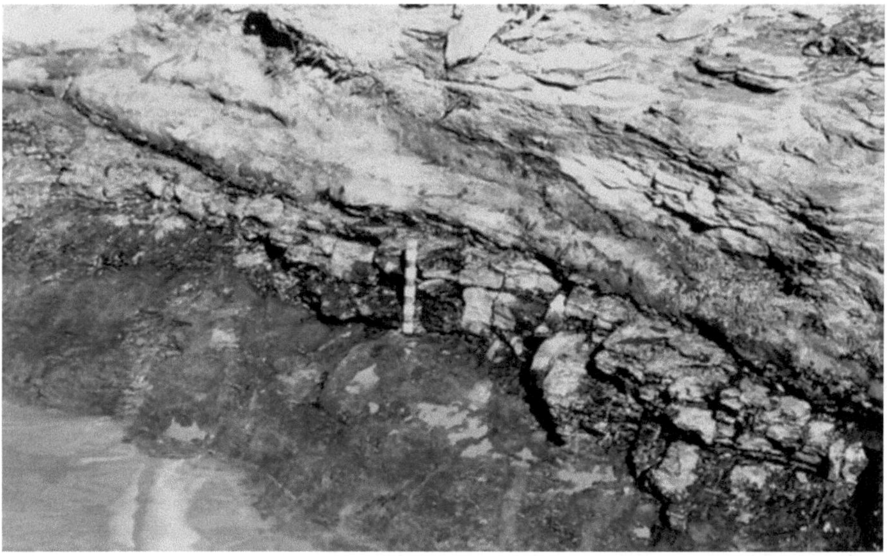

Fig. 4. The Cretaceous-Tertiary boundary at Moody Creek Mine, located within the 10 cm thick coal seam. (Photo by Vivi Vajda)

3.2 Palynological Processing

Standard palynological processing methods included treatment by hydrochloric acid, hydrofluoric acid, Schulze reagent, and an alkaline reagent (KOH) to disaggregate the organic material. The organic matter residue was sieved and retained on a 6 μm screen, and finally mounted on slides in glycerine jelly and sealed for examination under a transmitted light microscope. Relative abundance of spores and pollen is based on counts of >300 whole specimens/sample. The slides were further examined for rare taxa. Slides and macerated residues of the samples are deposited in the Paleontology Collection, Institute of Geological and Nuclear Sciences, Lower Hutt, New Zealand.

3.3 Samples

A total of 117 samples were studied palynologically from the three sections. At Moody Creek Mine, thirty-five palynological samples were selected from a 2.1 m vertical section through the uppermost Cretaceous and lowermost Paleocene (Fig. 5). At Compressor Creek, a vertical exposure of 7.8 m was sampled initially and later high resolution re-sampling targeted the 0.8 metres spanning the KTB: results from 3.5 m of section including the KTB are presented in Fig. 6. At mid-Waipara, forty-one samples, covering 24 m of the vertical section across the KTB were studied for palynology (Fig. 7).

4 Results

4.1 Palynology

A total of 105 pollen and spore taxa were identified in the three sections. The uppermost Cretaceous samples are characterised by a well-preserved and diverse palynoflora incorporating conifers, ferns and flowering plants. In all the sequences studied, the KTB is abruptly followed by a 20-95 cm interval characterized by a palynoflora dominated by fern spores (= a "fern spike"). Apart from the fern spike, the KTB is recognisable palynologically by extinction of some key taxa, e.g., *Tricolpites lilliei*, *Nothofagidites kaitangata* and *Quadraplanus brossus*, all produced by flowering plants (Raine 1984; 1989; 1994).

We consider the sediments at Moody Creek Mine and Compressor Creek to be practically "in situ" or very locally derived, and thus reflect local vegetation, while those at mid-Waipara were transported some distance from the land and therefore record regional vegetation. As the sediments have been transported to this near-shore marine setting, the palynological assemblages represent both inland and coastal vegetation. We have also taken in account, when interpreting the palynological assemblages, that the sediments of mid-Waipara have been affected by reworking, bioturbation, and weathering of the unconsolidated greensand.

The fern spike is most dramatic and abrupt in the Moody Creek Mine section (Fig. 5) where the palynological turnover is also coincident with the geochemical anomaly. This critical geochemical marker coincides with the palynological KTB, seen as an extinction of several miospore index taxa in the same horizon. The KTB shows no lithological signature.

The organic matter and palynoflora at this site is interpreted to be derived from *in situ* peat-forming vegetation, and consequently an undisturbed recovery succession is preserved. Fern spores make up 25% of the total miospore assemblage below the boundary, but reach 98% immediately above the KTB (Fig. 5).

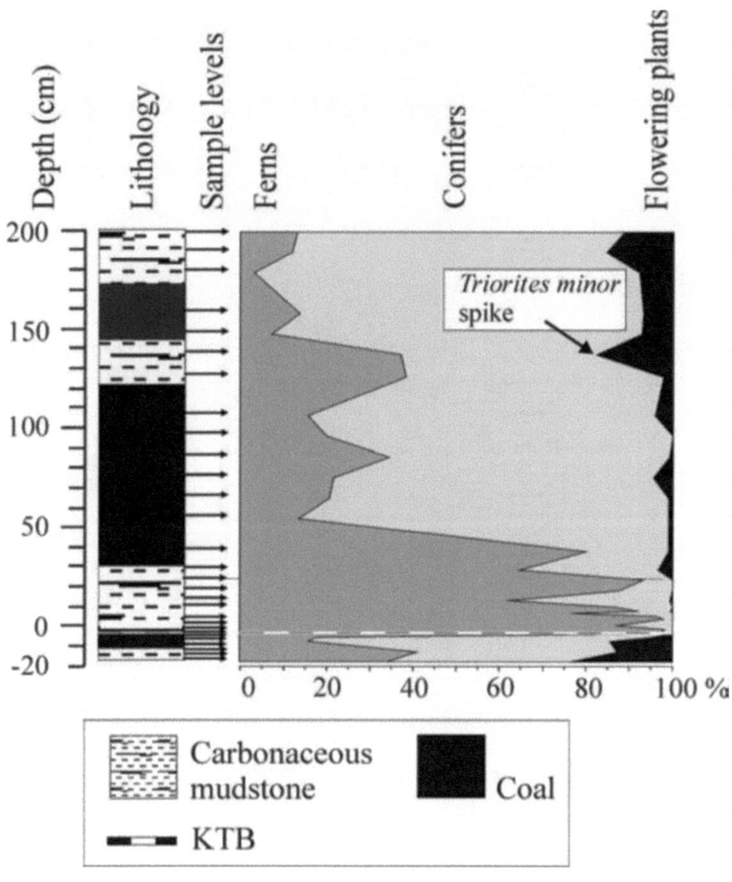

Fig. 5. Distribution and abundance of the major miospore groups - conifers, ferns and flowering plants (angiosperms) - in the terrestrial Moody Creek Mine section, also indicating the stratigraphical level of *Triorites minor* pollen peak.

At Compressor Creek, the sediments immediately overlying the KTB consist of sandy mudstone and the miospore assemblage shows evidence of short transportation within the local flood basin. Here, fern spore abundance rises from 34% below the KTB to 91% in the assemblage above (Fig. 6), with the same recovery pattern as at Moody Creek Mine.

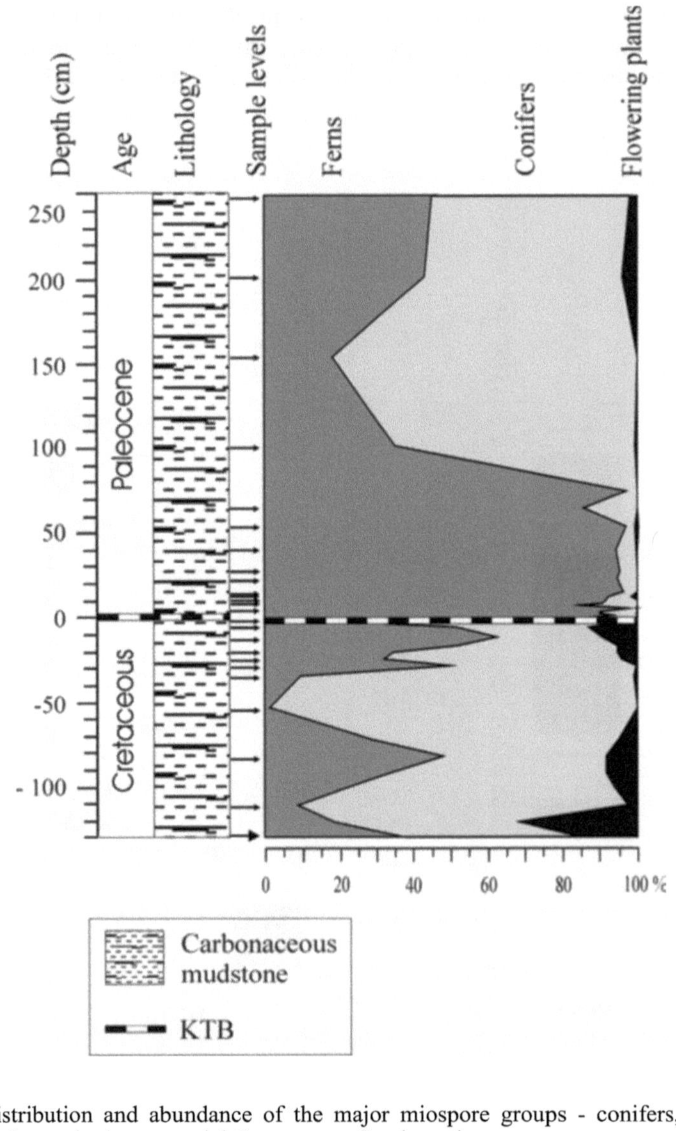

Fig. 6. Distribution and abundance of the major miospore groups - conifers, ferns and flowering plants - in the terrestrial Compressor Creek section.

The miospore assemblages of the near-shore marine section at mid-Waipara show a moderate increase in fern spores from 40% below the boundary to over 70% in the sediments above the KTB (Figs. 7 and 8; Vajda et al. 2001; Vajda and Raine 2003).

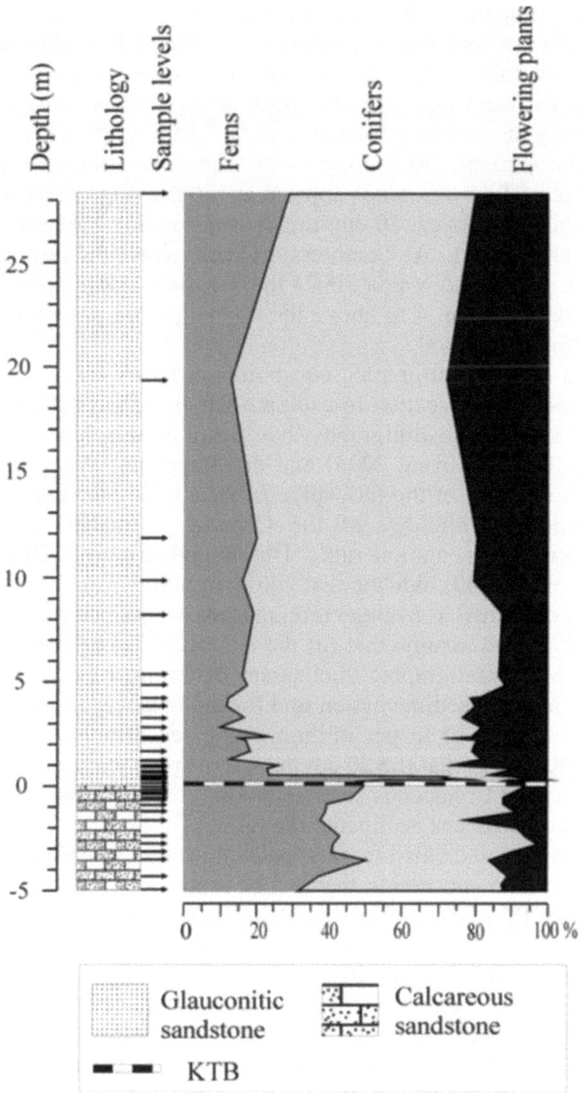

Fig. 7. Distribution and abundance of the major miospore groups - conifers, ferns and flowering plants - in the marine mid-Waipara section.

4.2 Timing of Vegetation Collapse and Recovery

The stratigraphic interval of the fern spike varies in thickness between the three studied sections. It spans c. 20 cm at mid-Waipara, c. 40 cm at Moody Creek Mine and c. 95 cm at Compressor Creek (Figs. 5, 6 and 8). Other features in the recovery succession follow the same pattern, suggesting more-or-less continuous deposition at the different sites, e.g. the peak of *Triorites minor* (an angiosperm within the recovery succession) coincides at the same relative distance from the KTB in two of the sections. At Moody Creek Mine where the fern spike is ca. 40 cm thick, the peak of *Triorites minor* appears at 140 cm (Fig. 5). At mid-Waipara, where the fern spike covers ca. 20 cm, the *Triorites minor* anomaly appears at 70 cm above the KTB (Fig. 8). At Compressor Creek, where the fern-spike is ca. 95 cm thick (Fig. 6; Raine and Vajda 2002) the *Triorites minor* anomaly would be expected at a distance of ca. 4 m above the boundary and, therefore, presumably lies above the sampled interval.

We have used the biostratigraphic constraints provided by marine microfossil events in the mid-Waipara section to estimate the duration of the fern spike. At mid-Waipara a refined biostratigraphy has been established on the basis of foraminifera (Hollis and Strong 2003) and dinoflagellates (Wilson 1984; 1987). Estimation of the duration of the fern spike is beset with uncertainties. However, the relatively uniform lithology of the Conway Formation at mid-Waipara suggests a uniform sedimentation rate. Dinoflagellate events (Roncaglia et al. 1999; Crampton et al. 2000) indicate that 200 m of uppermost Cretaceous strata in this section were deposited at average (compacted) sedimentation rate of ca. 25 m per million years. If we assume that (a) the Paleocene is complete in the section and that (b) the early stratigraphic thicknesses determined by Wilson (1963) are reliable, then the average sedimentation rate for the Paleocene is less than half the Cretaceous rate, i.e., ca. 10 m per million years, see Strong and Hollis (2003). These average rates imply that the 20 cm thick fern spike represents a minimum of 8000 years if based on Cretaceous sedimentation rates and a maximum of 20,000 years, based on the Paleocene sedimentation rates.

Examination of new and existing micropaleontological samples reveals an early Paleocene foraminiferal succession that can be tentatively related to international foraminiferal zones P0-P1c (P0, Pα, P1a, P1b, P1c). At 5-6 cm above the KTB in mid-Waipara, the Paleocene key taxon *Eoglobigerina eobulloides* was identified. This species ranges from Zone P0 to Zone P1b (Fig. 8). The P0 index, *Guembelitria cretacea,* was not recovered from this interval, which was considered too sparsely fossiliferous to adequately test for its presence/absence. The first occurrence of Pα zone index foraminifera is 23 cm above the KTB (Fig. 8), where *Globanomalina archaeoimitata* and probably *Globanomalina eugubina* are present. *Globanomalina archaeoimitata* has a reported stratigraphic range of Pα to P1a. This would indicate that the lowest 23 cm of Paleocene strata should be correlated with zone P0 and that the base of Zone Pα would roughly coincide with the top of the fern spike (Fig. 8). The Waipara situation is similar to a Japanese marine section where the fern-spike extends into the upper part of foraminifera zone P0 (Saito et al. 1986). From magnetostratigraphicbiostratigraphic data, the P0 zone had a duration of ca. 30,000 years (Berggren and Norris 1997), consistent in magnitude with the mid-Waipara estimates made above.

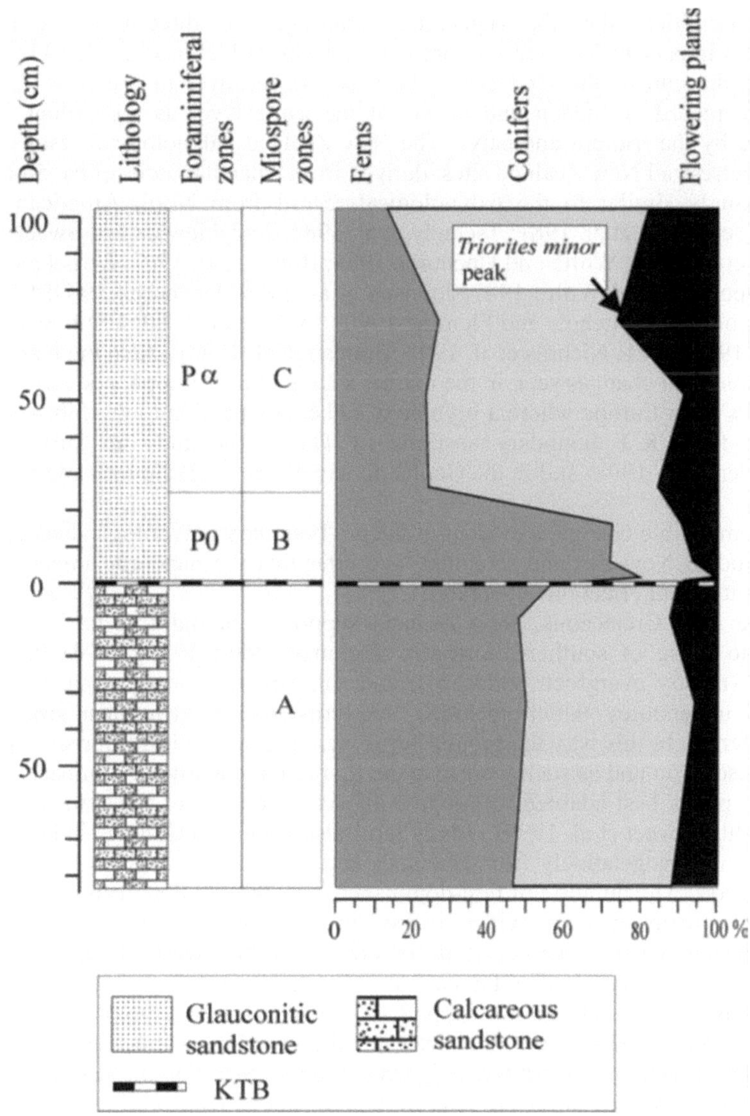

Fig. 8. K-T boundary at the marine mid-Waipara section (expanded vertical scale compared to Fig. 7), showing distribution and abundance of the major miospore groups - conifers, ferns and flowering plants, also the stratigraphical level of the *Triorites minor* abundance peak and zonation based on foraminifera.

5 Discussion

We are confident that the vegetation perturbation is directly linked to the Chicxulub impact in Yucatan (Alvarez et al. 1980; Hildebrand et al. 1991; Kyte 1998) at the end of the Cretaceous, because the turnover in the New Zealand miospore record is sudden and occurs at the same level as the fallout debris indicated by the iridium anomaly. The New Zealand palynological results from the two terrestrial New Zealand sites, derived from palaeolatitudes of ca. 50-60° S, are strikingly similar to the palynological record from North American KTB sections (e.g. Orth et al. 1981; Tschudy et al. 1984; Jerzykiewicz and Sweet 1986; Nichols et al. 1986; Wolfe and Upchurch 1986; Bohor et al. 1987; Lerbekmo et al. 1987; Upchurch and Wolfe 1987; Johnson et al. 1989; Upchurch 1989; Fleming and Nichols 1990; Nichols and Fleming 1990; Sweet et al. 1990, 1999; Sweet and Braman 1992, 2001; Nichols et al. 1992; Braman et al. 1993). Also the magnitude of palynological changes seen in the marine setting at mid-Waipara is very similar to records from Europe where a bryophyte spike, comparable to the fern-spike, is reported from K/T boundary sections in The Netherlands at Curf Quarry (Herngreen et al. 1998) and in the Geulhemmerberg caves (Brinkhuis and Schiøler 1996).

No comparable change is evident in the pre-boundary palynological succession in the studied New Zealand sections. We infer that the meteorite impact had a significant global effect on vegetation.

In the Late Cretaceous, New Zealand supported humid forests floristically similar to those of southern Australia (Dettman 1986; 1992; 1994) in which conifers (mostly evergreen podocarps) and angiosperms formed a canopy layer over an understorey which included tree-ferns, and a lowermost stratum of ground-ferns. In this way the ground-ferns were pre-adapted to flourish in places of low insolation and as such were also the first to recover after the impact as they were the plants best adapted to the post-impact conditions of low irradiance and high acidity (Sweet et al. 1999). Many ferns also have an advantage in being able to re-establish vegetatively from rhizomes and are thus not dependent on sexual reproduction. The duration of fern-dominance (ca. 8000-20,000 years) is orders of magnitude greater than that evident in modern landscapes following deforestation. This indicates not only the severe initial effects on the terrestrial vegetation, but also a continued period of unfavourable growing or regenerative conditions for seed plants. These data do not support the conclusions of Lomax et al. (2001), who postulated a 60-80 year time scale for the duration of the North American fern-spikes, based on process-based dynamic global vegetation modelling.

Vegetation recovery after the asteroid impact can be resolved in great detail in the two terrestrial sections. The recovery succession begins with abundant representation of the spore genus *Laevigatosporites,* produced by ground ferns related to extant *Blechnum.* This is followed by the spore genera *Baculatisporites* and *Gleichenidiites*, related to modern ground ferns of Osmundaceae and *Gleichenia*, a typical succession of pioneering plant communities on a landscape laid barren (Tryon and Tryon 1982). The extant *Gleichenia* is in New Zealand commonly seen in acidic swamps and bogs, and is tolerant of fire, achieving temporary local dominance in burnt areas until competing species recover (Wardle 1991).

The ground-fern succession is followed by tree-ferns represented by spores of *Cyathidites* and *Cibotiidites*, taxa related to modern Cyatheaceae and *Dicksonia*. The "old flora" then gradually returns, initially expressed by an increased relative abundance of conifers and, subsequently, by flowering plants. The conifers and flowering plants possibly persisted in refuge areas where, after surviving the relatively short impact winter, they grew in small numbers below the limit required for detection in the palynological record. Re-growth from the soil seed-bank succession are mainly represented by *Phyllocladidiites*. This is related to the extant Huon Pine (*Lagarostrobus franklinii*), a conifer thriving in cool-temperate climates (Gibson et al. 1991), and suggests cooler conditions than those of the late Maastrichtian.

Late Cretaceous and Paleocene vegetation-climates inferred from the miospore assemblages agree well with New Zealand leaf fossil physiognomic studies (Kennedy and Raine 2001; Kennedy et al. 2002; Kennedy 2003). These suggest mean annual temperatures of ca. 12-16°C and 7-11°C (at different sites) during the Campanian to Maastrichtian, i.e., cool-mild temperate climate with moderately high rainfall. Early Paleocene leaf assemblages showed lower dicotyledonous angiosperm species diversity, and cool- temperate conditions with mean annual temperatures of ca. 6-12°C.

Subsequent expansion of relatively few species of radiolarians and diatoms, together with a short-lived recovery of calcareous plankton during the early Palaeocene (Hollis et al. 1995; Hollis 1996; Strong 2000) coincides broadly with the recovery succession of land plants (Hollis et al. 2002). This indicates that the period with low insolation and photosynthesis cut-off was relatively short (Vajda et al. 2003) but was followed by a prolonged period of marine conditions which differed from those of the Late Cretaceous.

It took a considerable time for the forests to re-establish but land plants and mammals underwent an extremely rapid diversification in the Paleocene (Wolfe and Russell 2001) with first occurrences of new species directly following the KTB. In the New Zealand sections this is exemplified by the appearance of new pollen taxa such as *Nothofagidites waipawensis* directly above the boundary (Vajda and Raine 2003).

Based on species level of pollen-spore taxonomy, about 15 vol.% of the New Zealand palynoflora became extinct at the KT event. This number is well below the extinction rate of 45 vol.% described from North American palynological assemblages (Wolfe and Russell 2001). The fact that the New Zealand palynoflora at the time consisted mainly of conifer pollen and fern spores may explain this lower extinction rate compared to the North American record, which includes more numerous angiosperms (Upchurch 1989; Sweet and Braman 1992; Wolfe and Russell 2001). Extinction rates at the KTB were globally greatest among the flowering plants, but angiosperms have a generally low representation in the New Zealand pollen record.

6 Summary and Conclusions

Palynology serves as an excellent tool in pinpointing the KTB in New Zealand terrestrial and near-shore marine settings. The Cretaceous palynoflora is diverse and key Cretaceous indicator species have their highest stratigraphic occurrence at the KTB.

There is a massive change in palynofloral composition between adjacent samples (vertical separation of a few millimeters) indicating dramatic and geologically instantaneous vegetation change coincident with the geochemical anomaly at the KTB. In New Zealand KTB sediments, this is expressed as a sharp decline in the relative abundance of flowering plants and conifers and a rise in the relative abundance of fern spores from background levels of 25-40% to 76-98%. This proves that the KT mass-kill of plants was just as dramatic in the Southern Hemisphere as in the Northern Hemisphere, which strongly indicates that the turnover in the terrestrial vegetation and marine plankton is directly linked to the Chicxulub impact.

We calculate the time for the fern dominance following the K-T event to be ca. 8000-20,000 years, based on sedimentation rate and foraminiferal stratigraphy. Combined terrestrial and marine records indicate that initial recovery was in the form of low-diversity, pioneer communities that eventually gave way to more complex conifer-dominated vegetation on land and siliceous plankton-dominated communities in the marine realm.

Acknowledgements

Wolfgang Kiessling (Museum für Naturkunde, Berlin, Germany) and an anonymous referee are gratefully acknowledged for carefully reviewing and improving this paper with their comments and suggestions. Stephen McLoughlin (Queensland University of Technology, Brisbane, Australia) is acknowledged for helpful suggestions on this manuscript. This research was supported by the Swedish Research Council and the Crafoord Fund (20020547) (Vajda), and the New Zealand Foundation for Research, Science and Technology (Raine, Hollis and Strong).

References

Alvarez L, Alvarez W, Asaro F, Michel HV (1980) Extraterrestrial cause for the Cretaceous-Tertiary extinction. Science 208: 1095-1108

Askin RA (1988) The palynological record across the Cretaceous/Tertiary transition on Seymour Island, Antarctica. In: Feldmann RM, Woodburne MO (eds) Geology and Paleontology of Seymour Island, Antarctic Penninsula. Geological Society of America Memoir 16, pp 155-162

Askin RA, Jacobson SR (1996) Palynological change across the Cretaceous-Tertiary Boundary on Seymour Island, Antarctica: Environmental and depositional factors. In: Macleod N, Keller G (eds) Cretaceous-Tertiary mass extinctions: biotic and environmental changes, W.W. Norton and Co, London, pp 7-25

Bal A, Lewis DW (1994) A Cretaceous-early Tertiary macrotidal estuarine-fluvial succession: Puponga Coal Measures in Whanganui Inlet, onshore Pakawau Sub-basin, northwest Nelson, New Zealand. New Zealand Journal of Geology and Geophysics 37: 287-307

Berggren WA, Norris RD (1997) Biostratigraphy, phylogeny and systematics of Paleocene trochospiral planktic foraminifera. Micropaleontology 43: 1-116

Bohor BF, Modreski PJ, Foord EE (1987) Shocked quartz in the Cretaceous-Tertiary boundary clays - evidence for a global disruption. Science 236: 705-709

Braman DR, Sweet AR, Lerbekmo F (1993) Palynofloristic changes across the Cretaceous-Tertiary boundary and contiguous strata.[abs.]Canadian Geophysical Union, Joint Annual Meeting 18, p 12

Brinkhuis H, Schiøler P (1996) Palynology of the Geulhemmerberg Cretaceous/Tertiary boundary section (Limburg, SE Netherlands). Geologie en Mijnbouw 75: 193-213

Brooks RR, Hoek PL, Reeves RD, Strong CP (1986) Geochemical delineation of the Cretaceous/Tertiary boundary in some New Zealand rock sequences. New Zealand Journal of Geology and Geophysics 29: 1-8

Browne GH, Field BD (1985) The lithostratigraphy of Late Cretaceous to Early Pleistocene rocks of Northern Canterbury, New Zealand. New Zealand Geological Survey Record 6, pp 63

Couper RA (1960) New Zealand Mesozoic and Cainozoic plant microfossils. New Zealand Geological Survey Paleontological Bulletin 32, 87 pp

Crampton JS, Mumme TC, Raine JI, Roncaglia L, Schiøler P, Strong CP, Turner GM, Wilson GJ (2000) Revision of the Piripauan and Haumurian local stages and correlation of the Santonian-Maastrichtian (Late Cretaceous) in New Zealand. New Zealand Journal of Geology and Geophysics 43: 309-333

Dettmann ME (1986) Significance of the Cretaceous-Tertiary spore genus Cyatheacidites in tracing the origin and migration of Lophosoria (Filicopsida). Special Papers in Palaeontology 35: 63-94

Dettmann ME (1992) Structure and floristics of Cretaceous vegetation of southern Gondwana: implications for angiosperm biogeography. The Palaeobotanist 41: 224-233

Dettmann ME (1994) Cretaceous vegetation: the microfossil record. In: Hill RS (ed) History of the Australian vegetation. Cambridge, Cambridge University Press, pp 143-170

Fleming RF, Nichols DJ (1990) The fern-spore abundance anomaly at the Cretaceous-Tertiary Boundary: a regional bio-event in western North America. In: Kauffman EG, Walliser OH (eds) Extinction events in earth history. Lecture Notes in Earth Sciences 30, Springer Verlag, Heidelberg, pp 351-364

Gibson N, Davies J, Brown MJ (1991) The ecology of Lagarostrobos franklinii (Hook. f.) Quinn (Podocarpaceae) in Tasmania. 1. Distribution, floristics and environmental correlates. Australian Journal of Ecology 16: 215-222

Herngreen GFW, Schuurman HAHM, Verbeek JW, Brinkhuis H, Burnett JA, Felder WM, Kedves M (1998) Biostratigraphy of Cretaceous/Tertiary boundary strata in the Curfs Quarry, the Netherlands. Mededelingen Nederlands Instituut voor Toegepaste Geowetenschappen 61: 1-57

Hildebrand AR, Penfield GT, Kring DA, Pilkington M, Camargo Zanoguera A, Jacobsen SB, Boynton WV (1991) Chicxulub Crater; a possible Cretaceous/Tertiary boundary impact crater on the Yucatan Peninsula, Mexico. Geology 19: 867-871

Hollis CJ (1996) Radiolarian faunal change through the Cretaceous-Tertiary transition of eastern Marlborough, New Zealand. In: MacLeod N, Keller G (eds) Cretaceous-Tertiary

mass extinctions: Biotic and environmental changes, Norton and Company, New York, pp 173-204
Hollis CJ (2003) The Cretaceous-Tertiary boundary event in New Zealand: profiling mass extinction. New Zealand Journal of Geology and Geophysics 46: 307-321
Hollis CJ, Strong CP (2003) Biostratigraphic review of the Cretaceous/Tertiary Boundary transition, mid-Waipara River section, North Canterbury, New Zealand. New Zealand Journal of Geology and Geophysics 46: 243-253
Hollis CJ, Rodgers KA, Parker RJ (1995) Siliceous plankton bloom in the earliest Tertiary of Marlborough, New Zealand. Geology 23: 835-839
Jerzykiewicz T, Sweet AR (1986) The Cretaceous-Tertiary boundary in the central Alberta Foothills: stratigraphy. Canadian Journal of Earth Sciences 23: 1356-1374
Johnson KR, Nichols DJ, Attrep Jr M, Orth CJ (1989) High-resolution leaf-fossil record spanning the Cretaceous/Tertiary boundary. Nature 340: 708-711
Johnson KR, Raine JI (1991) A southern Hemisphere terrestrial Cretaceous/Tertiary boundary section: macro- and microfloral record from the Paparoa Trough, South Island, New Zealand. [abs.]Abstracts with Programs, Annual meeting, Geological Society of America 23: 358
Kennedy EM (2003) Late Cretaceous and Paleocene terrestrial climates of New Zealand: leaf fossil evidence from South Island assemblages. New Zealand Journal of Geology and Geophysics 46: 295-306
Kennedy EM, Raine JI (2001) Terrestrial Paleogene climate and leaf flora of New Zealand.(abs.) Abstracts, Climate and Biota of the Early Paleogene conference, Powell, Wyoming: 51
Kennedy EM, Spicer RA, Rees PM (2002) Quantitative paleoclimate estimates from Late Cretaceous and Paleocene leaf floras in the northwest of the South Island, New Zealand. Palaeogeography, Palaeoclimatology, Palaeoecology 184: 321-345
Kiessling W, Claeys P (2002) A geographic database approach to the KTB. In: Buffetaut E, Koeberl C (eds) Geological and biological effects of impact events, Springer, Berlin Heidelberg, pp 83-140
Kyte FT (1998) A meteorite from the Cretaceous/Tertiary boundary. Nature 396: 237-239
Lerbekmo JF, Sweet AR, St Louis RM (1987) The relationship between the iridium anomaly and palynological floral events at three Cretaceous-Tertiary boundary localities in western Canada. Geological Society of America Bulletin 99: 325-330
Lomax B, Beerling J, Upchurch Jr G, Otto-Bliesner B (2001) Rapid (10-yr) recovery of terrestrial productivity in a simulation study of the terminal Cretaceous impact event. Earth and Planetary Science Letters 192: 137-144
Nathan S (1978) Sheet S44 Greymouth. Geological map of New Zealand 1:63 360. Wellington, N.Z. Department of Scientific and Industrial Research
Nichols DJ, Fleming RF (1990) Plant microfossil record of the terminal Cretaceous event in the western United states and Canada In: Sharpton VL, Ward PD (eds) Global Catastrophes in Earth History; an Interdisciplinary Conference on Impacts, Volcanism, and Mass Mortality. Geological Society of America, Special Paper 247, pp 445-455
Nichols DJ, Brown J-L, Attrep Jr M, Orth CJ (1992) A new Cretaceous-Tertiary boundary locality in the western Powder River basin, Wyoming: biological and geological implications. Cretaceous Research 13: 3-30
Nichols DJ, Jarzen DM, Orth CJ, Oliver PQ (1986) Palynological and iridium anomalies at Cretaceous-Tertiary boundary, south-central Saskatchewan. Science 231: 714-717
Norris RD (2001) Impact of K-T Boundary events on marine life. In: Briggs DEG, Crowther PR (eds), Palaeobiology II. Blackwell Science, Oxford, pp 229-231
Orth CJ, Gilmore JS, Knight JD, Pillmore CL, Tschudy RH, Fasett JE (1981) An iridium abundance anomaly at the palynological Cretaceous-Tertiary boundary in northern New Mexico. Science 214: 1341-1343

Raine JI (1984) Outline of a palynological zonation of the Cretaceous to Paleogene terrestrial sediments in west coast region, South Island, New Zealand. Report, New Zealand Geological Survey 109, 82 p

Raine JI (1989) Palynology of outcrop upper Pakawau Group, northwest Nelson, New Zealand. New Zealand Geological Survey Report PAL 148, 18 p

Raine JI (1994) Terrestrial K-T boundary studies in New Zealand. Palaeoaustral 2: 9-12

Raine JI, Vajda V (2002) Vegetation change at the Cretaceous-Tertiary boundary in New Zealand, evidence for ecological disaster following the Chicxulub asteroid impact. [abs]. First International Paleontological Congress 2002, Geological Society of Australia, Abstracts 68, pp 132-133

Roncaglia L, Field BD, Raine JI, Schiøler P, Wilson GJ (1999) Dinoflagellate biostratigraphy of Piripauan-Haumurian (Upper Cretaceous) sections from northeast South Island, New Zealand. Cretaceous Research 20: 271-314

Saito T, Yamanoi K, Kaiho K (1986) End-Cretaceous devastation of terrestrial flora in the boreal Far East. Nature 323: 253-255

Strong CP (1984) Cretaceous-Tertiary boundary, Mid-Waipara River section, north Canterbury, New Zealand (Note). New Zealand Journal of Geology and Geophysics 27: 231-234

Strong CP (2000) Cretaceous-Tertiary foraminiferal succession at Flaxbourne River, Marlborough, New Zealand. New Zealand Journal of Geology and Geophysics 43: 1-20

Sweet R, Braman DR (1992) The K-T boundary and contiguous strata in western Canada: interactions between paleoenvironments and palynological assemblages. Cretaceous Research 13: 31-79

Sweet AR, Braman DR (2001) Cretaceous-Tertiary palynoflora perturbations and extinctions within the Aquilapollenites phytogeographic province. Canadian Journal of Earth Sciences 38: 249-269

Sweet R, Braman DR, Lerbekmo JF (1990) Palynofloral response to K/T boundary events: a transitory interruption within a dynamic system. Geological Society of America, Special Paper 247: 457-469

Sweet AR, Braman DR, Lerbekmo JF (1999) Sequential palynological changes across the composite Cretaceous-Tertiary (K-T) boundary claystone and contiguous strata, western Canada and Montana, USA. Canadian Journal of Earth Sciences 36: 743-768

Tryon R, Tryon A (1982) Ferns and allied plants with special reference to tropical America. New York, Springer-Verlag, pp 857

Tschudy RH, Pillmore CL, Orth CJ (1984) Disruption of the terrestrial plant ecosystem at the Cretaceous-Tertiary boundary, Western Interior. Science 225: 1030-1032

Upchurch Jr GR (1989) Terrestrial environmental changes and extinction patterns at the Cretaceous-Tertiary boundary, North America. In: Donovan SK (ed) Mass Extinctions. London, Belhaven Press, London, pp 195-216

Upchurch Jr GR, Wolfe JA (1987) Plant extinction patterns at the Cretaceous-Tertiary boundary, Raton and Denver basins. [abs.] Abstracts with Programs, Geological Society of America 19, 7: 874

Vajda V, Raine JI (2003) Terrestrial palynology of the Cretaceous/Tertiary boundary at mid-Waipara River, North Canterbury, New Zealand. New Zealand Journal of Geology and Geophysics 46: 255-273

Vajda V, Raine JI, Hollis CJ (2001) Indication of global deforestation at the Cretaceous-Tertiary boundary by New Zealand fern spike. Science 294: 1700-1702

Vajda V, Raine JI, Hollis CJ (2002) Global effects on the biota following the Chicxulub asteroid impact at the end Cretaceous - micropaleontological record from New Zealand K-T boundary. [abs.] In: Jakes P (ed) Abstracts, Workshop on Impacts: a geological and astronomical perspective – Prague (Czech Republic), Impact Programme, European Science Foundation, pp 63-64

Vajda V, Ocampo A, Buffetaut E (2003) Unmasking the KT catastrophe; evidence from flora, fauna and geochemistry. [abs.] In: Cockell C (ed) Abstracts, Workshop on Biological processes associated with impact events – Cambridge (United Kingdom), Impact Programme, European Science Foundation, 57

Ward S (1997) Lithostratigraphy, palynostratigraphy and basin analysis of the Late Cretaceous to early Tertiary Paparoa Group, Greymouth Coalfield, New Zealand. Ph.D. thesis, University of Canterbury, 200 pp

Wardle P (1991) Vegetation of New Zealand. Cambridge, Cambridge University Press, 672 pp

Wilson DD (1963) Geology of Waipara Subdivision. New Zealand Geological Survey Bulletin 64, 122 pp

Wilson GJ (1984) New Zealand Late Jurassic to Eocene dinoflagellate biostratigraphy - a summary. Newsletters on Stratigraphy 13: 104-117

Wilson GJ (1987) Dinoflagellate biostratigraphy of the Cretaceous-Tertiary boundary, mid-Waipara River Section, North Canterbury, New Zealand. New Zealand Geological Survey Record 20: 8-16

Wolfe JA, Upchurch Jr GR (1986) Vegetation, climatic and floral changes at the Cretaceous-Tertiary boundary. Nature 324: 148-152

Wolfe JA, Russell DA (2001) Impact of K-T boundary events on terrestrial life. In: Briggs DEG, Crowther PR (eds) Palaeobiology II. Blackwell Science, Oxford, pp 232-234

The Neugrund Marine Impact Structure (Gulf of Finland, Estonia)

Sten Suuroja[1,2] and Kalle Suuroja[1]

[1]Geological Survey of Estonia, Kadaka tee 82, Tallinn 12168, Estonia.
[2]Department of Mining, Tallinn Technical University, Kopli 82, Tallinn, Estonia.
(s.suuroja@egk.ee)

Abstract. The Early Cambrian (approximately 535 Ma) Neugrund marine impact structure is located on the southern side of the entrance to the Gulf of Finland, immediately eastward of Osmussaar Island, Estonia. The origin of the structure was noted already in 1995 - 1998, but data obtained during the expeditions of 2000 and 2001 have shed new light on its morphology. The impact structure is about 20 km in diameter and spatially delimited by a ring fault between dislocated rocks and mostly intact target rocks. The structure has a central depression (crater deep or crater proper) 5.5 km in diameter, surrounded by a 50 - 100 m high and anomalously wide (2.5 - 3 km) 3-ridge shaped rim wall. The crater deep is filled with post-impact Early Cambrian and Early Ordovician siliciclastic rocks and covered with Middle and Late Ordovician calcareous rocks. The slight (some metres) uplift of limestone beds in the centre of the crater suggests that a central uplift also exists. The Ordovician erosion-resistant limestone forms a circular Central Plateau (Neugrund Bank) above the crater proper about 4.5 km in diameter. The plateau is surrounded by a 200 - 500 m wide and 20 - 70 m deep canyon (Ring Canyon). A 3 - 5 km wide circular depression where the crystalline target rocks are dislocated lies outside the rim wall. Sedimentary target rocks are eroded in the northern part of the structure. Outside the ring fault (outer boundary of the structure), sedimentary target rocks are dislocated within about 10 km, obviously due to the Neugrund impact. The 1 - 2 m thick ejecta layer consists of sandstones with abundant shock-metamorphosed quartz grains with well-developed planar deformation features (PDFs). Erratics consisting of Neugrund Breccia, derived by glacial action from the exposed parts of the impact structure, spread in an area of more than 10 000 km^2.

1 Introduction

The approximately 535 Ma (the age of the structure was determined by using the reconstructions of Torsvik et al. 1992; Tucker and McKerrow 1995 and the International Stratigraphic Chart by IUGS 2000) Early Cambrian Neugrund impact structure (59° 20′ N and 23° 32′ E) is located on the southern side of the entrance to the Gulf of Finland (Figs. 1, 2) about 7 km NE of Osmussaar Island. The structure is about 20 km in diameter and is spatially delimited by a ring fault (Figs. 1, 5). The ring fault runs across Osmussaar Island and serves as a boundary between dislocated target rocks and mostly intact target rocks. It is not visible in recent relief but (Figs. 2, 3, 4) is quite well recognisable in seismic reflection profiles (Fig. 6).

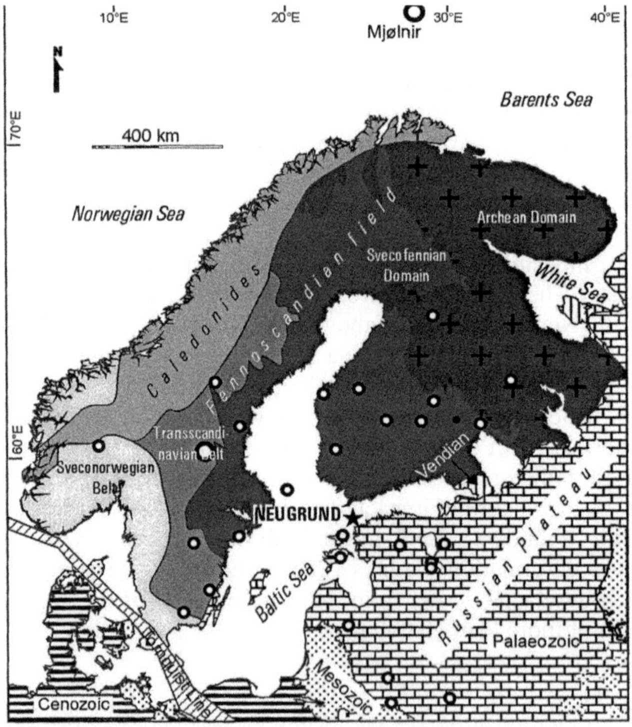

Fig. 1. Location of the Neugrund and other impact structures in the East European Platform. Black star – Neugrund impact structure; rings with a white core – other impact structures.

The impact structure has a central depression 5.5 km in diameter, surrounded by a 50 - 100 m high and 2.5 - 3 km wide 3-ridged rim wall (Fig. 4). A central uplift probably exists in the central depression, which is filled with the post-impact Early Cambrian siliciclastic rocks and covered by Middle and Late Ordovician limestones. There is no firm evidence yet, but slight (some metres) uplift of the infilling limestone in the central part of the crater proper supports this suggestion. The erosion-resistant Ordovician limestone forms a nearly 4.5 km wide circular Central Plateau (Neugrund Bank) over the central depression, which is surrounded by a 200 - 500 m wide and 20 - 70 m deep canyon, the so-called Ring Canyon (Figs. 7,8).

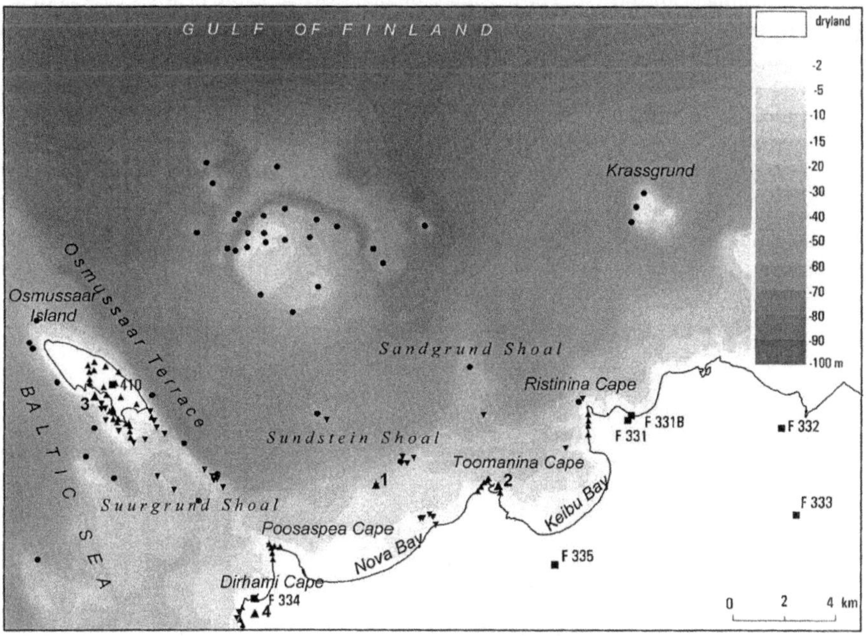

Fig. 2. Bathymetric map of the Neugrund impact structure area. Triangles: small – single, small to big erratic boulders made of Neugrund Breccia; big – gigantic erratic boulders made of Neugrund Breccia (1 = Toodrikivi, ca. 1200 m^3; 2 = Nõva Suurkivi, ca. 400 m^3; 3 = Skarvan, ca.400 m^3; 4 = remnants of Twins of Osmussaar. White marks land. Circles = drill hole and its number. Dots = diving and sampling sites.

The crystalline target rocks are strongly dislocated outside the rim wall in a 3 - 5 km wide circular area. These rocks are eroded, especially in the northern part, up to 10 km outside the impact structure. Beyond the ring fault, the sedimentary target rocks (Early Cambrian clay- and sandstones) are, in places, disturbed within 2 - 30 m of the upper part of the section, most likely due to the Neugrund impact event. Above the disturbed strata, a 1–2 m thick layer of post-impact silt- and sandstones occurs over thousands of square kilometres in Northwest Estonia. These sandstones contain abundant shock-metamorphosed quartz grains with well-developed planar deformation features (PDFs) and evidently represent the ejecta layer of the Neugrund impact.

The discovery of the Neugrund impact structure was based on the information obtained in the course of integrated geological and geophysical mapping at the NW Estonian coast in 1983–1999. The results of these investigations are presented in numerous reports (Malkov et al. 1986; Suuroja et al. 1987; Suuroja et al. 1998; Suuroja et al. 1999; Talpas et al. 1993) and papers (Suuroja et al. 2001b). The existence of an impact structure in this area was first suggested in 1995 (Suuroja and Saadre 1995). In 1996, this hypothesis was confirmed by seismic reflection profiling and in 1998 by direct submarine observations. The shock-metamorphic features in the brecciated crystalline rocks were studied some time later (Suuroja and Suuroja 2000).

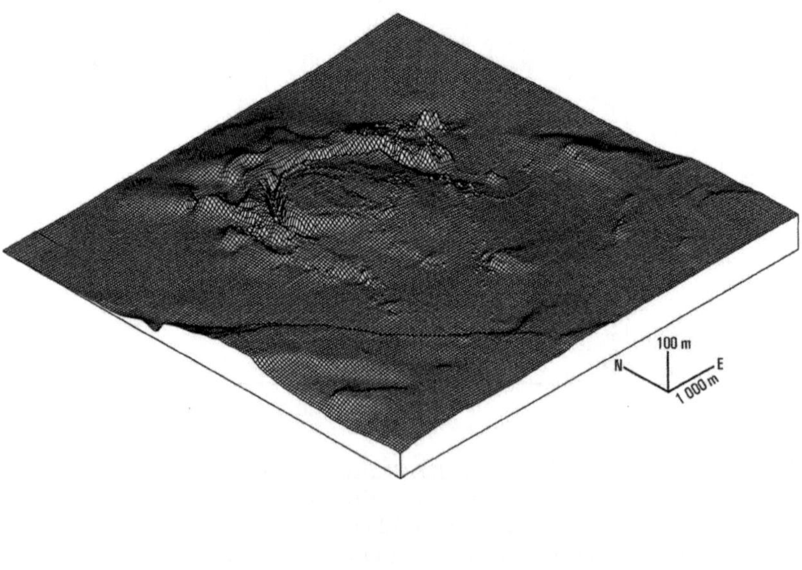

Fig. 3. 3D diagram of the topography of the Neugrund structure area.

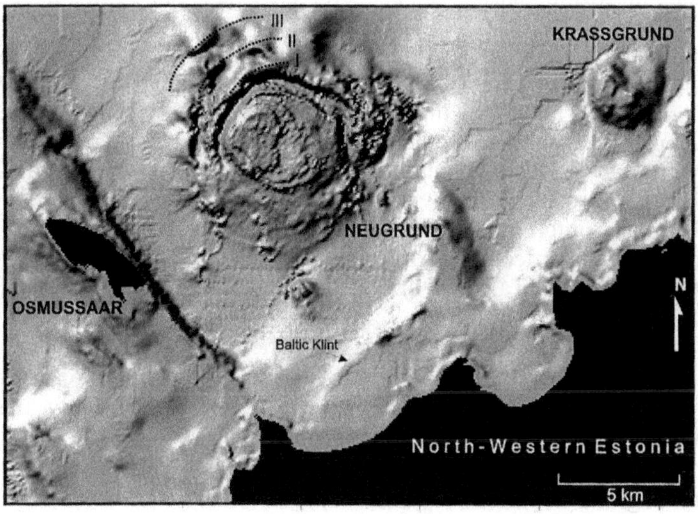

Fig. 4. Shaded relief of the Neugrund impact structure area. Ridges of several parts of the rim wall are marked by Roman numbers. Land is in black.

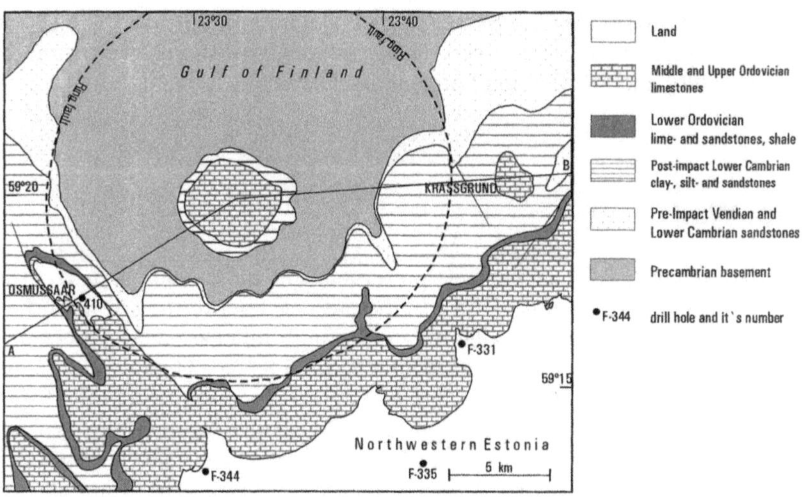

Fig. 5. Schematic bedrock geological map of the sea floor of the Neugrund impact structure area.

It was also established that the Neugrund and nearby Kärdla Palaeozoic marine impact structures had many similar features (Suuroja et al. 2001).

In 2000 and 2001 five marine expeditions to the Neugrund impact structure and its surroundings were carried out (four on board the r/v "Mare" and one on board the r/v "Skagerak"). The following methods were used: seismic reflection profiling – about 300 km; magnetometric profiling – about 200 km; observing submarine outcrops by a video robot; side scan sonar profiling – about 100 km; sampling of bottom deposits with gravity corer and scarp – at 52 sites; sampling of submarine bedrock outcrops during divings – 54 samples. Unpublished interpretations of some reflection profiles (about 120 km) have been provided earlier (Malkov et al. 1986; Talpas et al. 1993), and they are used also here (Figs. 6, 7, 8).

Based on the results of these and earlier expeditions, as well as mineralogical analyses, some concepts about the morphology of this old, but well-preserved and partially exposed submarine impact structure were revised. Additionally, the area of the distribution of erratics derived by the glacial action from the Neugrund impact structure was studied. In all, 1020 erratic boulders over 1 m in diameter, consisting of Neugrund Breccia, were observed over a land and sea area of some hundred square kilometres (Fig. 10).

The present paper gives a more detailed characterization of the morphology of the Neugrund impact structure (central depression, rim wall, circular depression, zone of farther dislocations, distal ejecta). The classification and nomenclature of shock-metamorphic rocks follows that proposed by Stöffler and Grieve (1996) to the IUGS Subcommission on the Systematic of Metamorphic Rocks. The shock-metamorphic stages were determined using the classifications elaborated by Grieve et al. (1996) and Stöffler and Langenhorst (1994).

2
The Rim Wall

The well-exposed rim wall is the best-preserved and most studied part of the Neugrund impact structure (Figs. 4,7,8). In the present relief of the sea floor there are 3 circular ridges of glacier-eroded 30 - 60 m high hillocks, where brecciated Precambrian metamorphic rocks crop out (Figs. 5,8,13). The preserved but strongly eroded rim wall is 60–120 m high and 2.5 - 3 km wide. The initial height of the rim wall might be 2 or 3 times higher. In the southern part of the structure the rim wall is more eroded and mostly buried under Quaternary (Pleistocene) deposits. The northern part of the impact structure, as well as the rim wall, was generally uplifted at post-impact time and the entire structure area has an incline of 3 m per km to the S or SW.

Nine hillocks were investigated in the rim wall area at depths of 16 - 34 m and sampled directly by diving (Figs. 2,13). The entire area, where the rim wall cropped out, was investigated by side-scan sonar, and seismic reflection profiling (Fig. 6). These investigations showed that the rim wall consisted of brecciated

Precambrian (Palaeoproterozoic) metamorphic rocks (amphibolites, migmatite granites, gneisses), penetrated by veins and bodies of clast- and matrix supported impact breccias containing shock-metamorphosed quartz with PDFs (Fig. 9d) and signs of partial melting.

The depressions between the hillocks or different parts of the rim wall are mostly buried under Quaternary (Pleistocene) glaciolacustrine deposits – varved clays. As suggested earlier (Suuroja and Suuroja 2000), the deformed Early Cambrian and Late Vendian siliciclastic pre-impact target rocks were not found in these depressions.

A more detailed characterisation of different parts (ridges) of the rim wall is given below.

1) The first, or so-called Inner Wall is 400–800 m wide at the base, 30 - 100 m high and has a rim-to-rim diameter of 6.5 km (Figs. 4,8) This is the best preserved and undisturbed part of the rim wall consisting of slightly brecciated Precambrian metamorphic target rocks penetrated by veins of clast- and matrix supported impact breccias, containing shock-metamorphosed quartz with PDFs. In the southern part of the structure, the inner wall is lower (30 - 60 m) and buried under Quaternary (Pleistocene) glaciolacustrine (varved) clays. In the south-eastern part of the structure, an approximately 0.5 km wide and 100 m deep gully cuts through the Inner Wall, as well as the other two walls (Figs 2, 4).

2) The second, or so-called Middle Wall is 0.5 - 1 km wide, 20 - 50 m high and has a rim-to-rim diameter of 9 km (Fig. 4). It has a common base with the Inner Wall but is separated from the latter by a 0.3 - 1 km wide irregular (largest in the western and narrowest in the eastern parts of the rim wall) depression. In the southern part of the structure, the Middle Wall as well is buried under varved clays. The Middle Wall is not as monolithic and intact as the Inner Wall. It consists of 0.5 - 2 km long and 20 - 80 m high hillocks of brecciated crystalline (metamorphic) rocks, cut of veins of clast- and matrix supported impact breccias containing shock-metamorphosed quartz with PDFs.

Most of the erratic boulders consisting of Neugrund Breccia, found so far in West Estonia, have been derived by glacial activity from this part of the rim wall. Generally, the Middle Wall is lower and more fragmentary than the Inner Wall. The reasons for this irregularity are not known; these can be primary (have contained soft siliciclastic rocks) or secondary (higher rate of erosion).

3). The third, or so-called Outer Wall, with a rim-to-rim diameter of about 12 km, consists of 1 - 1.5 km wide and 40 - 100 m high hillocks, which are separated from the Middle Wall by a 50 - 80 m deep and 1 - 1.5 km wide depression (Fig. 4). This part of the rim wall is most strongly eroded and deeply buried under the Quaternary deposits (varved clays). Thus, its remnants are observable only at four sites in the northern part of the structure. This allows us to suppose that the Outer Wall was primarily more monolithic and the irregularities appeared due to the prevalence of soft pre-impact siliciclastic rocks (silt- and sandstones) in the part of

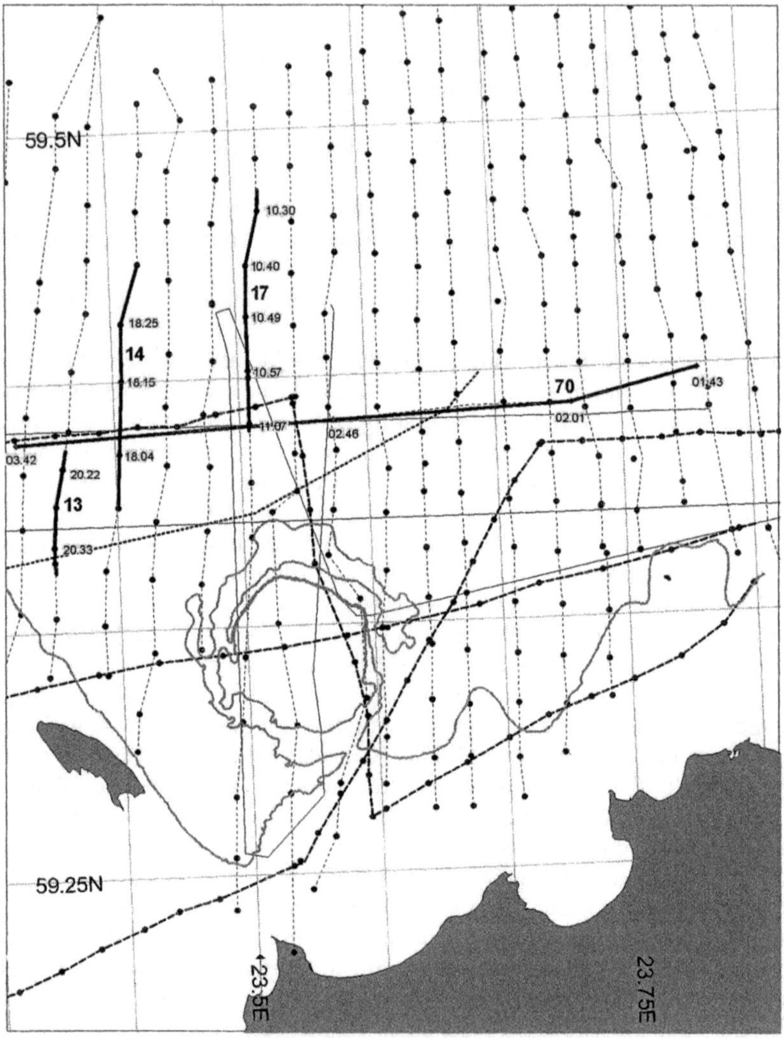

Fig. 6a. Location of the seismic reflection profiles on the Neugrund impact structure area. Bold lines are profiles given in Fig. 6 b.

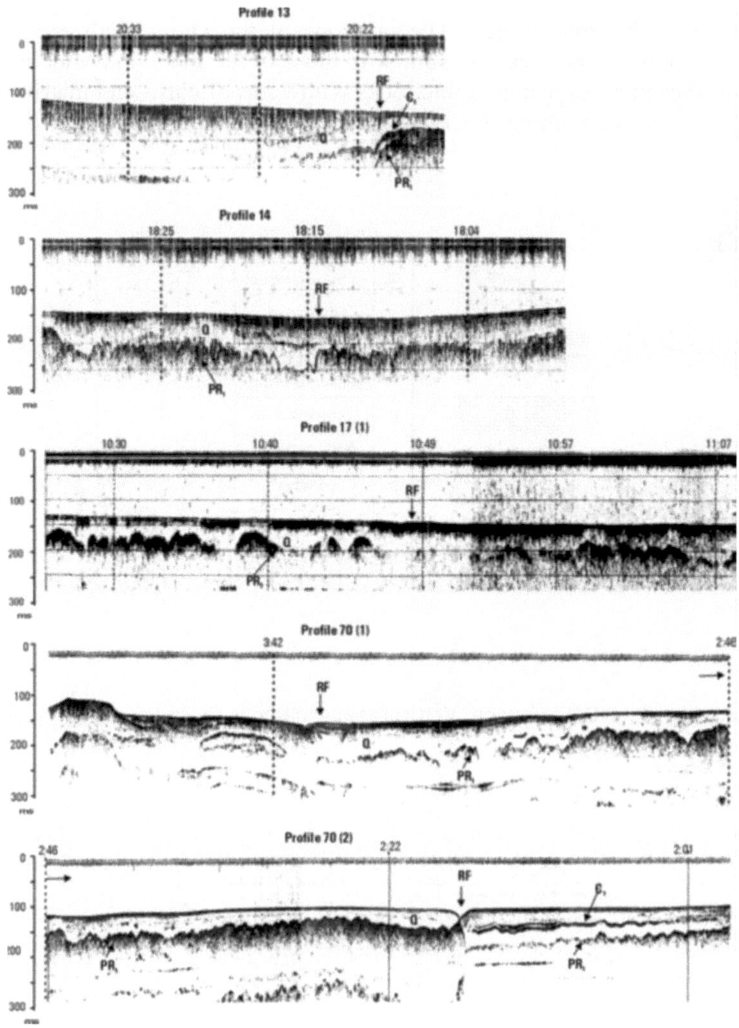

Fig. 6b. Fragments of the seismic reflection profiles crossing the Neugrund impact structure. Profiles were carried out by the r/v "Marina": profiles 13, 14, 20 - Talpas et al. (1993); profile 70 – Malkov et al. (1986). Space between horizontal lines 50 mm/sec or ca 36 m in water; space between vertical lines 220–260 m per minute. Record frequency 400–800 Hz. Q – Quaternary deposits; C_1 – pre-impact Early Cambrian; PR_1 – Precambrian (Palaeoproterozoic) basement; RF – ring fault.

the rim wall that later became eroded. The outcrops of the Outer Wall were sampled only at two points. The rocks are similar to those of the Inner and Middle walls – brecciated Precambrian metamorphic rocks with veins of clast- and matrix supported impact breccias containing shock-metamorphosed quartz, with PDFs.

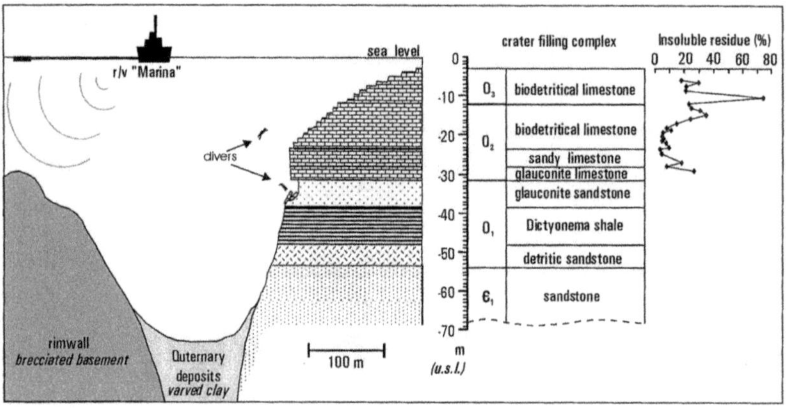

Fig. 7. The cross-section of the Ring Canyon. Vertical exaggeration is five times.

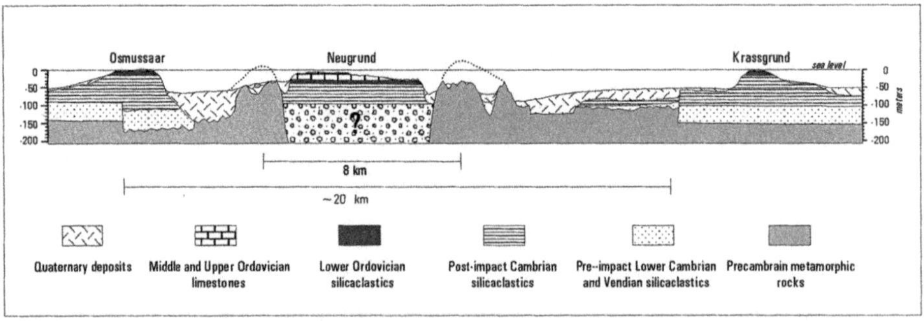

Fig. 8. West – east cross-section of the Neugrund impact structure.

3
The Circular Depression and Ring Fault

The target rocks, both sedimentary (siliciclastic) and crystalline, are dislocated in a 3 - 5 km wide circular area around the rim wall, called the circular depression. The bedrock relief of this area is strongly jointed due to uneven displacement (sinking) of huge (hundreds of metres in diameter) blocks of crystalline rocks during the impact. In the recent sea floor relief, these buried irregularities are not particularly expressed, but are well visible in seismic reflection profiles (Fig. 6). The dislocated bedrock is everywhere in this area buried under a 10 - 50 m thick layer of the Quaternary (Pleistocene) glacio-lacustrine deposits – varved clays, and in the southern part of the area also under the post-impact deposits. The crystalline target rocks lie at a depth of 160 - 180 m in the western part of the depression and at a depth of 100–130 m in its eastern part. The eastern part of the structure is uplifted by up to 60 m in relation to the western part (Fig. 8). The regional slope of the Precambrian basement in this region is 2–3 m per km south- or south-eastward (Suuroja et al. 1999).

The ring fault has been treated as the boundary between the disturbed rocks and mostly intact target rocks, but this does not mean that deformations do not occur farther away, especially in the sedimentary target rocks. In the recent bedrock topography, the ring fault is expressed as a 10 - 60 m high terrace, mostly buried under the Quaternary deposits and, therefore, not observable everywhere. In the southern part of the structure, the ring fault is also buried under the post-impact Lower Cambrian siliciclastic deposits. The crystalline basement in the north-western part of the structure, outside the ring fault, is uplifted by up to 50 m, but in the eastern part of the structure has sunk some 20 - 40 m. In the seismic reflection profiles, the ring fault in the crystalline basement is well observed only in the northern part of the structure and was identified in 15 profiles at 24 sites. The best results were obtained in areas where the thickness of the covering sedimentary rocks and Quaternary deposits is least (0 - 30 m) and depth of water is greater (30 - 110 m). In the southern part of the structure, the ring fault is not observable in seismic reflection profiles, due to shallow water (10 - 20 m) and a thicker (60 - 80 m) sedimentary rock cover.

4
The Zone of Distal Deformations

The metamorphic rocks of the Precambrian basement are disturbed within the area delimited by the ring fault, while the pre-impact sedimentary siliciclastic rocks are sporadically dislocated farther away, at radial distances of up to more than 20 km from the impact centre. In the section of drill hole F-332 (Vihterpalu), 20 km to

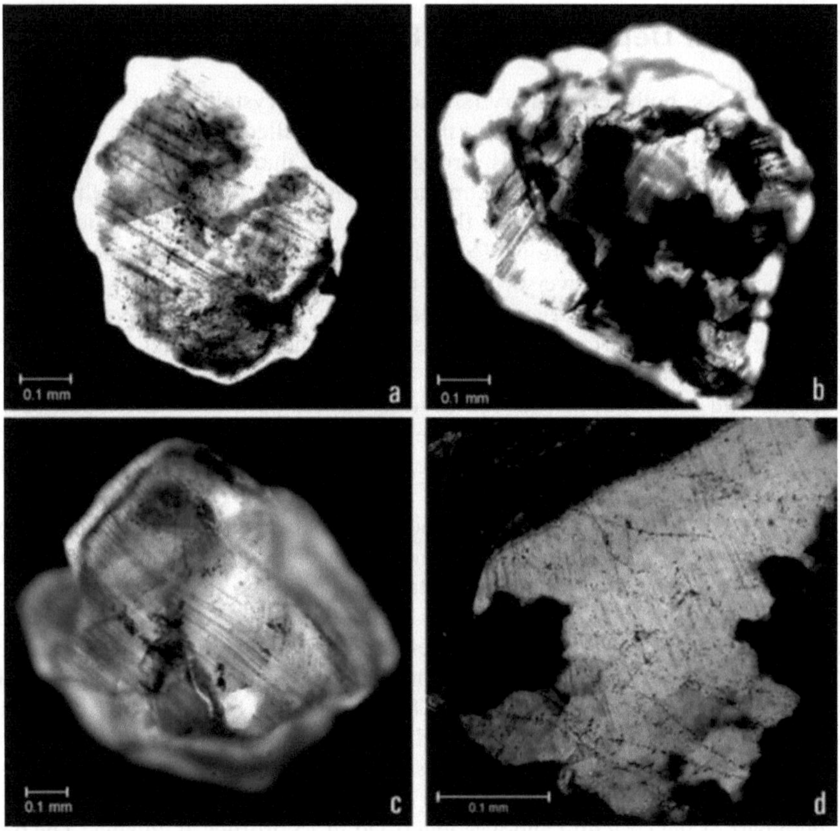

Fig. 9. Shock-metamorphosed quartz grains with multiple PDFs from the impact-related rocks of the Neugrund impact structure: a) Subrounded quartz grain with 1 set of decorated PDFs, separated from the brecciated target rocks (Early Cambrian clay- and sandstones) at a distance of 15 km from the impact centre. Sample 331-100 from drill hole F-331, at a depth of 100.0 m. In immersion liquid, cross-polarised light. b) Subrounded quartz grain with 3 sets of decorated PDFs, separated from the brecciated target rocks (Early Cambrian clay-and sandstones) at a distance of 15 km from the impact centre. Sample 331-40 (drill hole F-331 at a depth of 94.0 m). In immersion liquid, cross-polarised light. c) Subrounded quartz grain with 2 sets of decorated PDFs, separated from the distal ejecta at a distance of 15 km from the impact centre. Sample 331-08 (drill hole F-331 at a depth of 90.8 m). In immersion liquid, cross-polarised light. d) Quartz grain with 3 sets of slightly decorated PDFs. Sample N99-A1 from a vein of matrix supported impact breccia from the brecciated crystalline basement rocks of the rim wall at a depth of 24.2 m. Thin-section, cross-polarised light.

the south-east of the impact centre, the Early Cambrian pre-impact silt- and claystones of the Lontova Formation at depths of 100 - 110 m are slightly brecciated. Similar deformations at the same stratigraphical level are observed in the section of drill hole F-335, 22 km south of the impact centre. In the area of Ristna Cape, 3 km south-east of the ring fault or 12 km from the impact centre (Fig. 2), the pre-impact clay- and siltstones of the Lontova Formation are brecciated in the sections of drill holes F-331 and F-331A at depths of 92 -120 m. This breccia contains angular clasts of Vendian sandstones (5 - 20 cm in diameter) and Precambrian metamorphic rocks (0.5 - 5 cm in diameter).

Sandy fractions (0.063 - 0.5 mm) derived from this breccia matrix contain quartz grains (on average about 4% of total quartz) with well-developed PDFs (Fig. 9a, 9b). Up to four sets of PDFs with a frequency of 200 - 400 lamellas per 1 mm can be observed. The PDFs are most numerous (up to 8% of total quartz) in the 0.5 - 0.25 mm fraction and among subrounded grains (up to 5%). Planar deformation features are observed also in grains of apatite (up to 20% of the total apatite) and plagioclase.

The brecciated layer is absent in the section of drill hole F-331B, 200 m north-east of drill hole F-331 and closer to the impact centre. The corresponding rocks are only slightly crushed for some metres below the ejecta layer. Consequently, this breccia layer is not spread evenly. These deformations are connected with the Neugrund impact event. The crystalline basement in this area appears to be intact, but how should one explain the presence of the angular clasts of metamorphic rocks, undoubtedly derived from the crystalline basement of this area, more than 60 m higher, above the intact sedimentary rocks?

5
Distal Ejecta

The distal ejecta of the Neugrund impact event are not as well recognizable as those of the nearby Kärdla impact event (Fig. 10), where the sandy layer of impact-related deposits lies between limestone beds over thousands of square kilometres around the crater (Suuroja et al. 2000). Nevertheless, we have a reason to believe that the 1 - 2 m thick basal layer of the Early Cambrian siliciclastic deposits (silt- and sandstones of the Sõru Formation) may be the distal ejecta of the Neugrund impact event. This relation does not apply to the entire Sõru Formation, because the thickness of the Sõru Formation (1 - 2 m at a distance of 10 -20 km from the impact centre) increases farther away from the impact centre. In Hiiumaa Island, 80 km west of the Neugrund impact centre, the thickness of this layer is up to 20 m. In core sections of drill holes closest to the impact centre (F331and F331A- Ristna, distance between the drill holes is only 5 m, F332-Vihterpalu, 410-Osmussaar) the 1 - 2 m thick Sõru Formation, consisting of fine-grained microbedded quartzose sandstone and above-lying deformed pre-impact Early Cambrian claystones, contains abundant shock-metamorphosed minerals, especially quartz grains with well-developed PDFs.

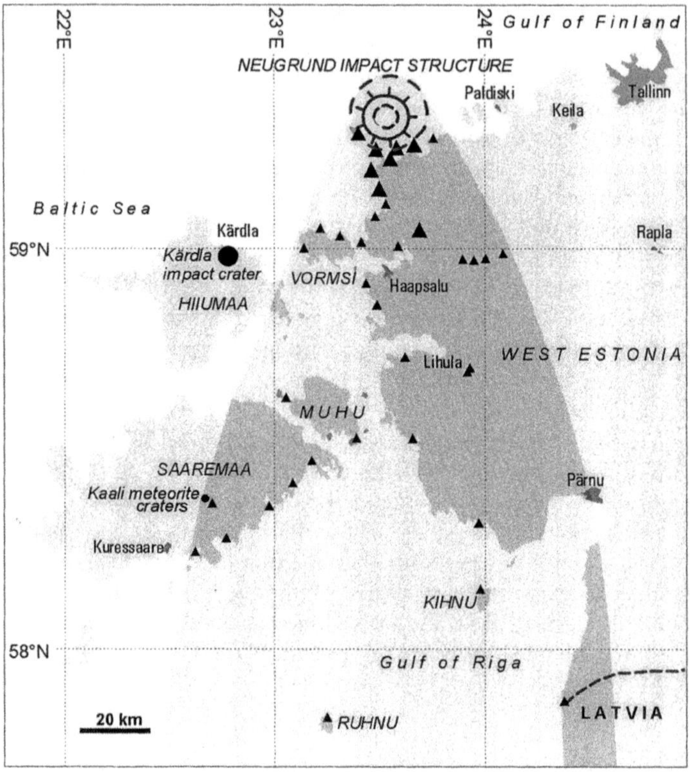

Fig. 10. Distribution of erratic boulders derived by glacial action from the exposed structures of the Neugrund Impact Structure. Large black triangles = numerous Neugrund Breccia boulders; small black triangles – rare Neugrund Breccia boulders.

For example, in drill core F331 at a depth of 90.8 m, 13 km to the south-east of the impact centre, up to 8 % of total quartz in the fraction 0.5 - 0.25 mm is represented by subrounded and rounded grains with PDFs. Here, up to 4 sets of PDFs with a frequency of 200 - 400 lamellas per 1 mm are also observed (Fig. 9c). Among fine fractions (0.063 - 0.25 mm) and subangular and well-rounded grains the shock-metamorphosed quartz grains are far less numerous (ca. 1 vol%) or may be completely absent just like amongst angular grains. The shape and fractional composition of the quartz grains with PDFs indicate that these are derived from the Early Cambrian or Vendian pre-impact siliciclastic target rocks and not from the crystalline basement of the target.

a

b

Fig. 11. Some erratic boulders of Neugrund Breccia derived by glacial action from the impact structure: a) "Skarvan" (ca 400 m^3) - on the west coast of Osmussaar Island; b) Nõva Suurkivi (ca 400 m^3) – on Põõsaspea Cape.

Farther from the impact centre, the distal ejecta form only a minor part of the Sõru Formation. The ejecta layer in the Neugrund case is deposited in water and consists mostly of pulverized pre-impact siliciclastic deposits, which at the time of the impact formed the about 100 m thick topmost part of the target (Suuroja et al.

2001b). As yet, no material derived from the crystalline target rocks has been detected in the ejecta. However, this does not mean that it is necessarily completely absent, because the ejecta were studied only in 3 drill core sections, where core recovery from this part of the core was very low (10 - 20%).

a

b

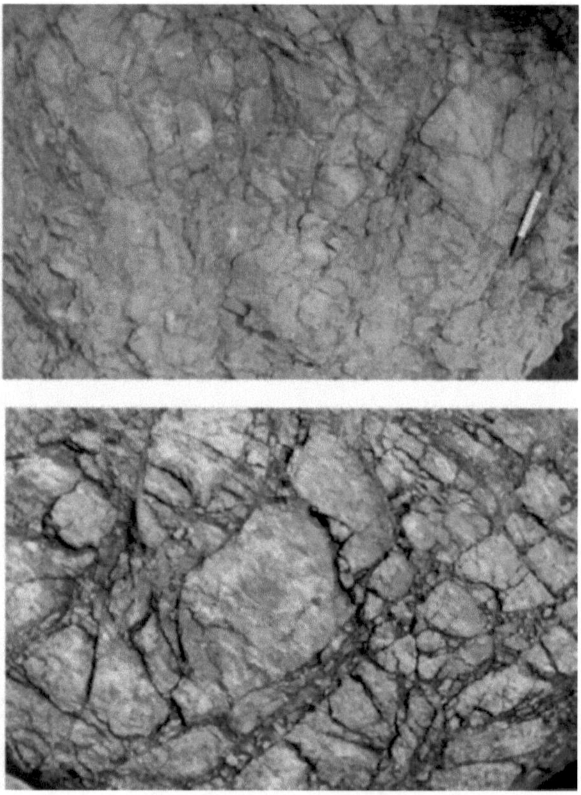

Fig. 12. Samples of the Neugrund Breccia from the erratic boulders of Toomanina Cape: a) brecciated granitoid; b) brecciated biotite gneiss. Pencil (13 cm) for scale in both a and b.

6 Distribution of Erratics of Neugrund Breccia

The distribution of erratics of Neugrund Breccia, derived from Pleistocene glacial action from the exposed areas of the impact structure, was studied in the course of the expeditions of 2000 and 2001. The following new sites were discovered: the shallow sea westward of Osmussaar Island, the coast between Cape Dirhami and

Riguldi village, Vormsi Island, surrounding of Rohuküla Harbour, Muhu Island, the coastal area southward of Haapsalu and Matsalu Bay, the coast of the Tõstamaa Peninsula and the coastal area of Northern Latvia (Fig. 10). A total of about 1000 erratics of Neugrund Breccia with a diameter of more than 1 m were observed in an area of about 5000 km^2. About 95% of these were found in the coastal areas. The erratics are easier to recognise on the coast than inland, so these numbers do not reflect the real pattern of distribution.

Fig. 13. Continental glacier eroded Inner Wall of the Neugrund impact structure, at a depth of 25 m.

The size of the erratics varied from small cobbles (smaller ones are not recognisable) to large boulders (diameter more than 10 m). Toodrikivi, the biggest erratic boulder (about 1200 m^3) in the Quaternary glaciation area of Eastern Europe (Fig. 2, 11), lying on the submarine Sandgrund Bank 5 km south of the rim wall, consists of Neugrund Breccia, as do gigantic boulders Skarvan (initial volume about 400 m^3) on the west coast of Osmussaar Island (Fig.12) and Nõva Suurkivi (about 400 m^3) on the coast of Toomanina Cape (Fig. 11). A total of 2120 erratic boulders, more than 1 m in diameter, were observed in a 6 km long and about 300-m-wide coastal zone between Cape Dirhami and Riguldi village. About 10% of these (210) were of Neugrund Breccia. Eleven boulders were larger than 5 m in a diameter.

By the composition of the clasts the following types of Neugrund Breccia erratic boulders were differentiated: amphibolitic (45 vol%); gneissic (25 vol%); granitoidic (18 vol%), migmatitic (12 vol%).

The distribution area of these erratics was limited by the Neugrund Breccia occurrences on Ruhnu Island and the coast of North Latvia, respectively 150 and 180 km from the impact structure. The north–south stretched shape of the area obviously indicates the movement direction of the last continental glacier sheet (Fig. 10).

7 Discussion

As a result of marine geological and geophysical investigations carried out in the Neugrund impact structure and the surrounding area in 2000 and 2001, the interpretation of the morphology and measurements of the structure has changed. Calculations of the impact energy and definitions of immeasurable parameters are almost always based on the rim-to-rim diameter of a structure (Dence 1973; Deutsch and Schärer 1994; French 1998; Grieve 1987; Melosh 1989; Melosh and Ivanov 1999; O'Keefe and Ahrens 1997; O'Keefe and Ahrens 1999; Pike 1985; Poag 1997; Tsikalas et al. 1998). The rim-to-rim diameter of the Neugrund structure is not easy to determine, because it has an extremely wide (up to 3 km) 3-ridged rim wall and a 20 - 21 km wide outer rim or ring fault outside it. Moreover, dislocations in the pre-impact sedimentary cover, obviously connected with the Neugrund impact event, can be detected outside the outer rim up to a distance of 20 km from the impact centre. The ring fault (limit of the outer wall) separates the area where crystalline target rocks are dislocated from the area where these are mostly intact.

What is the real rim-to-rim diameter of the Neugrund impact structure that could be used in all calculations: the 6.5 km wide ridge of the inner and best-preserved part of the rim wall (Inner Wall), the 9 km wide ridge of the Middle Wall, or the ring fault 20 km in diameter? In the cases where the inner rim is absent or not preserved, the diameter of the outer rim has been treated as the rim-to-rim diameter of an impact structure (Abels et al. 2000; Dypvik et al. 1996; Gibson and Reimold 2000; Kenkman et al. 2000; Koeberl and Anderson 1996; Lindström et al. 1996; Lilljequist 2000; Ormö and Lindström 2000). When there exists a well-developed rim wall and problems arise with the outer rim, the diameter of the former has been taken as the rim-to-rim diameter of a structure (Puura and Suuroja 1992; Suuroja et al. 2001b).

Also, the very wide distal distribution of the shock-metamorphosed material (mainly quartz grains with PDFs) and in deposits of different age is problematic in the case of the Neugrund structure. The size and roundness (subrounded and rounded grains dominate) of quartz grains bearing shock-metamorphic features indicate that these grains are derived mostly from Late Vendian and Early Cambrian siliciclastic deposits, which at the time of the impact formed an approximately 100 m thick layer. It is difficult to explain how shock-metamorphosed quartz got into the breccia-like deposits, or so-called sediment intrusions (Suuroja et al. 2002) of Osmussaar Island. These intrusions are about 60 Ma older than the Neugrund structure and the island is located on the outer boundary of the Neugrund impact structure. As other impact structures of the same age are absent in the nearest surroundings (Pesonen 1996; Pesonen and Henkel 1992; Ormö and Lindström 2000), the only probable explanation is that the shock-metamorphosed matter of these intrusions has been derived from the submarine outcrops of the Neugrund impact structure. This statement is not correct either, since the geological section of the crater proper shows that before penetration of sediment intrusions all structures of the crater must have been buried and siliciclastic deposits bearing shock-metamorphic matter were not accessible.

Acknowledgements

We are grateful to T. Flodén (Stockholm University) and the crew of the r/v "Skagerak" for help in organising the marine expedition and carrying out the geophysical investigations during summer 2001. We also thank marine archaeologists V. Mäss and A. Eero (Estonian Maritime Museum) for assistance in investigating and sampling submarine objects, inaccessible to geologists. Special thanks go to A. Talpas, our colleague from the Geological Survey of Estonia, for providing the authors unpublished geophysical information of the Neugrund structure area and for useful suggestions, and to A. Popova for mineralogical analyses. The Geological Survey of Estonia under the auspices of the State Geological Mapping Programme and the Impact Programme of the European Science Foundation is also thanked for financial support. We also thank the reviewers R. A. F. Grieve and H. Henkel for helpful comments and suggestions.

References

Abels A, Zumsprekel H, Bischoff L (2000) Basic remote sensing signatures of large, deeply eroded impact structures. Gilmour I, Koeberl C (eds) Impacts and the Early Earth. Lecture Notes in Earth Sciences 91, Springer Heidelberg, pp 309-326

French BM (1998) Traces of Catastrophe. A handbook of Shock-Metamorphic Effects in Terrestrial Meteorite Impact Structures. LPI Contribution No 954, Lunar and Planetary Institute, Houston, 120 pp

Dence MR (1973) Dimensional analysis of impact structures. Meteoritics 8: 343-344

Deutsch A, Schärer U (1994) Dating terrestrial impact events. Meteoritics 29: 301-322

Dypvik H, Gudlaugsson ST, Attrep M Jr, Ferrell RE, Krinsley DH, Mørk A, Faleide JI, Nagy J (1996) Mjølnir structure: An impact crater in the Barents Sea. Geology 24: 779-782

Gibson RL, Reimold WU (2000) Deeply exhumed impact structures: The case of the Vredefort Structure, South Africa. Gilmour I, Koeberl C (eds) Impacts and the early earth. Lecture Notes in Earth Sciences 91, Springer, Heidelberg, pp 249-277

Grieve RAF (1987) Terrestrial impact structures. Annual Reviews of Earth and Planetary Sciences 15: 245-270

Grieve RAF, Langenhorst F, Stöffler D (1996) Shock metamorphism in nature and experiment: II. Significance in geoscience. Meteoritics and Planetary Science 31: 6-35

Kenkmann T, Ivanov BA, Stöffler D (2000) Identification of ancient impact structures: Low-angle faults and related geological features of crater basements. In: Gilmour I, Koeberl C (eds) Impacts and the Early Earth. Lecture Notes in Earth Sciences 91, Springer, Heidelberg, pp 279-307

Koeberl C, Anderson RA (1996) Manson and company: Impact structures in the United States. In: Koeberl C, Anderson RR (eds) Manson Impact Structure, Iowa: Anatomy of an Impact Crater. Geological Society of America, Special Paper 302: 1-30

Lilljequist R (2000) The Gallejaur structure, Northern Sweden. In: Gilmour I, Koeberl C (eds) Impacts and the Early Earth. Lecture Notes in Earth Sciences, Springer Verlag, Berlin-Heidelberg 91, pp 363-387

Lindström M, Sturkell EFF, Törnberg R, Ormö J (1996) The marine impact crater at Lockne, central Sweden. GFF 118: 193-206

Malkov B, Kiipli T, Rennel G, Tammik P, Dulin E (1986) The regional geological-geophysical investigation of the Baltic Sea shelf area Estonian SSR at a scale of 1: 200 000. Report of Investigations (in Russian). Geological Survey of Estonian SSR, Tallinn, 199 pp

Melosh HJ (1989) Impact Cratering. A Geologic Process. Oxford University Press, Oxford 245 pp

Melosh HJ, Ivanov BA (1999) Impact crater collapse. Annual Reviews of Earth and Planetary Sciences 27: 385-425

O´Keefe JD, Ahrens TJ (1977) Meteorite impact ejecta: Dependence of mass and energy lost on planetary escape velosity. Science 198: 1249-1251

O'Keefe JD, Ahrens TJ (1999) Complex craters: relationship of stratigraphy and rings to impact conditions. Journal of Geophysical Research 104 (E 11): 27091-27104

Ormö J, Lindström M (2000) When a cosmic impact strikes the sea bed. Geological Magazine 137: 67-80

Peil T (1999) Settlement history and cultural landscapes on Osmussaar. Estonia Maritima 4: 5-38

Pike R (1985) Some morphologic systematic of complex impact structures. Meteoritics 20: 49-68

Pesonen L (1996) The impact cratering record of Fennoscandia. Earth, Moon and Planets 72: 377-393

Pesonen L, Henkel H (eds) (1992) Terrestrial impact craters and craterform structures, with a special focus on Fennoscandia. Tectonophysics 216: 1-234

Poag CW (1997) Structural outer rim of Chesapeake Bay impact crater: Seismic and and bore hole evidence. Meteoritics and Planetary Science 31: 218-226

Puura V, Suuroja K (1992) Ordovician impact crater at Kärdla, Hiiumaa Island, Estonia. Tectonophysics 216: 143-156

Stöffler D, Grieve RAF (1996) Classification and nomeclature of impact metamorphic rocks: a proposal to the IUGS Subcommission on the Systematics of Metamorphic Rocks [abs.]. Lunar and Planetary Science 25: 1347-1348

Stöffler D, Langenhorst F (1994) Shock metamorphism of quartz in nature and experiment: I. Basic observation and theory. Meteoritics 29: 155-181

Suuroja K (2001) Kärdla Meteorite Crater. Geological Survey of Estonia, Tallinn, 38 pp

Suuroja K, Saadre T (1995) Gneiss-breccia erratic boulders from Northwestern Estonia as witnesses of an unknown impact structure (in Estonian). Bulletin of Geological Survey of Estonia 5/1: 26-28

Suuroja K, Suuroja S (2000) Neugrund Structure – the newly discovered submarine early Cambrian impact crater. In: Gilmour I, Koeberl C (eds) Impacts and the Early Earth. Lecture Notes in Earth Sciences, 91, Springer Verlag, Berlin-Heidelberg, pp 389-416

Suuroja K, Koppelmaa H, Kivisilla J, Niin M (1987) The deep geological mapping of the Nõva-Haapsalu area at a scale of 1:200 000. Three maps with explanatory note (in Russian). Geological Survey of Estonian SSR, Tallinn, 220 pp

Suuroja K, Kadastik E, Ploom K (1998) Geological mapping of Northwestern Estonia at a scale of 1:50 000 (in Estonian with English summary). Three maps with explanatory note (in Estonian with English summary). Geological Survey of Estonia, Tallinn, 210 pp

Suuroja K, Suuroja S, Talpas A (1999) The marine geological investigations of the Neugrund structure area. Three maps with explanatory note (in Estonian). Geological Survey of Estonia, Tallinn, 180 pp

Suuroja K, Suuroja S, All T, Floden (2001a) Kärdla (Hiiumaa Island, Estonia) – the buried and well-preserved Ordovician marine impact structure. Deep Sea Research, Part II 982: 101-124

Suuroja S, All T, Plado J, Suuroja K (2001b) Geology and magnetic signatures of the Neugrund impact structure, Estonia. In: Plado J, Pesonen L (eds) Impacts in Precambrian Shields, Impact Studies vol. 2, Springer Verlag, Heidelberg, pp 277-294

Suuroja K, Kirsimäe K, Ainsaar L, Kohv M, Mahaney WC, Suuroja S (2002) The Osmussaar Breccia in Northwestern Estonia – Evidence of a 475 Ma Earthquake or an Impact? In: Koeberl C, Martinez-Ruiz F (eds) Impact Markers in the Stratigraphic Records, Impact Studies 3, Springer Verlag, Heidelberg, pp 333 - 347

Talpas A, Väling P, Kask J, Mardla A, Sakson M (1993) The geological mapping of the shelf area of the Baltic Sea at a scale of 1:200 000. Report of Investigations (in Russian). Geological Survey of Estonia, Tallinn, 152 pp

Torsvik TH, Smethurst MA, Van der Voo R, Trench N, Halvorsen E (1992) Baltica. A synopsis of Vendian-Permian palaeomagnetic data and their palaeotectonic implications. Earth-Science Reviews 33: 133-152

Tsikalas F, Gudlaugsson ST, Faleide JI (1998) The anatomy of a buried complex structure: The Mjølnir Structure, Barents Sea. Journal of Geophysical Research 103: 30469-30483

Tucker RD, McKerrow WS (1995) Early Paleozoic chronology: a review in light of new U-Pb zircon ages from Newfoundland and Britain. Canadian Journal of Earth Sciences 32: 368-379

Structure-filling Sediments of the Wetumpka Marine-target Impact Structure (Alabama, USA)

David T. King, Jr.[1], Thornton L. Neathery[2] and Lucille W. Petruny[3]

[1]Department of Geology, Auburn University, Auburn, AL 36849-5305, USA (kingdat@auburn.edu)
[2]Neathery and Associates, 1212-H Veteran's Memorial Parkway, Tuscaloosa, AL 35404, USA
[3]Department of Curriculum and Teaching, Auburn University, Auburn, AL 36849, USA and Astra-Terra Research, Auburn, AL 36831-3323, USA

Abstract. The Wetumpka impact structure is a deeply eroded, arcuate, 7.6-km diameter Late Cretaceous feature located within the inner Coastal Plain, Alabama, which was produced by a Late Cretaceous impact into shallow marine water. Wetumpka has distinctive exposed, geologic terrains produced by impact-related processes. These exposed terrains include Wetumpka's crystalline rim and two sedimentary terrains: (a) an interior unit (resurge unit), and (b) adjacent extra-structure unit (deformed unit) located outside the rim on the structure's southern side. Core drilling near the structure's geographic center revealed that Wetumpka's structure fill has two distinctive units: (1) an upper, resurge-deposited unit (~ 60 m thick; same as the "interior unit" above) and (2) a lower, structure-filling breccia unit comprised of fall-back ejecta layers, slumped target-rock blocks, and impact-related sandy breccias and sands (> 130 m thick).

1 Introduction

The Wetumpka impact structure, located in Elmore County, Alabama, USA (centered at N32° 31.3', W86° 10.4'; Fig. 1), is a locally prominent, semi-circular, rimmed feature with as much as 120 m of relief. This impact structure is composed of relatively highly indurated crystalline rock, which forms the impact-structure rim, and an unconsolidated mélange of resedimented and(or) deformed Upper Cretaceous sedimentary formations comprising two impact-structure related sedimentary terrains: (a) within the crystalline rim (interior unit); and (b) directly outside the crystalline rim on the southern side (extra-structure or deformed unit).

Wetumpka's impact-related surficial structures and stratigraphy were summarized initially by Neathery et al. (1976), who mapped and then named this structure "Wetumpka astrobleme," using the term astrobleme as suggested by Dietz (1961). However, lack of unequivocal petrographic evidence left open the question of Wetumpka's origin. In a comprehensive review of impact structures within the United States, Koeberl and Anderson (1996) recognized Wetumpka only as a "possible impact structure." Recent work has established unequivocal proof of impact origin using petrographic evidence (i.e., shocked quartz with abundant PDFs) and geochemical evidence (i.e., elevated Ir concentrations in impact breccias; King et al. 2002). In this paper, we refer to Wetumpka as an "impact structure" (*sensu* Stöffler and Grieve 1994), because it is deeply eroded and has a relatively degraded state of preservation.

Fig. 1. Location of the Wetumpka structure (dot at end of arrow) in the state of Alabama, USA. The main physiographic/geologic regions of Alabama are indicated.

2
Target Stratigraphy and Paleogeography

In the inner Gulf Coastal Plain near Wetumpka impact structure, three relatively soft, essentially unconsolidated Upper Cretaceous stratigraphic units lie unconformably upon harder crystalline, pre-Cretaceous Appalachian bedrock. These units comprise a broad monoclinal structure that dips south at approximately 8.5 m/km. In age order, these Upper Cretaceous stratigraphic units are: Tuscaloosa Group (consisting of approximately equally thick Coker Formation and overlying Gordo Formation; ~ 60 m total); Eutaw Formation (~ 30 m); and Mooreville Chalk (~ 30 m; Neathery et al. 1976; Neathery 1983; Fig. 2). Tuscaloosa is a Cenomanian fluvial and alluvial plain deposit, comprised of numerous 2- to 4-m thick, fining-upward sequences of clayey sands and sandy clays (Reinhardt et al. 1986). Eutaw comprises Santonian linear shoreline deposits of quartz-rich sands with small amounts of intercalated sandy marine clays (King 1994). Finally, Mooreville is comprised of Campanian marine marly clays and clayey marls (King 1994), a sediment type locally referred to as "chalk." All these units were target strata for the Wetumpka impact event and all occur as clasts within Wetumpka impact structure-filling units (King et al. 2002).

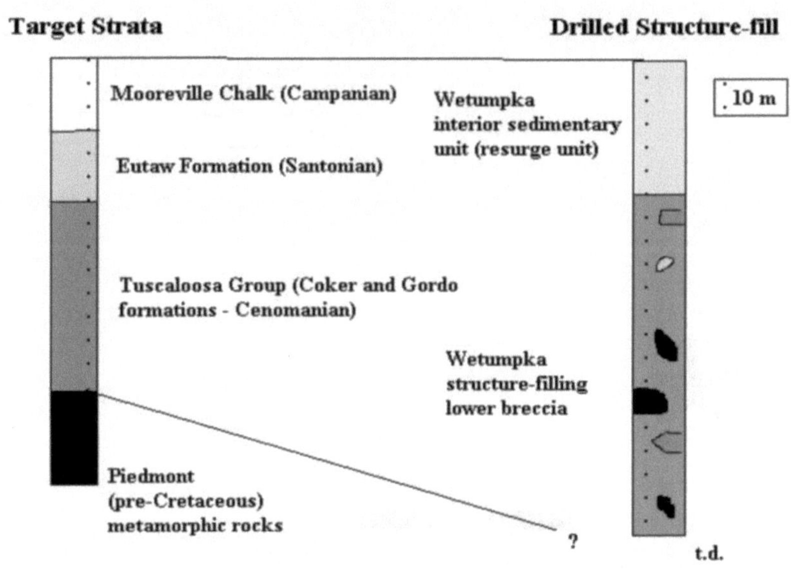

Fig. 2. Target stratigraphy versus stratigraphy in drilled core holes. Target strata occur as target-rock blocks in lower breccia and were resedimented together in the interior sedimentary (resurge) unit.

The Wetumpka impact structure is thought to be an early Campanian geologic feature (ca. 83 Ma) because the youngest layers involved in Wetumpka's impact event were derived from the lower part of the Mooreville Chalk, which contains early Campanian guide fossils (King 1997). At time of impact-structure formation, Wetumpka's target area was part of the Late Cretaceous continental shelf of southern Alabama, and the target site was likely shallow marine water within a few tens of kilometers of the paleo-shoreline (Fig. 3; King 1997; King and Neathery 1998). Judging from ostracode eye morphology (Puckett 1991) and supportive ichno-sedimentologic evidence within Campanian target strata (Rindsberg 1986), between ~ 35 and 100 m of seawater probably covered the sea floor at the target area.

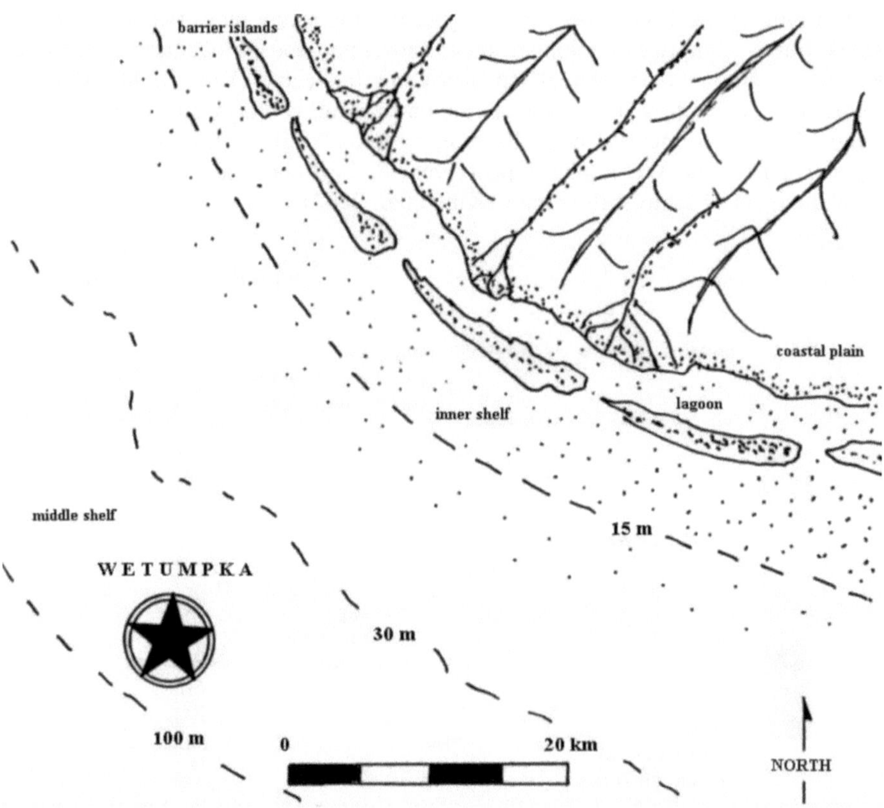

Fig. 3. Paleogeographic reconstruction of setting for Wetumpka impact event (Campanian, ca. 83 Ma; after King 1997). Hypothetical paleobathymetric contours show inferred depth for impact event (see text). Impact center marked by star.

3
Exposed Surficial Geology

Present surficial geology of the Wetumpka impact structure (Fig. 4a) consists of two main regions: (1) a heavily weathered, semi-circular crystalline rim terrain composed of post-orogenically deformed Appalachian Piedmont bedrock, which has up to 87 m of present relief, and (2) a contiguous tract of two relatively low-relief, stream-dissected sedimentary terrains (i.e., an "interior unit" comprising the floor of the structure and an extra-structure terrain or "deformed unit" located outside the rim on the rim's southern side). These sedimentary terrains are contiguous tracts of, respectively, resedimented and(or) deformed target strata and undisturbed to highly disturbed target strata (Fig. 4b).

The crystalline rim's outcrop has a northwest-southeast diameter of 6.1 km, however the total diameter of Wetumpka impact structure should be determined by measuring the maximum distance between concentrically disposed impact-related structures (Melosh 1989). In this instance, the present diameter is properly measured from a concentric fault and associated, small outlier of extra-structure terrain (on the southeast side Fig. 4a) to faulted crystalline rocks cropping out in the Coosa River (a part of the northwestern rim; at the Qal-crt contact in Fig. 4a). This distance yields a total structural diameter of 7.6 km.

The crystalline rim consists mainly of metamorphic pre-Cretaceous Appalachian piedmont units (e.g., Kowaliga Gneiss and Emuckfaw Group schists; Szabo et al. 1988), which crop out in a crescent-shaped arc (Fig. 4a,b). In contrast to exposures of the same and related piedmont units outside the Wetumpka impact structure; this terrain displays realigned (i.e., radially oriented) dips of foliation and concentrically striking normal faults (Neathery et al. 1976; Nelson 2000).

The interior unit consists mainly of two lithic types. One lithic type consists of resedimented and(or) deformed Tuscaloosa Group sediments, mostly red sands and clays of the Coker and Gordo formations, which enclose irregularly shaped and internally deformed outliers of younger target units (i.e., megablocks of Eutaw Formation and Mooreville Chalk; Fig. 4b). The other lithic type consists of blocks of internally deformed and undeformed Tuscaloosa Group strata. Unusual deformation features also characterize this interior-unit terrain, including outcrop-scale soft-sediment folding and shearing, clastic-dike injection, and imbricated stacking of slumped megablocks (Neathery et al. 1976; King 1998).

Within a small (~ 0.75 km^2), centrally located outcrop area, a rare surficial breccia unit crops out (Fig. 4a; King et al. 2002; Nelson 2000). This breccia contains numerous subrounded to subangular fragments, 5 cm to 30 m across, consisting of schists and gneisses like those present in the crystalline rim and sandstone clasts (from the Tuscaloosa Group and(or) Eutaw Formation), which are all supported by a fine sandy matrix. Associated with, and situated within, this breccia facies, are several lithic megablocks (15 to 25 m across) composed of micaceous schist.

The interior unit grades laterally toward the south into the extra-structure (deformed) terrain, which consists mainly of undisturbed to highly disturbed Tuscaloosa Group sediments. Some of these strata have experienced structural deformation of a probable tensional nature. For example, in several places, Mooreville Chalk fills what we interpret to be northeast-trending, graben-like features developed within Tuscaloosa Group and Eutaw Formation strata (Fig. 4a,b; Neathery et al. 1976; Nelson 2000). This terrain may be analogous to a partially developed outer 'terrace zone' attendant to typical complex impact structures (Melosh 1989).

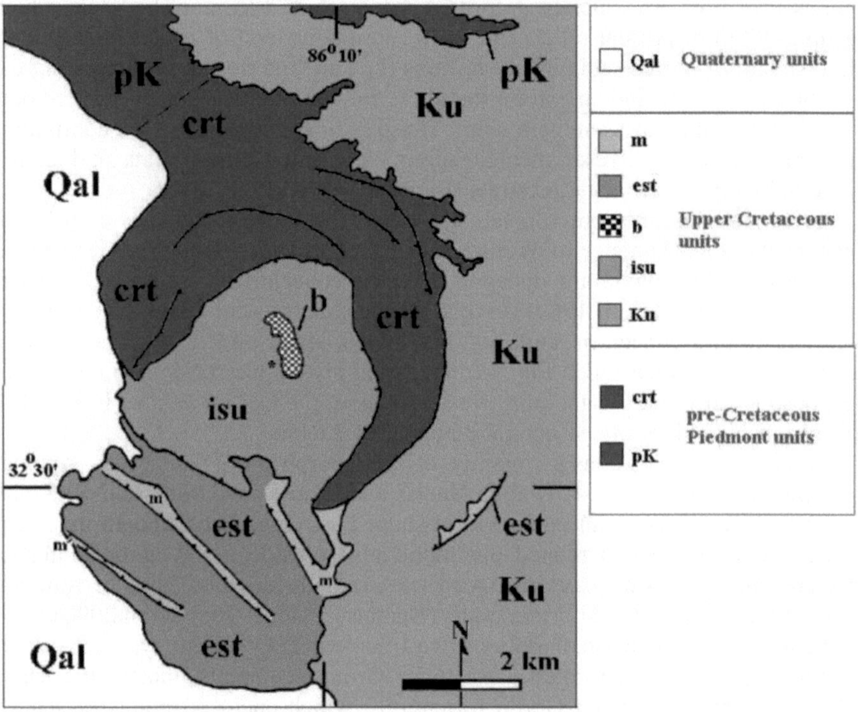

Fig. 4a. Simplified geological map of Wetumpka impact structure (after King et al. 2002) showing units **pK** (pre-Cretaceous metamorphic rocks), **crt** (crystalline rim terrain), **Ku** (Upper Cretaceous target strata), **isu** (interior sedimentary unit), **b** (surficial breccia unit), **est** (extra-structure terrain), **m** (Mooreville chalk inliers within extra-structure terrain), and **Qal** (Quaternary alluvium). Compare with Fig. 4b, below. Drill site marked by asterisk (*).

At Wetumpka, the interior unit and related extra-structure (deformed) terrain probably represent special types of 'broken formations' (a term modified from Raymond 1984) formed by marine-impact deformation and sedimentation processes. Target stratigraphic units, though recognizable as fragments or deformed elements within these terrains, cannot be mapped accurately because of their fine-scale complexity (Nelson 2000). These two terrains have been referred to previously as "Wetumpka mélange" (King 1998), and the interior unit has been referred to as "Wetumpka interior broken unit" (Nelson 2000), for lack of potentially more appropriate terminology. As Fig. 4a shows, these terrains comprise separate, mappable, lithostratigraphic units of impact-related origin, located in and around the Wetumpka impact structure (King 1998).

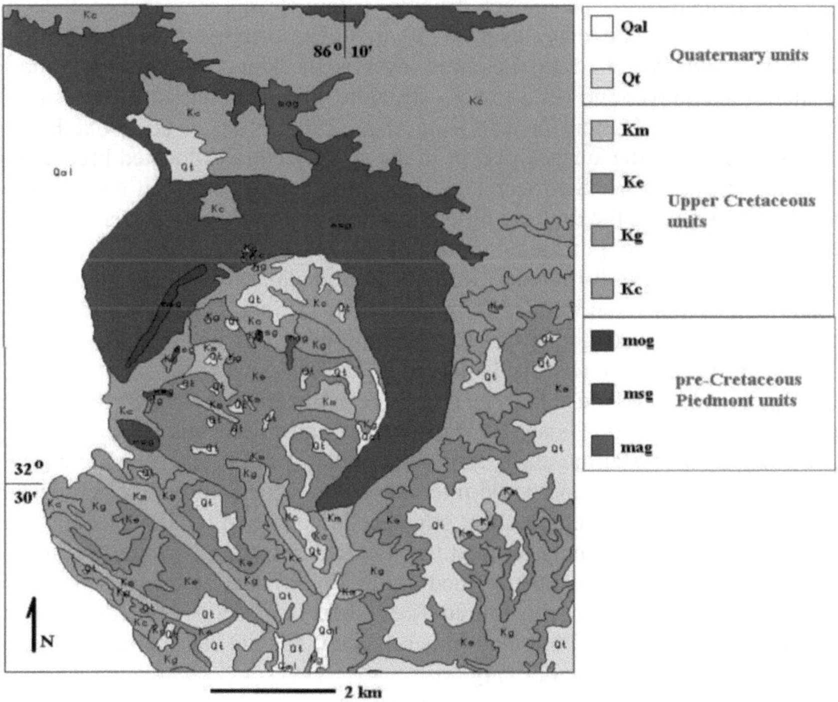

Fig. 4b. Detailed geological map of Wetumpka impact structure (after Neathery et al. 1976). Pre-Cretaceous units are **mag** (augen gneiss), **msg** (schist-gneiss), and **mog** (orthoquartzites and graphitic schists). Upper Cretaceous target stratagraphic units are **Kc** and **Kg** (respectively, Coker and Gordo formations of the Tuscaloosa Group), **Ke** (Eutaw Formation), and **Km** (Mooreville Chalk). Quaternary units are **Qt** (older high terrace) and **Qal** (present alluvium). Note that the **isu** (interior sedimentary unit) depticted in Fig. 4a has been mapped here as an amalgam of numerous irregular zones wherein large tracts (megablocks) of various target strata are recognizable. In the region mapped **est** (extra-structure terrain) on Fig. 4a, stratigraphic relations between units **Kc**, **Kg**, and **Ke** (shown above) appear mainly flat-lying, as in the undisturbed strata east of the structure, and northwest-striking grabens containing **Km** (Mooreville Chalk) characterize this terrain. Map areas of Figs. 4a and 4b are approximately the same.

4 Drilled and Interpreted Subsurface Geology

In 1998, drilling near Wetumpka's geographic center produced cored samples from two wells, the Schroeder and Reeves wells, located 122 m apart (Table 1). These drill cores display an impact structure-filling stratigraphy that includes (1) an upper, red sand and clay layer, ~ 60 m thick, comprised of resedimented and(or) disturbed target strata (i.e., this layer is the Upper Cretaceous "interior unit" described above) and (2) a lower, structure-filling breccia unit consisting of > 130 m of diverse breccia facies (King et al. 1999; King et al. 2002; Fig. 5), which is not seen at the surface except in the small, centrally located breccia outcrop noted above. In the Schroeder well core, the contact between the upper red sand and clay unit and the underlying breccia unit is very sharp (i.e., there is no intercalation as with other petrologic types in the cores).

The lower breccia unit includes, in order of abundance, three petrologic types: (1) sandy breccia and sand units; (2) impact breccias; and (3) target-rock blocks. Figure 6 shows the arrangement of these petrologic types within each well core, and the sequence of these units is displayed in Table 1 as well.

Sandy breccia and sand units are angular clast-bearing and matrix-supported. These green to grey deposits are typically clayey and micaceous. There is a varying abundance of crystalline clasts (mainly angular, 1 to 10 cm, schist and gneiss fragments) in these units, and some intervals contain rounded quartz pebbles and lignitic clasts. The main sedimentary structure is contorted bedding which is accentuated by laminae of finely divided lignitic material, clay, and micas (Fig. 7a). Sands in these deposits are mainly medium to fine. Schist and gneiss fragments are typically slightly-to-highly altered (i.e., sheet and framework silicates are wholly or partially converted to clay minerals) and thus the clasts commonly have a white ("bleached") appearance (Fig. 7b).

Impact breccias are monomict and polymict (i.e., shocked) clast-supported breccias (*sensu* Stöffler and Grieve 1994). Typically, these intervals are relatively well indurated. The constituent crystalline clasts, 0.5 to 4 cm across and angular to subangular, are typically slightly-to-highly altered like some of the clasts in the sandy breccias. The grey to green matrix is composed of fine, clayey micaceous sand (Fig. 7c). Grain support among clasts is common in most intervals. The upper and lower contacts of impact breccia layers are generally very sharp, but in some instances are gradational with sandy breccia and sand units over a few centimeters. Polymict impact breccias were the subject of intensive study by King et al. (2002), who reported on their shocked quartz content (dominant PDFs = $\omega\{10\bar{1}3\}$, r,z $\{10\bar{1}1\}$, and $\xi\{11\bar{2}2\}$) and slightly elevated Ir values.

In the lower breccia unit, sandy breccia and sand and impact breccia units are intercalated with 1- to 10-m thick, deformed and undeformed blocks of target strata and crystalline basement (King et al. 2002; Table 1). These target blocks are derived from the main stratigraphic units of the target sequence (i.e., the Tuscaloosa Group and Eutaw Formation) plus crystalline bedrock (schist and gneiss; Table 1). Upper and lower contacts with target blocks are typically sharp, lithic boundaries. Target blocks are internally deformed (plastically and by fracturing) in some places. Crystalline target blocks appear to be highly weathered because they are, in places, much softer than comparable fresh rock.

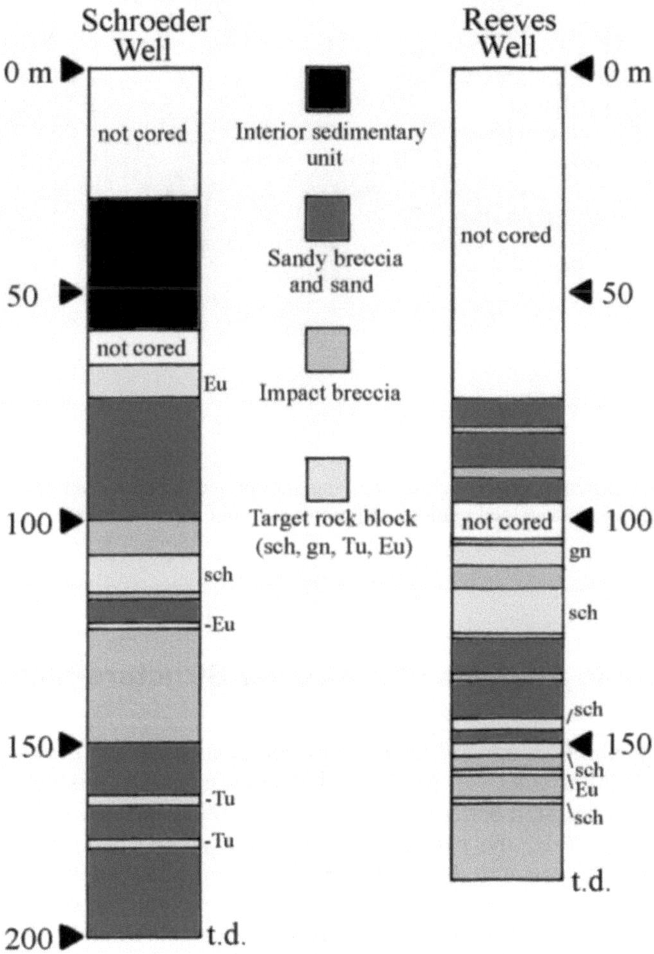

Fig. 5. Lithologic logs of drilled cores from Schroeder and Reeves wells (after King et al. 2002). Abbreviations for target-rock blocks are **sch** (schist), **gn** (gneiss), **Tu** (Tuscaloosa Group), and **Eu** (Eutaw Formation). See text for description of each lithic type in the logs.

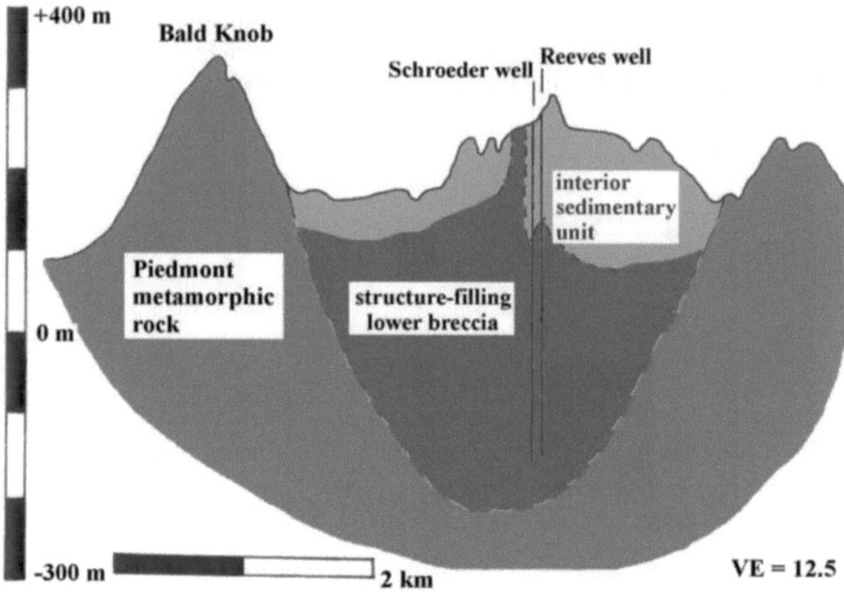

Fig. 6. Impact-structure cross-section showing interpreted relations between units based upon core drilling and surficial geologic relations, as explained in the text.

5 Stratigraphic Analysis of the Lower Structure-filling Units

A classical method for upward facies-transition analysis in sedimentary stratigraphy (Selley 1970; Miall 1973; McDonnell 1978) was used in this instance to help visualize and understand better the stratigraphic relationships among the structure-filling lithologies (i.e., the petrologic types given above) comprising the lower structure-filling breccia unit. Use of upward facies-transition analysis is applicable here because, unlike sedimentary sequences where recognition of upward transition from one facies to another is a statistical problem (Walker 1984), there is no problem of ambiguity in "facies recognition" in this situation. Further, the problems of sedimentary cyclicity, which raised questions about the usefulness of upward facies-transition analysis in papers published several years ago (see discussion in Miall 1996), are not an issue in these impact-structure filling units.

Lithologies from Table 1 were compiled in an "observed upward transition matrix" (Table 2), wherein the number in each row tallies the number of times that a petrologic type in the corresponding row's heading is overlain by the petrologic type in the corresponding column's heading. For example, in the "observed upward transition matrix" (Table 2), the most common upward transition is 'impact breccia' overlain by 'target-rock blocks' (i.e., this transition occurs 7 times). Cells in this observed matrix have only a small range of values, indicating a relatively even (but not random) distribution of upward transitions in these cores. Cells with zero value did not occur.

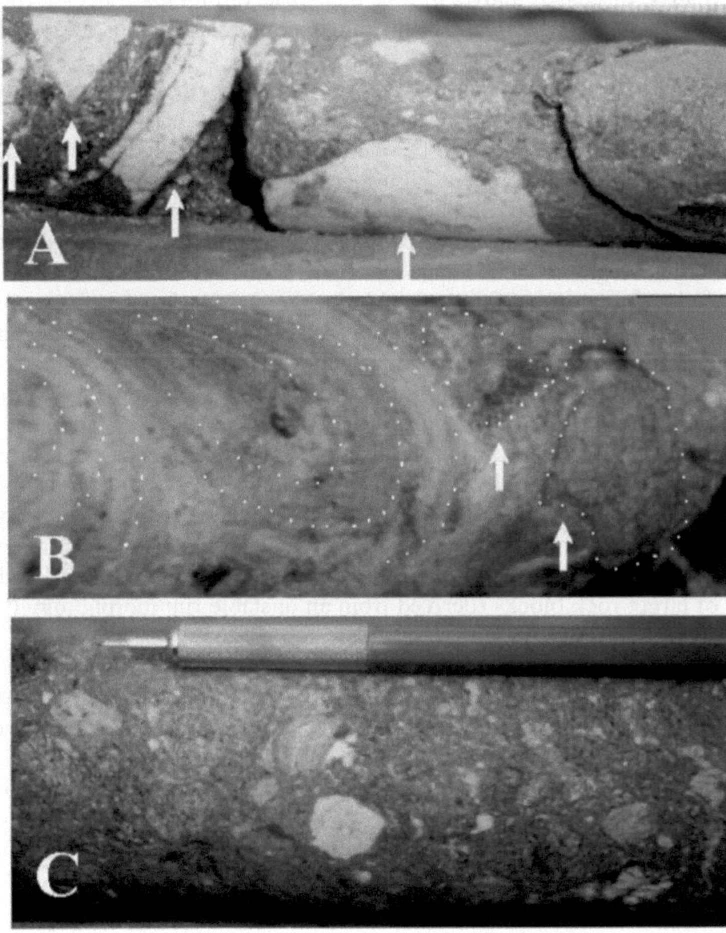

Fig. 7. Views of key petrologic types in drill core. A) Sandy breccia, which contains crystalline rock clasts (angular cobbles, arrows) that have been altered to a bleached condition in a matrix of micaceous medium to fine clayey sand (width of frame = 15 cm). B) Sandy breccia with contorted lamination (on left, outlined by dots), which contains relatively unaltered crystalline bedrock clasts (dark gneiss, on right, outlined by dots with arrows pointing; width of frame = 10 cm). C) Impact breccia (polymict, in this instance, as confirmed by thin-section analysis of shocked quartz), which is composed of clayey fine-sand matrix with subangular pebbles of crystalline bedrock (width of frame = 10 cm).

For comparative purposes, an "expected upward-transition matrix" (Table 2) was computed, wherein the expected value within each active cell is the row total multiplied by the column total, and this sum is divided by the total number of transitions in the matrix (Selley 1970). This expected matrix was subtracted from the observed matrix, discussed above, to give the "observed minus expected (or residual) matrix" (Table 2), which helps illustrate genetic relationships in an objective way.

The residual matrix shows that the 'impact breccia' petrologic type tends to be overlain by 'target-rock blocks' more commonly than 'sandy breccia and sand.' In turn, 'target-rock blocks' tend to be overlain by 'sandy breccia and sand' more commonly than by 'impact breccia.' Further, 'sandy breccia and sand' tend to be overlain more commonly by 'impact breccia' than by 'target-rock blocks.' The residual matrix shows that no upward transitions occur less often than expected (i.e., there are no negative values in cells).

The technique above is a "first-order Markov process" wherein "the probability of the process being in a given state at a particular time may be deduced from knowledge of the immediately preceding state" (Harbaugh and Bonham-Carter 1970). Therefore, inferences on preferred order of the lower breccia's units can give us a view of an 'ideal' or 'model' formative sequence (Miall 1996). As noted above, interpreting from these data, we obtain a 'model sequence' as follows: (1) impact breccia, (2) target-rock blocks, and (3) sandy breccia and sand units (see Reading 1986). The significance of this 'model sequence' cannot be established with certainty, but it could be interpreted feasibly as a sequence reflecting successive impact processes such as: (1) fall back of ejecta; (2) slump and related comminution of target-rock blocks (derived from an unstable rim morphology (?); and (3) centrifugal fluid flows within the structure.

An alternative explanation of the order of petrologic types within the lower breccia unit would be that they are entirely random or chaotic. We think if this were true, there would be no detectable 'ideal or model sequence' as above, or that the detected sequence would make no sense from a process point of view. Obviously, having more drill core and thus a larger number of matrix entries would be desirable in this instance. In order to test the usefulness of this technique, we encourage others who are working with petrologic sequences within drilled impact structures to look for similar "Markov" patterns.

6 Interpretation of Structure-filling Sequence

As noted in the descriptions above, Wetumpka impact structure-filling stratigraphy contains two distinct units. The 60-m thick red sand and clay unit drilled in the Schroeder well (corresponding to the "interior unit" of surficial geologic mapping), which, by virtue of its superior stratigraphic position, must represent deposition rather late during the modification stage of impact-structure development. The underlying succession of structure-filling impact breccias, target-rock blocks, and sandy breccia and sand units, therefore, must represent an earlier, structure-filling episode of the early modification stage (Figure 8).

Table 1. Lithologic synthesis of drill cores from Schroeder and Reeves wells. The ‘lithologic type’ headings correspond to those used in the text. Blank intervals were not cored.

SCHROEDER WELL			REEVES WELL		
N 32°31.368’ x W 086°10.369’			N 32°31.303’ x W 086°10.379’		
LITHOLOGIC TYPE	Top (m)	Base (m)	LITHOLOGIC TYPE	Top (m)	Base (m)
	0	29.5		0	75.7
Red sand and clay	29.5	60	Sandy breccia and sand	75.7	82.2
	60	67.5	Impact breccia	82.2	83.8
Target block (Eutaw Fm.)	67.5	75.3	Sandy breccia and sand	83.8	91.5
Sandy breccia and sand	75.3	103.7	Impact breccia	91.5	94.2
Impact breccia	103.7	111.6	Sandy breccia and sand	94.2	100
Target block	111.6	120.5		100	108.6
Impact breccia	120.5	122	Impact breccia	108.6	109.5
Sandy breccia and sand	122	127.4	Target block	109.5	114.4
Target block (Eutaw Fm.)	127.4	129.1	Impact breccia	114.4	119.9
Impact breccia	129.1	155.6	Target block	119.9	130.3
Sandy breccia and sand	155.6	167.3	Impact breccia	130.3	131.5
Target block (Tuscaloosa)	167.3	170	Sandy breccia and sand	131.5	139.5
Sandy breccia and sand	170	177.6	Impact breccia	139.5	141.3
Target block (Tuscaloosa)	177.6	179.8	Sandy breccia and sand	141.3	149.9
Sandy breccia and sand	179.8	200	Target block	149.9	153
			Sandy breccia and sand	153	156
			Target block	156	159
			Impact breccia	159	162.1
			Target block (Eutaw Fm.)	162.1	163.5
			Impact breccia	163.5	168.7
			Target block	168.7	170.1
			Impact breccia	170.1	187.5

The 60-m thick, upper red clay and sand unit (or “interior unit”) is interpreted to be a catastrophic deposit from a late modification-stage resurge event related to large-scale wall collapse (King et al. 2002; cf. Ormö and Lindström 2000). Wetumpka’s incomplete (~ 270°) rim may have something to do with this event, as this area has experienced no significant post-impact orogenic modification of the rim. Such a catastrophic resurge event may explain most of this interior unit's peculiar physical characteristics, which includes some deformational and some sedimentary features, noted above, and, of course, its stratigraphic position.

Table 2a-c. Upward facies-transition matrices (see text for discussion). Underlying lithologic types are row headings, overlying lithologic types are column headings. Data from drill core descriptions in Table 1. See text for description of matrix derivation.

A. Observed upward-transition matrix

	Target-rock blocks	Sandy breccia/Sand unit	Impact breccia
Target-rock blocks	0	5	5
Sandy breccia/Sand unit	4	0	6
Impact breccia	7	4	0

B. Expected upward-transition matrix

	Target-rock blocks	Sandy breccia/Sand unit	Impact breccia
Target-rock blocks	0	2.90	3.55
Sandy breccia/Sand unit	3.55	0	3.55
Impact breccia	3.90	3.19	0

C. Observed minus Expected (residual) matrix

	Target-rock blocks	Sandy breccia/Sand unit	Impact breccia
Target-rock blocks	0	2.10	1.45
Sandy breccia/Sand unit	0.45	0	2.45
Impact breccia	3.10	0.81	0

The lower breccia units are interpreted as the effects of fall-back (i.e., the impact breccias), wall slumping and disintegration (i.e., target-rock blocks), and centrifugal fluid flows within the structure (i.e., the sandy breccias and related sands). These effects occurred very early on in the structure's modification process. These deposits may have been strongly influenced by the flow of seawater which probably surged back over the rim and partially filled the early structure prior to the postulated later rim collapse (Figure 8; King et al. 2002; cf. von Dalwigk and Ormö 2001). As the drill cores did not penetrate to the base of the structure-filling sequence (i.e., drilling did not reach a "structure bottom"), we do not have an entirely complete picture of when and how lower breccia unit deposition began.

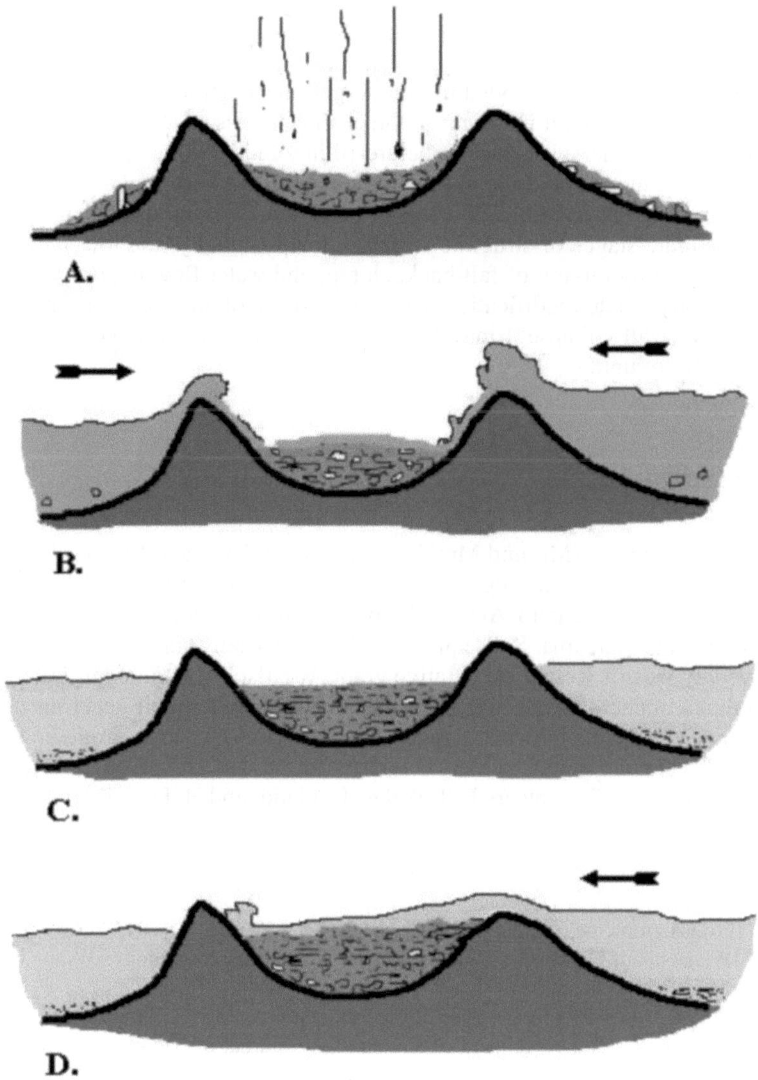

Fig. 8. Sequential diagrams showing inferred modification-stage history of Wetumpka impact structure. A) Fall-back of ejecta during early modification, before return of sea water. B) Post-impact (early modification) resurge of sea water, which washes over crystalline rim thus causing sea water to enter and help deposit sandy breccias and sands in lower breccia unit (such resurge may have occurred several times). C) Stasis, during which sea water is excluded by crystalline rim. D) Late-modification stage rim-collapse event resulting in catastrophic resurge sedimentation of interior sedimentary unit.

7 Conclusions

Wetumpka is a rather deeply eroded marine-target impact structure, which reveals some important details about its origin in the characteristics of its interior unit, extra-structure terrain, and subsurface structure-filling units. Although further work is needed to better understand the exact origin(s) of all of Wetumpka's structure-filling stratigraphy, the two-fold division of subsurface stratigraphy clearly suggests two, separate stages of structure filling. Perhaps initially, the lower breccia unit was produced by mixing of fall-back, slump, and water flow-deposited material. Subsequently, a late modification-stage event of involving a catastrophic secondary resurge (wall collapse?) may have deposited the complex red sands and clays of the interior unit.

Acknowledgments

We thank local residents (Mr. and Mrs. P. Schroeder and Mr. and Mrs. G. Reeves) for access to their property during well drilling. Grant support for drilling at Wetumpka was an in-kind gift to Auburn University from Vulcan Materials Company, Birmingham, Alabama. Subsequent work was partially supported by an Auburn University Dean's Research Initiative grant. We also thank all contributors to the "Wetumpka Impact Crater Fund" at Auburn University for their generous help. We appreciate the help of Mr. J. Graves, Auburn University, in producing the digital geological map figure (4a). The manuscript was improved due to the thoughtful comments of reviewers F. Tsikalas, L. Plado, and H. Dypvik.

References

Dietz RS (1961) Astroblemes. Scientific American 205: 50-58

Harbaugh JW, Bonham-Carter G (1970) Computer simulation in geology. Wiley Interscience, New York, 98 pp

King Jr DT (1994) Upper Cretaceous depositional sequences in the Alabama Coastal Plain: their characteristics and constituent clastic aquifers. Journal of Sedimentary Research B64: 258-265

King Jr DT (1997) The Wetumpka impact crater and the Late Cretaceous impact record. Alabama Geological Society Guidebook 34c: 25-56

King Jr DT (1998) Wetumpka melange, a new stratigraphic unit in Alabama. Gulf Coast Association of Geological Societies Transactions 48: 151-158

King Jr DT, Neathery TL (1998) The Wetumpka asteroid impact structure in Alabama, U.S.A. [abs.] American Association of Petroleum Geologists Annual Convention Abstracts 7: abstract no. 358, 6 p (CD ROM)

King Jr DT, Neathery TL, Petruny LW (1999) Impactite facies within Wetumpka impact crater, Alabama. [abs.] Lunar and Planetary Science 30: abstract no.1634 (CD ROM)

King Jr DT, Neathery TL, Petruny LW, Koeberl C, Hames WE (2002) Shallow marine-impact origin for the Wetumpka structure (Alabama, USA). Earth and Planetary Science Letters 202: 41-549

Koeberl C, Anderson RR (1996) Manson and company: Impact structures of the United States. In: Koeberl C, Andersom RR (eds) Manson Impact Structure, Iowa; Anatomy of an Impact Crater. Geological Society of America Special Paper 302: 1-30

McDonnell KL (1978) Transition matrices and the depositional environments of a fluvial sequence. Journal of Sedimentary Petrology 48: 43-48

Melosh HJ (1989) Impact Cratering, a Geologic Process. Oxford University Press, New York, 245 pp

Miall AD (1973) Markov chain analysis applied to an ancient alluvial plain succession. Sedimentology 20: 347-364

Miall AD (1996) The geology of fluvial deposits: sedimentary facies, basin analysis, and petroleum geology. Springer-Verlag, Belin Heidelberg, 582 pp

Neathery TL (1983) Description (of stops 15A, B, and C). Alabama Geological Society Guidebook 20: 48-52

Neathery TL, Bentley RD, Lines GC (1976) Cryptoexplosive structure near Wetumpka, Alabama. Geological Society of America Bulletin 87: 567-573

Nelson AI (2000) Geological mapping of Wetumpka impact crater area, Elmore County, Alabama [unpublished M.S. thesis]. Auburn University, Auburn, Alabama, 187 pp

Ormö J, Lindström M (2000) When a cosmic impact strikes the seabed. Geological Magazine 137: 67-80

Puckett TM (1991) Absolute paleobathymetry of Upper Cretaceous chalks based on ostracodes: Evidence from the Demopolis chalk (Campanian-Maastrichtian) of the northern Gulf Coastal Plain. Geology 19: 449-452

Raymond LA (1984) Classification of mélanges. In: Raymond LA (ed) Melanges, Their Nature, Origin, and Significance, Geological Society of America Special Paper 198, pp 7-20

Reading HG (1986) Facies. In: Reading HG (ed) Sedimentary Environments and Facies, 2nd. ed., Oxford, Blackwell Scientific, pp 4-19

Reinhardt J, Smith LW, King, Jr. DT (1986) Sedimentary facies of the Upper Cretaceous Tuscaloosa Group in eastern Alabama. Geological Society of America, Centennial Field Guide-Southeastern Section, pp 363-367

Rindsberg AK (1986) Cretaceous trace fossils in Alabama chalks. Alabama Geological Society Guidebook 26: 111-119

Selley RC (1970) Studies of sequence in sediments using a simple mathematical device. Geological Society of London Quarterly Journal 125: 557-581

Stöffler D, Grieve RAF (1994) Classification and nomenclature of impact metamorphic rocks: A proposal to the IUGS subcommission of the systematics of metamorphic rocks. In: Montanari A, Smit J (eds) Post-Östersund Newsletter, European Science Foundation (ESF) Network on Impact Cratering and Evolution of Planet Earth, Strasbourg, pp 9-15

Szabo MW, Osborne WE, Copeland Jr CW, Neathery TL (1988) Geologic map of Alabama. Geological Survey of Alabama, Special Map 220, scale 1:250,000

von Dalwigk I, Ormö J (2001) Formation of resurge gullies at impacts at sea: the Lockne crater, Sweden. Meteoritics and Planetary Science 36: 359-369

Walker RG (1984) Facies, Facies Sequences, and Facies Models. In Walker RG (ed) Facies Models 2nd. ed., Toronto, Geological Association of Canada, pp 1-10

Krk-breccia, Possible Impact-Crater Fill, Island of Krk in Eastern Adriatic Sea (Croatia)

Tihomir Marjanac*[1], Ana Marija Tomša[1], and Ljerka Marjanac[2]

[1] Department of Geology, Faculty of Science, University of Zagreb, Kralja Zvonimira 8, 10000 Zagreb, Croatia
[2] Institute of Quaternary Paleontology and Geology, Croatian Academy of Sciences and Arts, Ante Kovačića 5, 10000 Zagreb, Croatia
* Corresponding author: tmarjan@public.srce.hr

Abstract. We propose an impact origin of a polymict breccia of presumed Eocene-Oligocene age, which occurs in a continuous blanket (ca. 150 km^2) and many smaller patches scattered around the island of Krk in Eastern Adriatic Sea. This breccia is not differentiated in published geological maps as one lithological unit, but as two stratigraphically different units; namely of Early Cretaceous age and of Late Eocene to Early Oligocene age. The age of this breccia was attributed after the age of youngest clasts found, but the debris of Middle–Late Eocene age was recently also found within "Early Cretaceous" breccia, thus indicating its much younger age. Our map shows that previously differentiated breccias belong to a single lithological unit which we informally call the Krk-breccia.

The Krk-breccia is generally massive, with chaotic fabric and a variable amount of matrix. Its bedding is unclear, except where matrix-rich breccia underlies matrix-poor variety, and soft-sediment injection structures mark the contact. The Krk-breccia debris is stratigraphically varied; the oldest debris is of Jurassic age, and the youngest clasts are represented by Middle Lutetian Flysch sandstones. The debris is of pebble- to cobble-size, but in the lower part of the breccia there occur very coarse clasts of Early Eocene Foraminiferal limestones, some of which exceed 64 m across. The clasts are angular and occasionally *in situ* fractured with matrix-filled fissures, what indicates very rapid deposition. Approximately in the centre of the mapped Krk-breccia area there occurs a ca. 5 km wide field of scattered large limestone blocks, which herein are referred to as "megablock facies".

The Krk-breccia is 1500 m thick, as revealed by the Krk-1 deep exploration well, which reached the Triassic/Jurassic boundary at a depth of 3100 m. Comparison with "ideal" stratigraphic succession reveals that the drilled stratigraphic units are uplifted by ca. 810–1360 m at the well location.

The extraordinary thickness of the Krk-breccia, its lensoid geometry, unsorted-chaotic fabric, unselective large-scale erosion which provided the debris and cartographic appearance in an elliptical unit, which is restricted to a depression in the central part of the Krk Island, may be explained in terms of an impact-crater fill.

1 Introduction

The Eastern Adriatic islands and coast (Fig. 1) are localities of "classical" carbonate platform deposits (e.g., Grubić 1980). This area was also studied because of its seismic activity (e.g.,Aljinović et al. 1987; Prelogović et al. 1982; Prelogović et al. 1995) and tectonical framework (e.g., Aubouin et al. 1970; Jamičić et al. 1995; Miljuš 1973; Oluić et al. 1972). However, only a few researchers attempted to interpret the origin of the clastic sequences occurring in this region (e.g., Babić and Zupanič 1983, 1998; Marjanac 1989, 1996).

The authors of geological maps of the northern Adriatic islands and coastal mountains (e.g., Šušnjar et al. 1970; Grimani et al. 1973) differentiated various carbonate breccias on the Krk Island, some of which were attributed to Early Cretaceous age, and some were identified as “carbonate breccias (E_3Ol_1)”.

The age of breccias is always a problem, especially when the matrix is unfossiliferous, which is common in this case. Thus, the Early Cretaceous age was attributed to some breccias only because no younger clasts were found. In this way stratigraphically different breccias were identified on the geological map of the Island of Krk (Šušnjar et al. 1970) as of Lower Cretaceous and Eocene–Oligocene age, respectively. However, some research students (Boljat 1981; Milinović 1981) found Paleogene debris in many outcrops of the “Lower Cretaceous” breccia, which suggested that the extent of the Eocene–Oligocene breccia must be much wider than shown on the geological map. The exploration well Krk-1 drilled almost in the centre of the island (Fig. 2), revealed that the first 1500 m of the well were drilled through a breccia. This breccia is informally called the Krk-breccia. Along the coast, on the Velebit Mountain and, locally, in its hinterland occurs extensive, lithologically similar breccia which is known as Jelar-beds (Bahun 1974; Herak and Bahun 1979), also of presumed Eocene-Oligocene age (Fig. 1).

We have attempted to map in detail the extent of the Krk-breccia, because it has not been differentiated on existing maps, and tried to interpret its genesis.

2 Location and Morphology

The Island of Krk is the largest island in the Adriatic Sea (410 km^2) (Fig. 1), and topographically comprises two distinct morphological units (Fig. 2). The northwestern part of the island is characterized by relatively low topography. The hills are ca. 230–250 m high, and form an arcuate ridge around depressions that are close to the sea-level (reservoir lake bottom at +16 m, Punat Bay bottom at -9 m). This part of the island is characterized by karst morphology, with numerous carbonate solution pits – dolines. The southeastern part of the island is a ca. 400 m high plateau with about 500 m high peaks (Fig. 2).

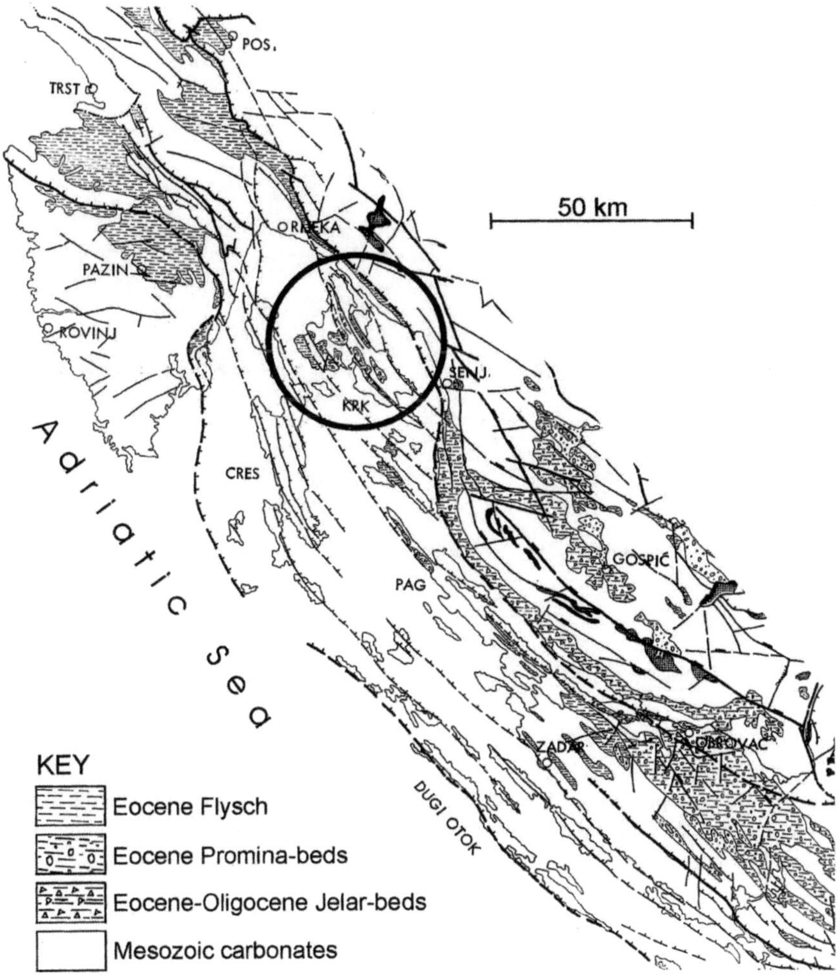

Fig. 1. Position of the studied area. Tectonical map is from Oluić et al. (1972).

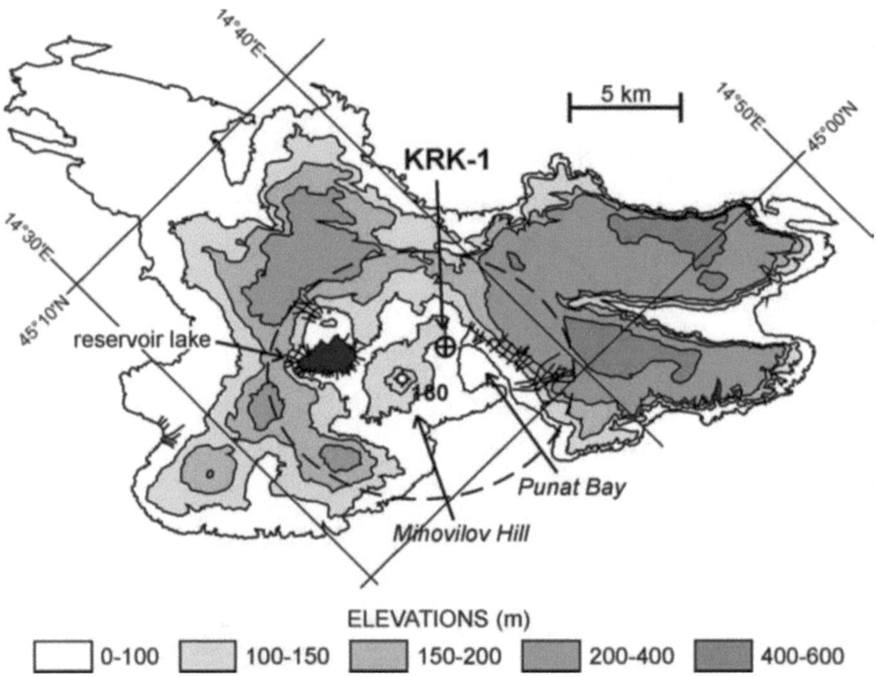

Fig. 2. Simplified topography of Krk Island. The dashed line outlines an annular depression, that hosts a reservoir lake, several smaller valleys and Punat Bay. Mihovilov Hill is almost in the centre of the depression.

Two centripetal drainage systems are developed in the central part of the island, which feed a reservoir lake on the northwestern side of the Mihovilov Hill and the Punat Bay on its southeastern side.

3 Geology

Geotectonically, the Island of Krk is located in the Dalmatian Zone (Aubouin et al. 1970) of the Karst Dinarides, which is composed of 8000 m thick (Velić 2000), deeply karstified carbonates (Kranjec 1981).

Crystalline basement occurs at 10–13 km depth, as revealed by reflection seismics (Aljinović 1981). The island of Krk (Figs. 3 and 4) consists of Mesozoic platform carbonates and Tertiary shallow marine carbonates and flysch clastics (Mamužić et al. 1969; Šikić et al. 1969; Šušnjar et al. 1970). The oldest strata exposed are lower Cretaceous limestones and dolomites, with thin breccia intercalations. Paleogene deposits are represented by Early–Middle Eocene Alveolina Limestones and Middle Eocene Nummulite Limestones and flysch. The contact of Upper Cretaceous and Tertiary carbonates is locally marked by an erosional paleokarstified surface (Ibrahimpašić and Gušić 2000), or by bauxite (Grimani et al. 1973; Šinkovec and Sakač 1981). Locally, there are also erosional remnants of coarse carbonate breccia of presumed Late Eocene–Early Oligocene age (Šušnjar et al. 1970; Grimani et al. 1973).

Structurally, the island geology is characterized by NW–SE trending folds (Mamužić et al. 1969; Šikić et al. 1969; Šušnjar et al. 1970; Grimani et al. 1973; Mamužić and Milan 1973) (Fig. 4). Some synclines are bordered by thrust faults with opposite vergence, and it is interesting to note deviation of structural axes which is unusual on Adriatic islands.

Šušnjar et al. (1970) have differentiated Lower Cretaceous and Upper Eocene–Lower Oligocene breccias because of their respective clast populations. The former age was attributed to grey breccia which comprises only Lower Cretaceous and older debris, whereas the latter age was attributed to variegated breccia which also comprises Early–Middle Eocene debris. Our research showed that these two breccias represent variants of a single breccia unit (the Krk-breccia), with significant variation in clast composition over short distances.

In 1965–1966 the Naftaplin oil company drilled the 3715 m deep exploration well Krk-1, which was located centrally at a structure that was in those days interpreted to be the core of the anticline. The exploration target were uplifted Permian clastics which were expected to be petroliferous. This well revealed that first 1500 m of the drilled sequence are built of breccia, which was assigned a Paleogene age, because it comprises fragments of Early-Middle Eocene Alveolina Limestone.

Our mapping showed that what was previously regarded as a Lower Cretaceous anticlinal structure in the central part of the island is actually the Krk-breccia, whose central part is covered by large, randomly scattered blocks of well-bedded Cretaceous limestones which are herein referred to as the "megablock facies" (Fig. 4).

4 Krk-breccia

At the surface, the Krk-breccia disconformably overlies strongly fractured Upper Cretaceous carbonates, whereas in the Krk-1 well it overlies Upper Malmian limestones and dolomites (Đurasek et al. 1981). The Krk-breccia covers an area of ca.

50 km² (Fig. 4) in a continuous sheet, but there are also many patches of identical breccia scattered around the island, which probably represent erosional remnants

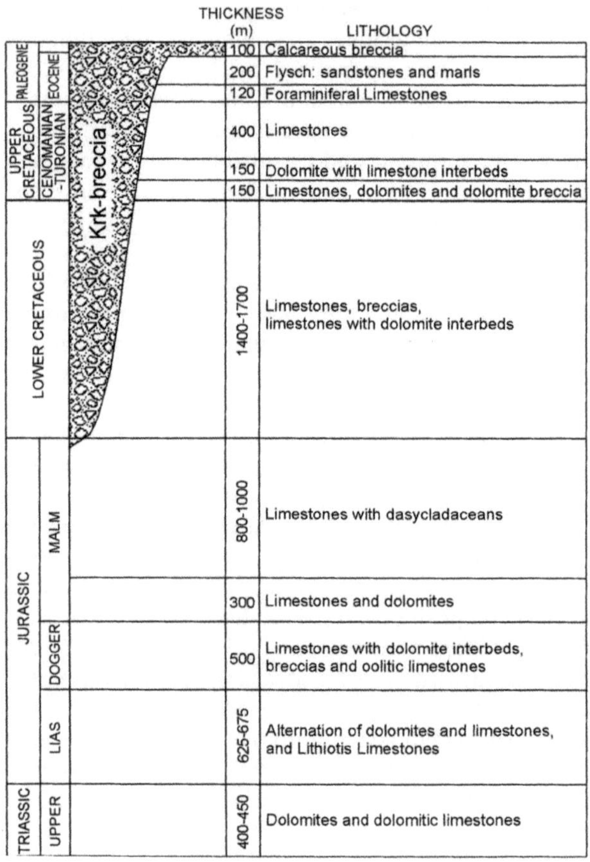

Fig. 3. Stratigraphic column of the Krk Island. The depth of excavation in the base of the Krk-breccia is illustrated only approximately. Simplified after Šušnjar et al. (1970).

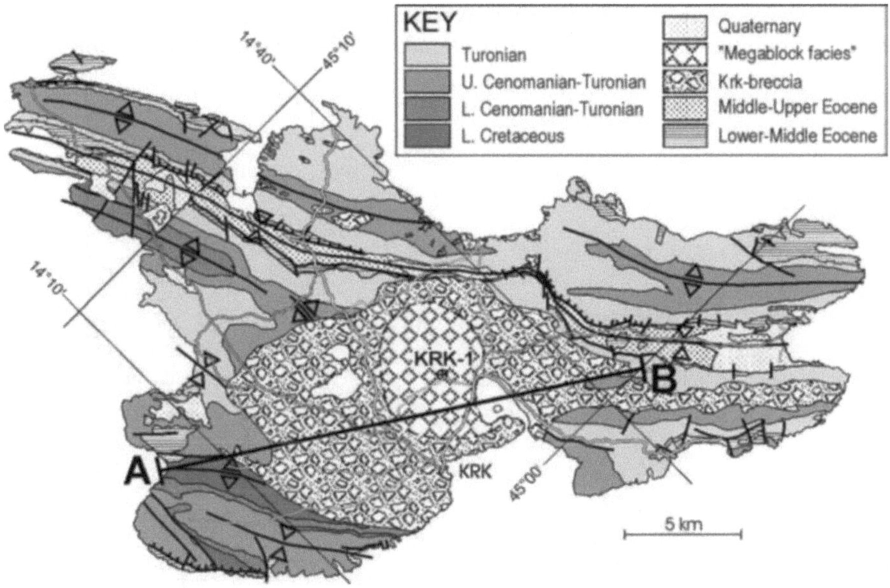

Fig. 4. Simplified geological map of the Krk Island (partly after Mamužić et al. 1969, Šikić et al. 1969, Šušnjar et al. 1970). The illustrated extent of the Krk-breccia is the result of our mapping. Note that structural axes apparently deviate around the Krk-breccia body. Conceptual cross-section in Fig. 12 is based on A–B section line.

of a once more extensive breccia layer. There are also a few narrow breccia "zones" which extend from the main breccia body (Fig. 4). Here the breccia is only several meters thick, and the underlying Cretaceous limestones, which also border these breccia "zones" laterally, protrude in erosional "windows".

The breccia seems to have chaotic fabric; angular clasts are surrounded by a variable amount of calcareous matrix (Fig. 5), so that locally the breccia may be grain-supported and matrix-supported. The Krk-breccia is composed of carbonate debris that ranges from small gravel to large boulder size; the boulders can be as large as 64 m across. The largest clasts occur near the base of the Krk-breccia, as observed at the outcrops located around the breccia periphery, but they disappear upwards after a few tens of meters. Thus, in the upper part of the breccia package, predominates debris with a maximal clast size that seldom exceeds 1 m.

In the lower part of the breccia some clasts are fractured *in situ* and their fissures are filled with matrix (Fig. 6). The matrix is reddish to brownish in colour, locally grayish, and composed of sand-size carbonate debris, rare fine-grained quartz grains, and microcrystalline calcite cementing the grains. It is unfossiliferous, and also lacks fossil debris. For the purpose of determination of the matrix/clast ratio, the matrix is defined herein as material that is finer than 2 mm, and the ratio was determined as a cumulative clast cross-section along a random oriented 0.5 m long line. Thus, the matrix abundance only approximates volume

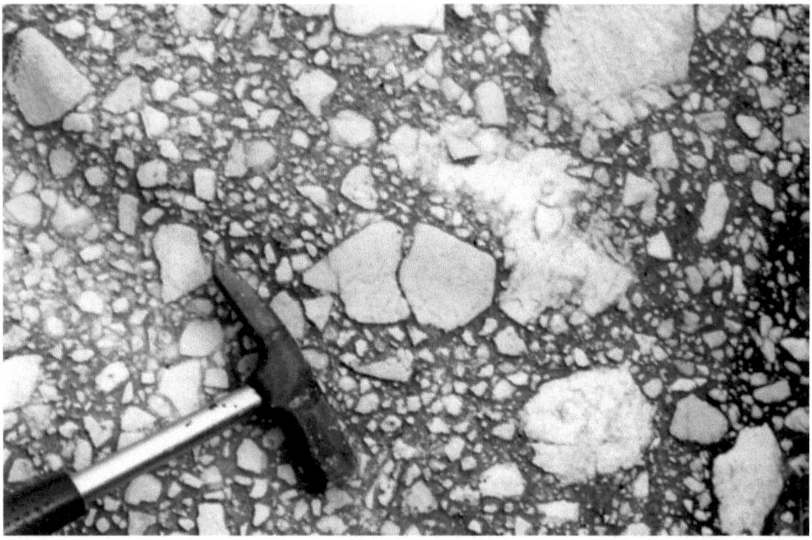

Fig. 5. Relatively fine-grained variety of the Krk-breccia with reddish carbonate matrix.

Fig. 6. *In situ* fractured Flysch sandstone clast with matrix-filled open fissure indicates heavy loading during deposition, probably caused by very rapid deposition of a large volume of debris. *In situ* fracturing and matrix-filled fissures were also observed in limestone clasts of various sizes.

percentage, because it was measured on two-dimensional exposures. Based on the clast/matrix ratio, two end-members can be differentiated, i.e. matrix-rich breccia, which contains ca. 53 % of matrix, and matrix-poor breccia, in some cases with 13.5 % of matrix. The contact between these two breccia types is generally gradual, but locally it is characterized by soft-sediment injection of the underlying matrix-rich breccia (Fig. 7).

Fig. 7. Contact of matrix-rich (below) and matrix-poor (above) breccia (dashed line) is characterized by loading and upwards injection of the underlying matrix-rich breccia, which suggests rapid deposition, which exerted high pressure over the underlying deposit. Field of view is ca. 3 m.

The provenance of the Krk-breccia debris can be recognized already in local stratigraphic units (Mesozoic carbonates and Tertiary clastics, Fig. 3). In its lower part, the Krk-breccia comprises abundant clasts of Early to Late Cretaceous age, scattered clasts of Eocene age (Alveolina Limestone and Flysch sandstone), and occasional small bauxite clasts. The clasts of Eocene age are the largest; Alveolina Limestone clasts commonly reach 10 m (exceptionally even 64 m) in length, and Flysch clasts can be as large as 10 m. However, the Flysch clasts are poorly exposed, so they may be much larger (possibly up to > 100 m across) as inferred from the subtle topography that is commonly developed on a marly substratum. The abundance of large Eocene debris decreases upwards, and in the rest of the Krk-breccia Cretaceous debris predominates. The debris lithologies are varied and unevenly distributed giving the breccia locally a variegated, but also greyish appearance. The variegated breccia is composed of Upper Cretaceous white and pink limestones, and Eocene yellowish to greyish limestones and brownish sandstones encased in a reddish matrix. The greyish breccia is composed of Jurassic, Lower and Middle Cretaceous dark grey limestones and dolomites. These two breccia variants have puzzled previous researchers (e.g., Šušnjar et al. 1970) who attributed them to different stratigraphic units. Locally, however, it is possible to see rare clasts of variegated breccia, incased into grey breccia. The breccia cored in the Krk-1 well (Fig. 8) is lithologically identical to the breccia exposed at the surface.

Fig. 8. Krk-1 well core shows that the breccia drilled at depth of 964—966 m does not differ from the rocks exposed at the surface.

The thickness of the Krk-breccia can be estimated from the deep exploration well Krk-1 which drilled through 1500 m of breccia. The well composite log (Fig. 9) shows that breccia interval is characterized by relatively uniform, but low natural radioactivity (GR log), and high neutron porosity (Nphi log) in the first 350 m compared to the breccia below which shows higher oscillations in porosity. The GR-log is lithologically controlled; thus a uniform pattern may indicate limited difference in lithology, or "homogenized" rock, such as breccia, whereas higher excursions of the GR curve generally suggest lithological heterogeneity, or brecciated rocks with the cement (or matrix) that is mineralogically different from the debris. The Nphi-log suggests that the shallowest, 350 m thick breccia interval is apparently made of water-saturated porous rock, which is underlain by alternating water-saturated porous rocks interbedded with rocks of significantly lower porosity (possibly partly cemented) down to a depth of 1500 m. The interval 625–750 m in the Nphi-log shows markedly lower porosity which may be due to large limestone clasts that were drilled in that interval. The composite well log (see also

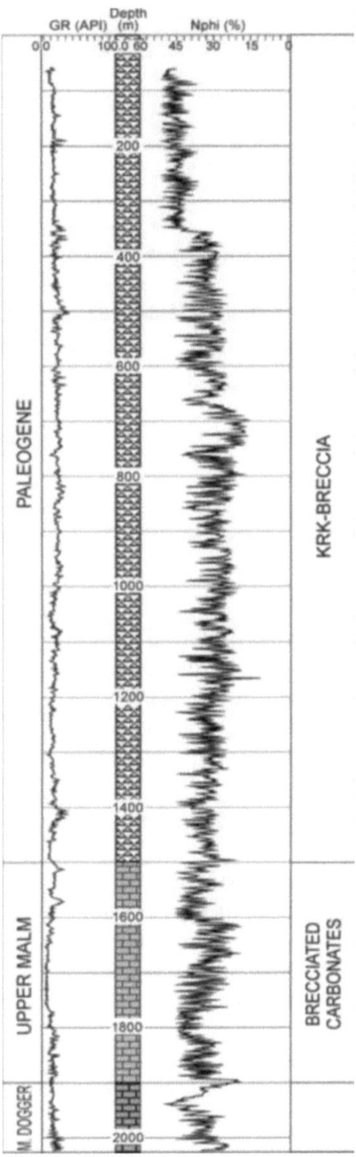

Fig. 9. Krk-1 well composite log with original stratigraphic attribution.

Đurasek et al. 1981) shows that Krk-breccia overlies 400 m thick Upper Malmian carbonates.

It must be noted that nowhere in the coastal Dinarides occur breccias of comparable thickness. The lithologically similar breccia is known as Jelar-beds which occur along the coast and on coastal mountains in thickness of ca. 300 m (Bahun 1974; Herak and Bahun 1979). This breccia comprises abundant very coarse (commonly 10s m size) debris of Dinaric provenance, but its exact age and genesis are still a matter of debate.

5 "Megablock Facies"

Approximately in its central part, the Krk-breccia is overlain by a "megablock-facies" (Fig. 4) which is represented by large blocks of well-bedded limestone of Early Cretaceous age. The size of these blocks commonly exceeds 10 m, and are surrounded by *terra rossa* matrix. They are randomly oriented – scattered, and commonly have different, even opposite bedding at small distances of several meters (Fig. 10). The Early Cretaceous age of the blocks probably mislead previous researchers to interpret this as an anticline core (e.g., Grimani et al. 1973).

The stratigraphic position of the "megablock facies" relative to the Krk-breccia is not visible at the surface, but the Krk-1 well revealed that the breccia underlies these blocks, because drilling was started in this "facies", but the well soon entered the breccia. The thickness of the "megablock facies", and its contact with the Krk-breccia, are however, unknown, because the first 56 m of the well were not logged.

Fig. 10. "Megablock facies" is made of scattered large limestone blocks that typically protrude from soil in a random orientation. White lettering indicates measured position of beds in individual blocks.

6
Age of the Krk-breccia

Grimani et al. (1973), in a discussion on the Krk-island geology, referred to the Krk-1 well and speculated on a possible Early Cretaceous age of the breccia. However, Paleogene age of the Krk-breccia was first proposed by the petroleum industry paleontologists (Fig. 9), because they found fragments of Eocene limestone in well-cuttings and cores. Šušnjar et al. (1970) and Grimani et al. (1973) differentiated Eocene–Oligocene breccia in their maps, also according to the stratigraphic ages of the apparently youngest clasts. However, they considered the grey-coloured breccia as being of Early Cretaceous age.

No fossils are found in breccia matrix, and its age can only be estimated on the basis of the age of the youngest clasts observed. These are clasts of Flysch sandstone (Fig. 6) and marls of early–middle Eocene age as indicated by their foraminiferal fauna (Grimani et al. 1973), indicating that the breccia is somewhat younger. Consequently, the oldest possible age might be late Eocene (Priabonian), but it could be of Oligocene or even younger age.

The age of the "megablock-facies" is, however, unknown, because we could not find any age indicators in the field. This "facies" is apparently younger than the Krk-breccia.

7
Paleomagnetism

Márton et al. (1990) published paleomagnetic data obtained on the islands of Cres, Krk, Rab, and Pag. The largest data base was obtained for the Krk island, where they took samples at 15 localities, representing all lithostratigraphic units. Natural remanent magnetization (NRM) was generally weak, and in two samples it was too low to yield data. Paleomagnetic declinations and inclinations for the Krk, Cres, and Rab samples have a wide dispersal, as shown in Figure 11. Both counter-clockwise as well as clockwise rotations of NRM, relative to modern field are found in the Krk and Rab samples, whereas on the Island of Cres (Márton et al. 1990) and in central Istria (Márton et al. 1995) paleomagnetism shows better-ordered declination vectors with a smaller scatter.

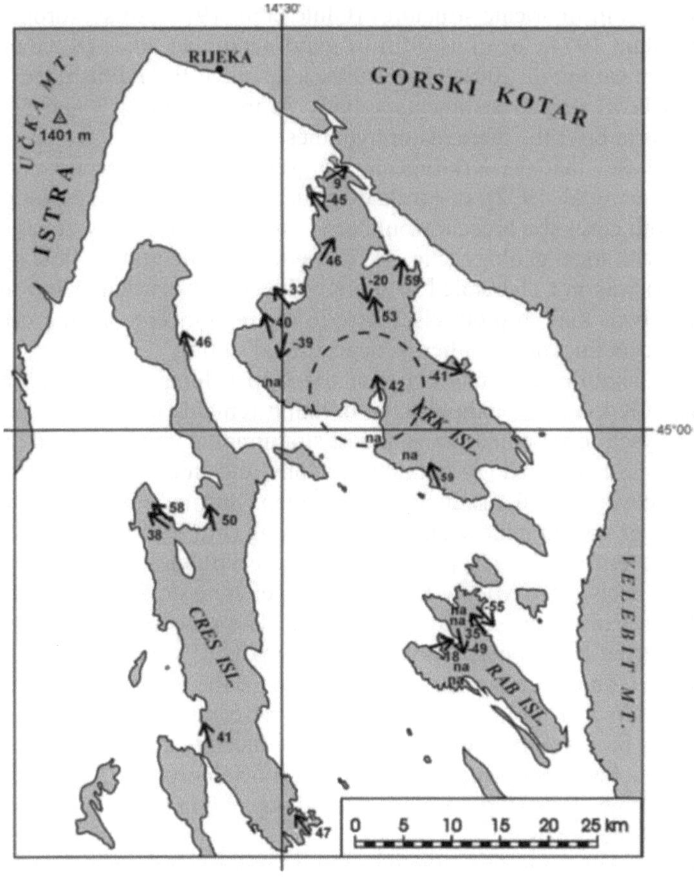

Fig. 11. Paleomagnetic data of Márton et al. (1990) show a strong dispersal of NRM deviation on the islands of Krk and Rrab, unlike on the neighbouring island of Cres. The dashed line outlines the proposed impact structure.

8 Discussion

The origin of the Krk-breccia has not been treated as an individual issue in the published literature. It was interpreted as a syn-orogenic formation by Bahun (1974) and Herak and Bahun (1979), who did not differentiate it from the coarse Jelar-deposits (breccia) in coastal Dinarides. The Krk-breccia is differentiated only as narrow zones in geological maps of Mamužić et al. (1969) and Šušnjar et al. (1970), which inspired some researchers to interpret the breccia variably as: a) stacked breccias in imbricate structures (Oluić et al. 1972), b) syn-orogenic gravity flows (Bahun 1974), or c) as infill of giant neptunian dikes (S. Grandić, personal communication in 2001), but Đurasek et al. (1981) illustrated the Krk-breccia as a bowl-shaped sedimentary body in their Figures 3 and 5. We will briefly discuss in brief the three major hypotheses.

The hypotheses that the Krk-breccia represents stacked breccias in imbricate structures (Oluić et al. 1972), or infills of giant neptunian dikes, are not plausible, because in both cases the breccia would crop out in rather narrow zones, which is inconsistent with local geology (Fig. 4). The syn-orogenic gravity flow hypothesis (Bahun 1974) was not elaborated to the Krk-breccia in particular but to coarse-grained Jelar-beds that crop out elsewhere in the region (Fig. 1), and can not explain Krk-breccia thickness neither its depositional setting.

There is a significant hiatus between the Upper Jurassic deposits, and Krk-breccia in the Krk-1 well, although the duration remains unclear. Assuming that the Eocene flysch represented the youngest stratigraphic unit subjected to erosion, we can estimate from analogy with stratigraphy and thicknesses of stratigraphic units in area covered by the same geological map (Šušnjar et al. 1970), that about 2500–3000 m of stratigraphic sequence are missing (Fig. 3).

Unselective erosion, which produced the large volume of Krk-breccia debris (96.8 km^3) by incision of Middle Eocene to Upper Jurassic deposits, is difficult to explain in terms of normal erosional processes. The process was clearly unselective, as it affected limestones, sandstones, and marls of various ages and degrees of cementation. The large sizes of Eocene clasts in the Krk-breccia indicate very intensive erosion and rapid transport, so that these relatively ductile lithologies could not have been significantly rounded and reduced in size. Soft sediment loading at the contact between matrix-poor and matrix-rich breccia (Fig. 7), and *in situ* fracturing of some clasts (Fig. 6) suggest rapid deposition and relatively high pressure applied to the debris.

Observed varieties in clast lithologies and stratigraphies, between variegated and grey-coloured breccias in particular, can be attributed to debris supply from different sources.

The transport mechanism for matrix-rich conglomerate is commonly inferred to have been a debris flow (Lowe 1982). Such flows may have generated rather thick coarse-grained deposits with the largest clasts floating in finer-grained "matrix". The large thickness of the Krk-breccia may be the result of amalgamation of several thinner breccia units, as may be indicated by the presence of breccia-clasts within the breccia. The Krk-breccia seems to be thickest near the locality of the Krk-1 well, thus its depositional environment must have been a ca. 1500 m deep depression, bordered by steep walls that could have provided debris surges by retrogressive slope failures. The initial depression was thus filled, at least to the present ground-level, but also significantly widened. Narrow "zones" of thin breccia, that extend from the main breccia body, may represent a) erosional remnants of wider, but relatively thin breccia sheet, and b) infill of paleo-gullies. The breccia in these "zones" is indeed very thin but seems to be laterally bordered by higher elevated "basement" rocks what gives impression of breccia-filled channels.

The "megablock facies" must be genetically related to the Krk-breccia, because it is not exposed elsewhere. The piles of large blocks give the impression of a rock-fall rubble, which accumulated almost in the centre of the island, far from any steep rock faces.

The Krk-1 well drilled the Triassic/Jurassic boundary at the depth of 3100 m (Đurasek et al. 1981). However, the known thicknesses of stratigraphic units (Fig. 3) indicate that the boundary should be much deeper, namely 3910–4460 m deep. It appears that the boundary is uplifted by about 810–1360 m at the well site.

A centripetal drainage system (Fig. 2) indicates active modern subsidence at both sides of the Mihovilov Hill, approximately in the Krk-breccia centre.

The interpretation of the Krk island paleomagnetic data by Márton et al. (1990) was essentially a tectonic one; they interpreted a counter-clockwise rotation of the islands relative to a stabile Europe as a consequence of the advancement of nappes from the SE to NW, but doubted whether the locality with the clockwise rotation belongs to the Dalmatian zone at all.

Below, we present an alternative hypothesis on the Krk-breccia genesis which can account for most of its observed characteristics.

9
Impact Hypothesis

The exceptional thickness and volume of the breccia, unselective large-scale erosion, rapid deposition with relatively high stress applied on debris and underlying sediments, deposition in an at least 1.5 km-deep circular to oval-shaped depression, and very wide dispersal of NRM values in all stratigraphic units on the island of Krk, might be explained in terms of an impact crater infill. The semi-circular arrangement of the hills around the Mihovilov Hill (Fig. 2), may mark the rim of a 13 x 11 km structure. In this interpretation, the main sedimentary body of the Krk-breccia represents crater-fill (Fig. 12), whereas narrow breccia zones may represent infills of resurge gullies.

The coarse polymictic debris probably represents debris surge deposit (possibly with a contribution from fall-back breccia), which filled the cavity soon after the excavation of the transient crater. The youngest debris in the breccia (clasts of Flysch sandstone and marl) indicates the probable target lithology. The abundance of Flysch debris in the Krk-breccia is low, compared to the thickness of Flysch elsewhere in the Adriatic. This can indicate that a large part of the Flysch target rocks was melted and/or evaporized, and ejected out of the crater.

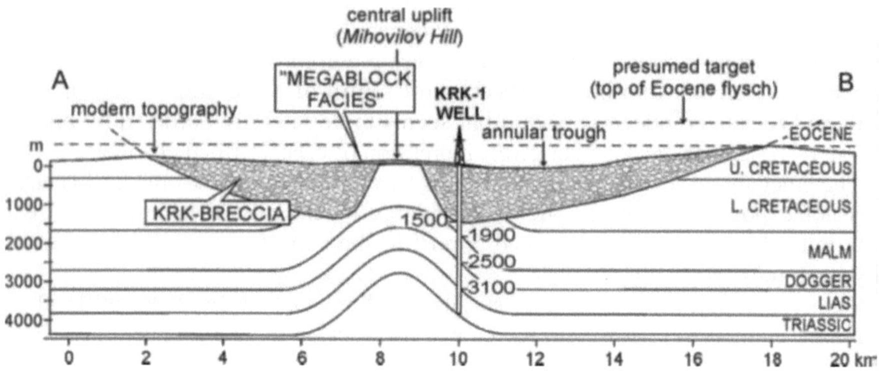

Fig. 12. Conceptual cross-section of proposed Krk impact structure. Stratigraphic thicknesses are adopted from geological map of Šušnjar et al. (1970). In this reconstruction, the top of modern topography is taken as the approximate level of the Eocene base. The thickness of Eocene flysch is very speculative; on the Krk Island it reaches (post-erosional) 200 m, whereas in the neighbouring areas it commonly attains ca. 600—700 m. The Krk-1 well depths are given on the left, whereas depths of drilled stratigraphic contacts are labelled next to the well.

The thickness of the Krk-breccia exceeds the thickness of all other impact crater-fill breccias developed from a limestone target (85 km Chesapeake Bay: 600–1200 m, Poag 1997; 20–24 km Haughton Dome: 110 m, Osinski and Spray 2001; 45 km Montagnais crater: 552 m, Jansa et al. 1989), which presents a problem to explain. Assuming that the impact hypothesis is correct, two possible causes can be considered: 1) the crater diameter is actually much larger than recognized at the moment, and 2) the depth to diameter ratio is different for a deeply karstified limestone target than for a sedimentary target. Regarding the first possibility, it is unlikely that the crater could be as wide as, say, 100 km, because on neighboring islands and on the mainland there are no indications of a possible impact structure. Regarding the second possibility, the exceptional thickness of the Krk-breccia crater-fill could be the result of an impact into strongly karstified target rocks. We hypothesize that an impact into karstified target could create an unusually deep crater, and an enormous volume of crushed rocks, because the karstified carbonate rocks have a large volume of caverns which decrease the target density.

The lack of any *in situ* impact melt may be a consequence of intensive erosion, which has lowered the northwestern part of the island by at least 300 meters, as indirectly indicated by the thicknesses of the missing stratigraphic units (see Fig. 4).

The top of the Mihovilov Hill is located almost in the centre of 5-km-wide zone of the "megablock facies", and surrounded by an annular set of valleys (Fig. 2). The "megablock facies" is far from any steep slopes (it actually builds a hill), it is surrounded by the Krk-breccia (Fig. 4) and may have been formed by central peak collapse. Indirect evidence suggesting that below the "megablock facies" is a buried central uplift can be derived from depth of the drilled stratigraphic contacts which are shallower than expected. Figure 12 shows a conceptual geological section through the Krk Island. The distribution of the "megablock facies" is over 5 km wide, twice as predicted central-peak diameter of an 11 km impact structure (Melosh 1989).

As regards the timing of possible impact, it seems that it occurred before folding of the coastal Dinarides, because the structural axes deviate around the Krk-breccia body (Fig. 4). If the central part of the structure already hosted a central uplift, it must have acted as a rigid "plug" in folding, and could not be folded like the remaining part of the area.

As the age of the assumed impact is unknown and no fossils are found in the Krk-breccia matrix, it is not possible to reconstruct the environment prior to impact. However, we favor a shallow marine interpretation because it can explain the lack of impact melt and occurrence of breccia-filled resurge gullies.

10 Conclusions

Polymictic Krk-breccia on the northern Adriatic Krk Island may be interpreted as an impact crater infill that was created by transient crater slope failures and debris back-fall soon after the impact. The breccia appears to be largely confined to a 14 x 11 km complex crater, which is presently about 1.5 km deep.

The age of the Krk-breccia is poorly confined and it can be (at present) only estimated as post-Lutetian, but very likely pre-dated final folding of the Dinarides.

However, the impact interpretation still remains a hypothesis, because none of the impact-characteristic indicators (microscopic or macroscopic evidence of shock metamorphism, or traces of meteoritic matter) have been found in the studied area.

Acknowledgements

The data presented are the outcome of research project 119-303, funded by the Ministry of Sciences and Technology of the Republic of Croatia, and in part, of the unpublished graduate thesis of the second author. Well log data are courtesy of

Naftaplin petroleum Co. We are indebted to Chris Koeberl and Uwe Reimold for their thorough reviews that helped us to improve many of our ideas.

References

Aljinović B (1981) Depths of basement of sediments along the line Brač-Palagruža. Geološki vjesnik 34: 121–125

Aljinović B, Prelogović E, Skoko D (1987) Novi podaci o dubinskoj geološkoj građi i seizmotektonski aktivnim zonama u Jugoslaviji (*New data on deep geological structure and seismotectonic active zones in region of Yugoslavia*). Geološki vjesnik 40: 255–263

Aubouin J, Blanchet R, Cadet J-P, Celet P, Charvet J, Chorowicz J, Cousin M, Rampnoux J-P (1970) Essai sur la géologie des Dinarides. Bulletin de la Société géologique de France 7: 1060–1095

Babić Lj, Zupanič J (1983) Paleogene clastic formations in northern Dalmatia. In: Babić Lj, Jelaska V (eds) Contributions to sedimentology of some carbonate and clastic units of the Coastal Dinarides, Excursion Guide Book of 4th International Association of Sedimentologists Regional Meeting Split, Croatia, pp 37–61

Babić Lj, Zupanič J (1998) Nearshore deposits in the middle Eocene clastic succession in northern Dalmatia (Dinarides, Croatia). Geologia Croatica 51: 175–193

Bahun S (1974) Tektogeneza Velebita i postanak Jelar-naslaga (*The tectogenesis of Mt. Velebit and the formation of Jelar deposits*) (in Croatian).Geološki vjesnik 27: 35–51

Boljat E (1981) Stratigrafsko-tektonski odnosi okolice Krka na otoku Krku (*Tectonostratigraphic relations of Krk environs on the island of Krk*). Unpublished graduation thesis. University of Zagreb

Đurasek N, Frank G, Jenko K, Kužina A, Tončić-Gregl R (1981) Prilog poznavanju naftnogeoloških odnosa u sjeverozapadnom dijelu jadranskog primorja (*Contribution to the understanding of oil-geological relations in NW Adriatic area*). Symposium Proceedings "Complex oil-geological aspects for offshore and coastal Adriatic areas", Split, Croatia 1981 (1) (in Croatian). Radovi znanstvenog savjeta za naftu 8: 201–213

Grimani I, Šušnjar M, Bukovac J, Milan A, Nikler L, Crnolatac I, Šikić D, Blašković I (1973) Osnovna geološka karta SFRJ 1:100.000. Tumač za list Crikvenica L33-102 (*General geological map of SFRJ 1:100.000. Explanation notes for sheet Crikvenica L33-102*), Institut za geološka istraživanja Zagreb, Savezni geološki zavod Beograd, pp 5–43

Grubić A (1980) An outline of geology of Yugoslavia. In: Grubić A (ed) Yugoslavia, An Outline of Geology of Yugoslavia Excursions 201A-202C, 26th International Geological Congress Paris, Guide-book 15, pp 5–49

Herak M, Bahun S (1979) The role of the calcareous breccias (Jelar Formation) in the tectonic interpretation of the High Karst Zone of the Dinarides. Geološki vjesnik 31: 49–59

Ibrahimpašić H, Gušić I (2000) Biostratigraphical Correlation of the Deposits of Southeastern Part of the Krk Island. In: Proceedings of 2nd Croatian geological congress Cavtat-Dubrovnik, Croatia 2000, Institute of geology, Zagreb: pp 213–217

Jamičić D, Prelogović E, Tomljenović B (1995) Folding and deformational style in overthrust structures on Krk Island. In: Rossmanith HP (ed) Mechanics of Jointed and Faulted Rock, Balkema, Rotterdam, pp 359–362

Jansa LF, Pe-Piper G, Robertson PB, Friedenreich O (1989) Montagnais: A submarine impact structure on the Scotian Shelf, eastern Canada. Geological Society of America Bulletin 101: 450–463

Kranjec V (1981) Neke značajke naftoplinonosnosti naslaga i moguća daljnja nalazišta ugljikovodika u predjelima vanjskih Dinarida i Jadranskog podmorja (*Some characteristics of oil and gas-bearing deposits and possibilities of future hydrocarbon finds in external Dinarides and Adriatic offshore*) (in Croatian). Pomorski zbornik 19: 385–410

Lowe DR (1982): Sediment gravity flows: II depositional models with special reference to the deposits of high-density turbidity currents. Journal of Sedimentary Petrology 52: 279–297

Mamužić P, Milan A (1973) Osnovna geološka karta SFRJ 1:100000, Tumač za list Rab L33-114, (*General geological map of SFRJ 1:100.000. Explanation notes for sheet Rab L33-114*) Institut za geološka Istraživanja Zagreb, Savezni geološki zavod Beograd, pp 5–39

Mamužić P, Milan A, Korolija B, Borović I, Majcen Ž (1969) Osnovna geološka karta SFRJ 1:100.000, List Rab L 33-114 (*General geological map of SFRJ 1:100.000, Sheet Rab L33-114*), Institut za geološka Istraživanja Zagreb, Savezni geološki zavod, Beograd

Marjanac T (1989) Ponded megabeds and some characteristics of the Eocene Adriatic basin (middle Dalmatia, Yugoslavia). Memorie della Societa Geologica Italiana 40: 241–249

Marjanac T (1996) Deposition of megabeds (megaturbidites) and sea-level change in a proximal part of Eocene-Miocene flysch of central Dalmatia (Croatia). Geology 24: 543–546

Márton E, Milicevic V, Veljovic D (1990) Paleomagnetism of the Kvarner islands, Yugoslavia. Physics of the Earth and Planetary Interiors 62:70–81

Márton E, Drobne K, Cimerman F, Ćosović V, Košir A (1995) Paleomagnetism of latest Maastrichtian through Oligocene rocks in Istria (Croatia), the Karst Region, and S of the Sava Fault (Slovenia). In: Proceedings of 1st Croatian geological congress Opatija, Institute of geology, Zagreb, Croatia 1995: 355–360

Melosh HJ (1989) Impact cratering – A Geologic Process. Oxford University Press, Oxford, 245 pp

Milinović M. (1981) Stratigrafsko-tektonski odnosi područja Kornića, Muraja, Sv. Dunata, okolice Punta na otoku Krku (*Tectono-stratigraphic relations of Kornić, Muraj, St. Dunat, and Punat environs on the island of Krk*). Unpublished graduation thesis, University of Zagreb, 2 maps , 30 pp

Miljuš P (1973) Osnovne crte geološko-tektonske građe Dinarida (Geological-tectonic structure and evolution of Dinarides and review on hydrocarbon explorations of Sava-Vardar zone). Nafta 24/7-8: 1–15

Oluić M, Grandić S, Haček M, Hanich M (1972) Tektonska građa vanjskih Dinarida Jugoslavije (*Tectonic structure of external Dinarides in Yugoslavia*). Nafta 23/1–2:3–16

Osinski GR, Spray JG (2001) High shocked low density sedimentary rocks from the Haughton impact structure, Devon Island, Nunavut, Canada. [abs.] Lunar and Planetary Science 32, # 1908, CD-ROM

Poag CW (1997) The Chesapeake Bay bolide impact: a convulsive event in Atlantic Coastal Plain evolution. Sedimentary Geology 108: 45–90

Prelogović E, Cvijanović D, Aljinović B, Kranjec V, Skoko D, Blašković I, Zagorac Ž (1982) Seizmotektonska aktivnost duž priobalnog dijela Jugoslavije (*Seismotectonic activity along the coastal area of Yugoslavia*). Geološki vjesnik 35: 195–207

Prelogović E, Kuk V, Jamičić D, Aljinović B, Marić K (1995) Seizmotektonska aktivnost Kvarnerskog područja (*Seismotectonic activity in Kvarner area*). In: Proceedings of 1st Croatian geological congress Opatija, Institute of Geology, Zagreb, Croatia 1995: 487–490

Šikić D, Polšak A, Magaš N (1969) Osnovna geološka karta SFRJ 1:10000, List Labin L 33-101, (*General geological map of SFRJ 1:100.000. Sheet Labin L33-101*) Institut za geološka istraživanja Zagreb, Savezni geološki zavod Beograd

Šinkovec B, Sakač K (1981) Boksiti starijeg paleogena na otocima sjevernog Jadrana (*The Early Eocene bauxites on north Adriatic islands*). Geološki vjesnik 33: 213–225

Šušnjar M, Bukovac J, Nikler L, Crnolatac I, Milan A, Šikić D, Grimani I, Vulić Z, Blašković I (1970) Osnovna geološka karta SFRJ 1:10000, List Crikvenica L 33-102 (*General geological map of SFRJ 1:100.000. Sheet Crikvenica L33-102*), Institut za geološka istraživanja Zagreb, Savezni geološki zavod Beograd

Velić I (2000) The Karst Dinarides: the Adriatic carbonate platform or carbonate pie up to 8.000 m thick, built through the period of more than 200 milions of years (abstract). Riassunti delle comunicazioni orali e dei poster, 80. riunione estiva Trieste, Societá Geologia Italiana, pp 452–455

Did the Puchezh-Katunki Impact Trigger an Extinction?

József Pálfy

Hungarian Natural History Museum, Department of Geology and Paleontology, POB 137, Budapest, H-1431 Hungary. (palfy@paleo.nhmus.hu)

Abstract. The 80 km diameter Puchezh-Katunki impact crater is the only one of the six largest known Phanerozoic craters that has not been previously considered as a factor in a biotic extinction event. The age of impact is currently regarded as Bajocian (Middle Jurassic), on the basis of palynostratigraphy of crater lake sediments, but there is no significant extinction in the Bajocian. Earlier K-Ar age determinations of impactites compared with a current Jurassic time scale permit that either the end-Triassic or the Early Jurassic (Pliensbachian-Toarcian) extinction was coeval with the Puchezh-Katunki crater. The stratigraphical and paleontological record contains clues that suggest that an impact may have occurred at these horizons. The age of the Puchezh-Katunki crater needs reevaluation through $^{40}Ar/^{39}Ar$ dating of impact rocks and/or revision of the palynology of the oldest crater fill. A definitive age determination will help constrain the impact-kill curve.

1 Introduction

The putative link between extraterrestrial impacts and mass extinction events has been the focus of much interdisciplinary research for over 20 years. The terminal Cretaceous bolide impact that created the Chicxulub crater is now widely regarded as the main cause of the Cretaceous/Tertiary boundary extinction, confirming the original hypothesis of Alvarez (1980). Building on the Cretaceous/Tertiary example, large body impacts have been considered as potential causal agents in other extinction events (Rampino and Haggerty 1996). Such a research agenda was clearly formulated by Raup (1992): "... it is appropriate, even obligatory, to entertain the possibility that impacts could have been responsible for extinctions other than the K-T event." While most known large impact craters have been evaluated in this context, the Puchezh-Katunki crater represents a notable exception.

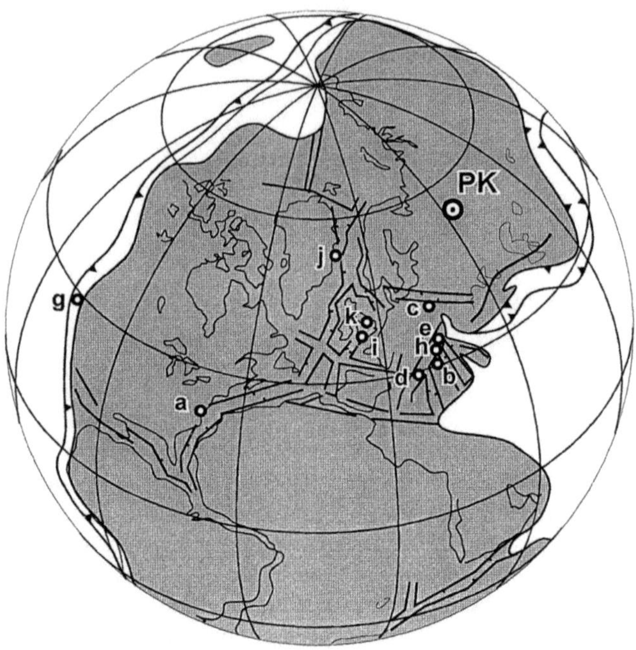

Fig. 1. Early Jurassic paleogeographic map showing the Puchezh-Katunki (PK) impact crater (circle, not to scale) and location of reported possible impact indicators (dots) from the Triassic-Jurassic boundary and Early and Middle Jurassic. For key to labels (location and reference), see Table 3. Base map from Ziegler (1990). See text for discussion.

The Puchezh-Katunki structure is a large impact crater at the Volga River, approximate 400 km northeast of Moscow in Russia (Fig. 1). Measuring 80 km in diameter, it is the fifth largest known terrestrial impact crater in the Phanerozoic (Grieve et al. 1995; Grieve 1997) (Table 1). According to Raup's (1992) "impact-kill curve", an impact of that magnitude might have produced a noticeable extinction in the paleontological record. The predicted magnitude of species extinction is approximately 40%; taking into account the uncertainties, a range between 20 to 70% is suggested.

The results of multidisciplinary scientific investigations of the Puchezh-Katunki crater, including studies of the 5374 m deep Vorotilovskaya borehole drilled at the crater's center, were recently summarized in a book in Russian (Masaitis and Pevzner 1999). The age of the impact is regarded as Bajocian. It is stratigraphically bracketed by the youngest target rocks, Early Triassic in age, and the overlying lake sediments, thought to be Bajocian (Middle Jurassic) in age. Here I present arguments that the age cannot be regarded as definitively determined. I consider the uncertainties of the crater age and give a literature review of possible impact indicators from the latest Triassic to Middle Jurassic, in order to investigate if the Puchezh-Katunki impact is recorded in the global sedimentary

Table 1. The largest known Phanerozoic impact craters (after Grieve 2001)

Crater name	Diameter	Age	Location
Chicxulub	170	65.0±0.1 Ma (K-T boundary)	Yucatán, Mexico
Manicouagan	100	214±1 Ma (Late Triassic)	Quebec, Canada
Popigai	100	35.7±0.2 (Late Eocene)	Siberia, Russia
Chesapeake Bay	85	35.5±0.6 (Late Eocene)	Virginia, USA
Puchezh-Katunki	***80***	***see discussion***	***Russia***
Morokweng	70	146±2 (latest Jurassic)	South Africa

record and to explore the possibility that it may be related to either of the two extinction events in this interval, the end-Triassic or the Pliensbachian-Toarcian (Early Jurassic). Notably, there is no significantly elevated extinction rate registered in the Bajocian fossil record (Sepkoski 1996). A definitive dating of the Puchezh-Katunki crater will help constrain the validity and shape of the impact-kill curve.

2
The Puchezh-Katunki Impact Structure

The 80-km-wide Puchezh-Katunki impact structure is nearly completely buried under Neogene and Quarternary sediments. The only natural exposures of impactites are found along the banks of Volga River. Geophysical surveys revealed the crater morphology that features a central dome, ring depression and ring terrace (Masaitis et al. 1996).

The target stratigraphy consists of Archean crystalline basement rocks and overlying uppermost Proterozoic to lowermost Mesozoic sedimentary rocks. In the area adjacent to the crater, the crystalline basement occurs at a depth of ~2 km. The sedimentary sequence typically consists of 500 m Vendian clastics, 800 m Devonian limestone and shale, 450 m Carboniferous carbonates and marl, 250 m Lower Permian carbonates, evaporites and clay, 160 m Upper Permian clastics, and 80 m Lower Triassic clay and siltstone (Masaitis et al. 1996; Masaitis and Pevzner 1999).

Impact rocks and crater lake sediments were penetrated by nearly 180 drill holes, including the super-deep Vorotilovskaya borehole at the center of the crater (Masaitis and Pevzner 1999). The lithologic column of the uppermost part of the crystalline basement, the sequence of impact rocks and the overlying crater lake deposits are shown on Fig. 2.

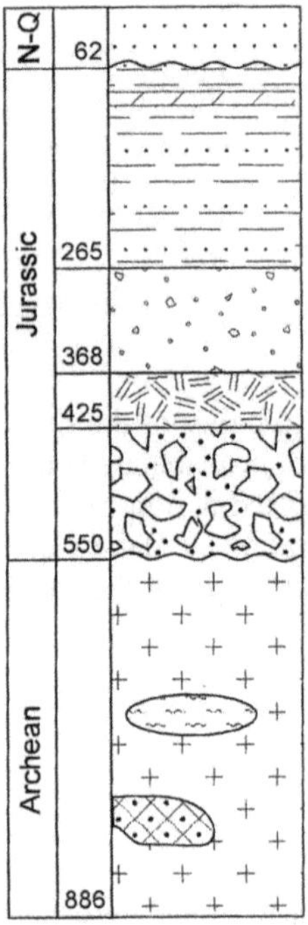

Fig. 2. Stratigraphic column penetrated by the uppermost part of the Vorotilovskaya borehole drilled through the center of the crater (after Vorontsov in Masaitis and Pevzner 1999).
Legend: 0-62 m – Neogene and Quarternary sand and clay; 62-265 m – Crater lake sediments: clay, siltstone and sandstone, subordinate carbonate (Kovernino Formation); 265-368 m – Koptomict gravel; 368-425 m – Suevite; 425-550 m – Allogenic polymict breccia; 550-886 m – Crystalline basement rocks: gneiss, amphibolite, peridotite.

3 Dating the Puchezh- Katunki Crater

Raup (1992) succinctly pointed out that "the dating problem will have to be investigated before we have definitive answers to the impact-extinction question." The prevailing view on the age of the Puchezh-Katunki impact regards it as Bajocian (Middle Jurassic), on the basis of palynostratigraphy from the oldest crater lake sediments (Masaitis et al. 1996; Masaitis and Pevzner 1999). The lacustrine Kovernino Formation, thought to have deposited in the lake that filled the crater, contains palynomorphs that are said to range from the Bajocian to the Bathonian. However, several problems call the validity of this age determination into question.

It is difficult to demonstrate that crater lake sedimentation immediately followed the impact. This argument, however, is weakened if one accepts that pollens suggesting the same age were also recovered from the matrix of impact breccias.

The palynological data is presented as a taxonomic list and abundances (Masaitis and Pevzner 1999), using a taxonomic terminology and methodology preferred by Russian workers which is different from practices followed by western palynologists. Therefore comparison and independent evaluation of data are difficult. I am not aware of published illustration of the pollens and spores recovered from the Kovernino Formation. The latest comprehensive work (Masaitis and Pevzner 1999) quotes the list of pollen taxa by referring only to the explanatory notes of the geological map of the area. A modern revision of the palynostratigraphy is warranted.

Independent K-Ar radiometric dating of impactites yielded scattered, ambiguous results, ranging from 200±3 Ma to 183±5 Ma (Masaitis and Pevzner 1999) (Fig. 3). Due to these discrepancies, radiometric ages were deeemed unreliable and not considered further in the estimation of crater age. Various compilations of the cratering record cite different numeric ages for the Puchezh-Katunki structure, according to the time scale used to convert the Bajocian biostratigraphic age (Grieve et al. 1995; Grieve 2001).

In general, K-Ar dating of impact rocks is often fraught with problems. Older apparent ages can result from the lack of complete resetting (i.e., retention of older ^{40*}Ar from the target lithology), whereas younger apparent ages may reflect ^{40}Ar* loss through devitrification or post-impact hydrothermal processes (Deutsch and Schärer 1994). However, as the youngest unit in the target rocks is Early Triassic in age, the stratigraphic brackets permit that any one of the radiometric age determinations could in fact be accurate. A comparison with the recently revised Jurassic numeric time scale (Pálfy et al. 2000) (Fig. 3) reveals that either the end-Triassic (~200 Ma) or the Early Jurassic (Pliensbachian-Toarcian, ~183 Ma) extinctions could hypothetically be coeval with the Puchezh-Katunki impact. Therefore, a definitive age determination of the Puchezh-Katunki crater is desirable.

To this end, new radiometric dating employing the more accurate and precise ^{40}Ar/^{39}Ar method is planned. This dating technique has been successfully employed in the age determination of other large impact craters, such as Chicxulub (Swisher et al. 1992) and Popigai (Bottomley et al. 1997). The lack of a coherent impact melt sheet and widespread hydrothermal alteration makes the radiometric dating challenging. Alternatively, the age of the impact may be verified if its distal ejecta were discovered in well-dated stratigraphic successions.

4
Reported Possible Impact Signatures in the Lower and Middle Jurassic and near the Triassic-Jurassic Boundary

In order to evaluate the feasibility of different ages for the Puchezh-Katunki crater, here I briefly review the literature records of possible geological evidence for a large impact near the Triassic-Jurassic boundary or in the Early to Middle Jurassic (Fig. 3, Table 2). The location of the reported possible impact signatures are shown on an Early Jurassic paleogeographic map (Fig. 1). Anomalously high Ir concentration and the presence of shocked quartz and/or microspherules are considered possible direct impact indicators. Sharp negative carbon anomalies and a paleontological record of catastrophic changes in ecosystems may be regarded tentatively as indirect impact indicators.

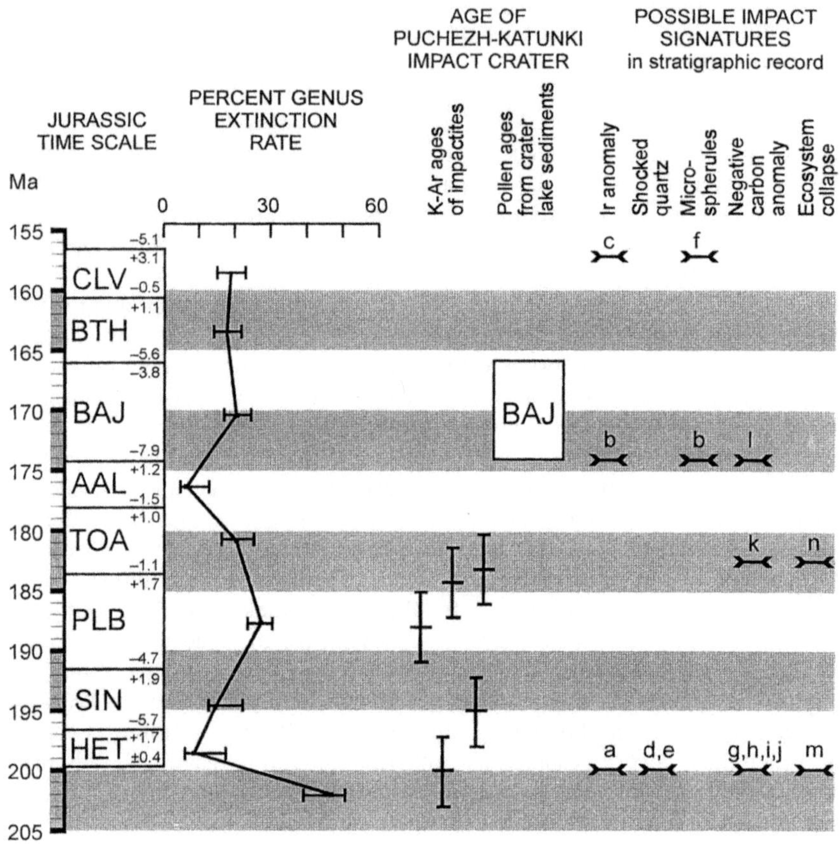

Fig. 3. Summary of age determinations of the Puchezh-Katunki crater (Masaitis and Pevzner 1999), possible impact signatures in the terminal Triassic and Early to Middle Jurassic stratigraphic record, and extinctions rates (Sepkoski 1996). Numeric time scale from Pálfy et al. (2000). For key to labels (location and reference), see Table 3. Stage abbreviations: HET – Hettangian; SIN – Sinemurian; PLB – Pliensbachian; TOA – Toarcian; AAL – Aalenian; BAJ – Bajocian; BTH – Bathonian; CLV – Callovian.

None of the three direct indicators is fool-proof, but their combined presence may make the strongest case for impact. The interpretation of the stable isotope stratigraphic record is more contentious. A negative carbon anomaly may reflect different kinds of disruption in the carbon cycle, including that through an impact-induced shutdown of marine primary productivity. However, this postulation requires additional support from other lines of evidence. Similarly, ecologic collapse in itself cannot be uniquely associated with impact.

Table 2. Possible impact signatures in the terminal Triassic and Early to Middle Jurassic stratigraphic record

Label	Type	Locality	Age	Reference	Remarks
a	Ir anomaly	Newark Basin	Tr/J boundary	(Olsen et al. 2002a, b)	Preliminary results, moderately elevated Ir values (Also note that no shocked quartz has been found in the Newark Supergroup, despite repeated search (Mossman et al. 1998; Olsen et al. 2002b))
b	Ir, micro-spherules	Southern Alps	Bajocian	(Jéhanno et al. 1988)	From ferruginous hardground; two spherule populations, one consists of micrometeorites, the other may have derived from ablation of a larger meteorite with an estimated minimum D of 50 m
c	Ir	Poland	Callovian	(Brochwicz-Lewinski et al. 1986)	Occurs in a condensed stromatolitic layer, where organic fixation may account for geochemical peculiarities
d	shocked quartz	Northern Apennines	Tr/J boundary	(Bice et al. 1992)	Grains occur at three levels, planar deformation is not convincing for impact origin, biostratigraphic constraints are loose
e	shocked quartz	Northern Alps	Tr/J boundary	(Badjukov et al. 1987)	Impact origin of grains has been seriously doubted
f	micro-spherules	Poland	Callovian	(Brochwicz-Lewinski et al. 1984)	Same locality as in (c)
g	C spike	Queen Charlotte Is.	Tr/J boundary	(Ward et al. 2001)	Short-lived negative anomaly recorded in marine organic matter; authors consider productivity collapse as most likely explanation
h	C spike	Hungary	Tr/J boundary	(Pálfy et al. 2001)	Negative anomaly recorded in marine carbonate and organic matter; authors consider productivity collapse or CH_4 release as potential causes
i	C spike	England	Tr/J boundary	(Hesselbo et al. 2002)	Negative anomaly recorded in organic matter
j	C spike	East Greenland	Tr/J boundary	(Hesselbo et al. 2002)	Negative anomaly recorded in terrestrial plant material
k	C spike	England	Toarcian	(Hesselbo et al. 2000)	Methane release is implied beacuse parallel change in terrestrial and marine organic matter observed
l	C spike	England	Bajocian	(Hesselbo et al. 2001)	From terrestrial organic material
m	fern spike	Newark Basin	Tr/J boundary	(Fowell and Olsen 1993; Fowell et al. 1994)	Invokes similarity to K-T fern spike, implies large-scale terrestrial ecosystem disruption
n	extinction	NW Europe	Toarcian	(Little 1996)	Benthic species extinction linked to Oceanic Anoxic Event

A perusal of the compilation presented in Fig. 3 and Table 2 permit the following observations. The Triassic-Jurassic boundary, marked by one of the „big five" mass extinctions, has been repeatedly linked to an impact event. Existing direct evidence for shocked quartz is disputed but a moderate Ir anomaly was recently reported (Olsen et al. 2002a, b). The fossil record of the extinction event is compatible with forcing by short-term, drastic environmental change, as is the disruption of the global carbon cycle. Scenarios that implicate intense and brief flood basalt volcanism of the Central Atlantic magmatic province appear better substantiated (Pálfy et al. 2002, Pálfy in press), but the role of an impact cannot be excluded.

Direct evidence for impact that would correlate with the Toarcian or Pliensbachian-Toarcian extinction is lacking. The paleontological record and the geochemical anomalies provide only hypothetical and circumstantial clues. Impact causation for this extinction has never been proposed, and the well-documented oceanic anoxic event and coeval volcanism of the Karoo-Ferrar province provide more plausible alternative forcing mechanisms (Pálfy et al. 2002).

From the Bajocian an iridium anomaly, microspherules and a carbon isotope excursion were reported, although none have been observed in more then one section. This stage is not noted as a time of elevated extinction rates but it represents the presently accepted age of the Puchezh-Katunki crater.

5 Prospects for Search of Puchezh – Katunki Ejecta in the Sedimentary Record

The paleogeography of the Russian Platform, as inferred from the distribution of marine strata, did not favor the preservation of a proximal ejecta blanket. No Lower or lower Middle Jurassic marine sediments are known in the vicinity of the Puchezh-Katunki crater, i.e., within $5R_{crater}$=200 km.

In most parts of the Russian Platform, marine sedimentation did not start until the later part of the Middle Jurassic, with the exception of the Donets folded area (Krymholts et al. 1988). In this area, marine strata were deposited from the Early Toarcian onward (Krymholts 1972). This depocenter, however, is some 800 km to the southwest from the Puchezh-Katunki crater. Therefore only a relatively thin distal ejecta layer may be expected, but so far no thorough search has been carried out.

6 Discussion

Whether or not the Puchezh-Katunki crater, the fifth largest in the Phanerozoic, is related to any extinction event, bears directly on the proposed relationship

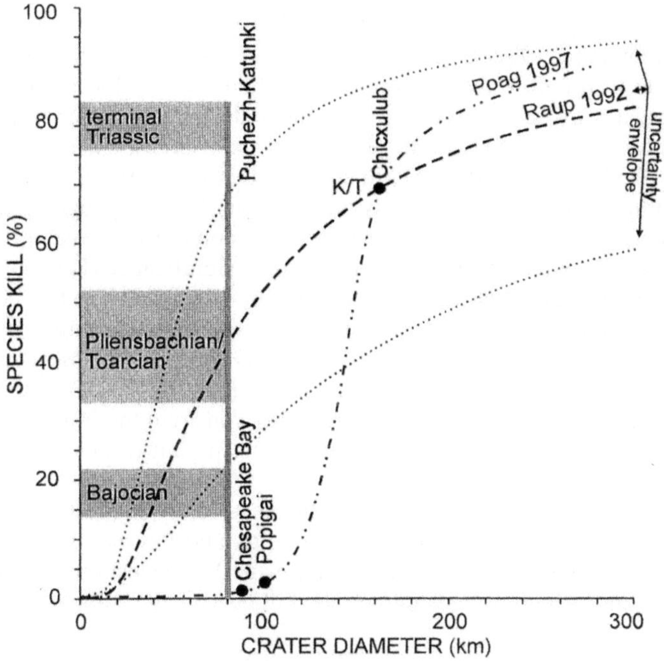

Fig. 4. Potential constraints of the impact-kill curve from the Puchezh-Katunki crater. Predicted extinction levels corresponding to the size of of the crater shown for the original impact-kill curve of Raup (1992) and the modified version of Poag (1997). Percent species extinction estimates for possible ages of the Puchezh-Katunki are shown based on Sepkoski (1996). Note that the percent estimates are for entire stages.

between large impacts and extinction. The impact-kill curve concept of Raup (1992) attempts to correlate the size of the impact (expressed by the crater diameter) and its biotic effect (measured by percent species kill) (Fig. 4).

Raup's (1992) original, sigmoidal kill curve was based solely on the extinction time series data and estimates of the impact flux. Remarkably, the discovery of Chicxulub revealed a crater size that matched the prediction of the curve for the K/T event. However, modifications were suggested subsequently on the basis of other craters. Jansa (1993) used the early Eocene Montagnais crater, which is 45 km in diameter, to place a lower threshold on the killing effect of an impact and argued for a hyperbolic, rather than sigmoidal kill curve.

The kill curve was recently revisited by Poag (1997), using data from two large, nearly coeval Late Eocene impact craters (Popigai, D=100 km and Chesapeake Bay, D=85 km). The lack of a significant extinction event directly related to these impacts contradicts the predictions of the original Raup curve, requiring at least its modification (Poag 1997) (Fig. 4).

Equal in size to Popigai and thus the second or third largest in the Phanerozoic (Table 1), the Manicouagan crater (D=100 km) was suggested to be linked to the end-Triassic extinction (Olsen et al. 1987). However, its U-Pb age of 214 Ma (Hodych and Dunning 1992) corresponds to the early Norian, a time of no significant extinction well before the Triassic-Jurassic boundary at 200 Ma (Pálfy et al. 2000). It postdates by several million years the end-Carnian, a time of a disputed vertebrate extinction event (Benton 1991). The sixth largest Phanerozoic crater, Morokweng (D=70 km) is 146±2 Ma in age (Koeberl et al. 1997), thus the impact is coeval within error with the Jurassic-Cretaceous boundary age (Pálfy et al. 2000). The end-Jurassic extinction, apparently a second-order peak in Sepkoski's data (1996), was disputed as a biotic event of major significance (Hallam 1996).

Reliable dating of the Puchezh-Katunki structure is important in this context, as it may provide critical evidence for possible biological effects of impacts that produce craters in the 80-100 km diameter range. As long as the link between the Chicxulub crater and the K/T event remains the only firmly established impact-extinction link, it is difficult to confirm or further constrain the impact-kill curve. Dating uncertainties of the Puchezh-Katunki crater permit discussion of the following possibilities.

(1) If the Puchezh-Katunki impact is Bajocian, as currently suggested by the palynostratigraphic data (Masaitis et al. 1996; Masaitis and Pevzner 1999), its biotic effects may be negligible. This is in agreement with the suggestion of Poag (1997) that the minimum extinction threshold on the kill curve is well above 100 km crater diameter.

(2) If the Puchezh-Katunki impact is of Triassic-Jurassic boundary age, it would require a major revision of the previously accepted crater age. Furthermore, its biotic effect would exceed the predictions of the original kill curve of Raup (1992).

(3) If the Puchezh-Katunki impact is of early Toarcian age, a significant revision of the previously accepted crater age would still be required. The relationship between extinction magnitude and crater size would be consistent with the original kill curve, falling within the uncertainty band of Raup (1992), and would be comparable with that of the Morokweng crater and the Jurassic-Cretaceous bioevent.

Should the Puchezh-Katunki impact be related to an extinction but the somewhat larger Popigai and Chesapeake Bay impacts are not, it lends support to complex extinction models, e.g. the multiplicative multifractal model of Plotnick and Sepkoski (2001). This model suggests that the extinction magnitude is not exclusively controlled by the external perturbation, i.e., the impact size alone, but it also depends on the state of the biota at the time of perturbation. Furthermore, that size alone does not determine the biotic effect of an impact, is proposed for Chicxulub and the K/T event, where the carbonate- and evaporite-rich target stratigraphy may have played an important role in unleashing the environmental catastrophe.

7 Conclusions

The age of the Puchezh-Katunki impact structure is not known with certainty. The cited Bajocian palynostratigraphic age needs better documentation before it can be accepted. Existing K-Ar radiometric ages are scattered between the Triassic-Jurassic boundary and Early Jurassic (Pliensbachian-Toarcian).

The possibility cannot be ruled out that the Puchezh-Katunki impact is coeval with either the end-Triassic or the Early Jurassic (Pliensbachian-Toarcian) extinction.

Much of the Russian Platform was subaerial during the Early and early Middle Jurassic. Some Puchezh-Katunki ejecta may be preserved in the Donets folded structure, some 800 km southwest of the crater, where marine sedimentation prevailed from the early Toarcian onward.

Possible distal ejecta and direct or indirect geochemical impact signatures are known from several localities and stratigraphic horizons within the suggested possible age range of the Puchezh-Katunki crater.

A more conclusive crater age determination is expected from $^{40}Ar/^{39}Ar$ dating of impact melts and/or glass – such a project is planned.

Better radiometric and stratigraphic dating will help either prove or disprove the presently highly hypothetical impact-extinction link. The results should provide important constraints on the impact-kill curve (Raup 1992), as the biological effects of an impact that produced a crater 80 km in diameter can be assessed.

Acknowledgements

I thank V. Masaitis (St. Petersburg) for discussion and providing copies of some of the essential Russian literature, and S. Feist-Burkhardt (London) for advice and comments on the palynostratigraphy. A critical review by S. Hesselbo helped improve the manuscript. The reserach was carried out during a Humboldt Research Fellowship and a Bolyai Research Fellowship. My conference participation was sponsored by the Alexander von Humboldt Foundation and the ESF Impact Programme. This is a contribution to IGCP 458.

References

Alvarez L, Alvarez W, Asaro F, Michel H (1980) Extraterrestrial cause for the Cretaceous-Tertiary extinction. Science 208: 1095-1108

Badjukov DD, Lobitzer H, Nazarov MA (1987) Quartz grains with planar features in the Triassic-Jurassic boundary sediments from the Northern Limestone Alps, Austria. [abs.] Lunar and Planetary Science 28: 38 - 39

Benton MJ (1991) What really happened in the Late Triassic? Historical Biology 5: 263-278

Bice DM, Newton CR, McCauley S, Reiners PW, McRoberts CA (1992) Shocked quartz at the Triassic-Jurassic boundary in Italy. Science 255: 443-446

Bottomley R, Grieve R, Masaitis V (1997) The age of the Popigai impact event and its relation to events at the Eocene/Oligocene boundary. Nature 388: 365-368

Brochwicz-Lewinski W, Gasiewicz A, Krumbein WE, Melendez G, Sequeiros L, Suffczynski S, Szatkowski K, Tarkowski R, Zbik M (1986) Anomalia irydowa na granicy jury srodkowej i gorney. Przeglad Geologiczny 33: 83-88 (in Polish)

Brochwicz-Lewinski W, Gasiewicz A, Suffczynski S, Szatkowski K, Zbik M (1984) Lacunes et condensation a la limite Jurassique moyen-supérieur dans le sud de la Pologne: manifestation d'un phénomene mondial? Comptes Rendus de l'Académie des Sciences, Paris, Series II 299: 1359-1362

Deutsch A, Schärer U (1994) Dating terrestrial impact events. Meteoritics 29: 301-322

Fowell SJ, Olsen PE (1993) Time calibration of Triassic/Jurassic microfloral turnover, eastern North America. Tectonophysics 222: 361-369

Fowell SJ, Cornet B, Olsen PE (1994) Geologically rapid Late Triassic extinctions: Palynological evidence from the Newark Supergroup. In: Klein GD (ed) Pangea: Paleoclimate, Tectonics, and Sedimentation During Accretion, Zenith, and Breakup of a Supercontinent. Geological Society of America Special Paper 288, pp 197-206

Grieve R, Rupert J, Smith J, Therriault A (1995) The record of terrestrial impact cratering. GSA Today 5: 189-196

Grieve RAF (1997) Extraterrestrial impact events: The record in the rocks and the stratigraphic column. Palaeogeography, Palaeoclimatology, Palaeoecology 132: 5-23

Grieve RAF (2001) Impact Crater website. http://gdcinfo.agg.nrcan.gc.ca:80/crater/index_e.html/. Address of 2003:http://www.unb.ca/passc/ImpactDatabase

Hallam A (1996) Major bio-events in the Triassic and Jurassic. In: Walliser OH (ed) Global Events and Event Stratigraphy in the Phanerozoic. Springer, Berlin, pp 265-283

Hesselbo SP, Robinson SA, Surlyk F, Piasecki S (2002) Terrestrial and marine mass extinction at the Triassic–Jurassic boundary synchronized with major carbon-cycle perturbation: A link to initiation of massive volcanism? Geology 30: 251-254

Hesselbo S, Morgans-Bell H, McElwain J, Rees PM, Stuart R (2001) A major carbon-cycle perturbation in the Middle Jurassic and accompanying climatic change adduced from the land plant record. [abs.] EUG XI, Strasbourg, Abstracts, http://www.campublic.co.uk/ EUGXI/CC03.pdf

Hesselbo SP, Gröcke DR, Jenkyns HC, Bjerrum CJ, Farrimond P, Morgans Bell HS, Green OR (2000) Massive dissociation of gas hydrate during a Jurassic oceanic anoxic event. Nature 406: 392-395

Hodych JP, Dunning GR (1992) Did the Manicouagan impact trigger end-of-Triassic mass extinction? Geology 20: 51-54

Jansa LF (1993) Cometary impacts into ocean: their recognition and the threshold constraint for biological extinctions. Palaeogeography, Palaeoclimatology, Palaeoecology 104: 271-286

Jéhanno C, Boclet D, Bonté P, Castellarin A, Rocchia R (1988) Identification of two populations of extra-terrestrial particles in a Jurassic hardground of the southern Alps. Proceedings of Lunar and Planetary Science Conference 18, pp 623-630

Koeberl C, Armstrong RA, Reimold WU (1997) Morokweng, South-Africa - a large impact structure of Jurassic-Cretaceous boundary age. Geology 25: 731-734

Krymholts GY (1972) Stratigraphy of the USSR. Vol. 10: The Jurassic System (in Russian). Gosgeoltechizdat, Moscow, 524 pp

Krymholts GY, Mesezhnikov MS, Westermann GEG (1988) The Jurassic ammonite zones of the Soviet Union. Geological Society of America Special Paper 288, Boulder, Colorado, 116 pp

Little CTS (1996) The Pliensbachian-Toarcian (Lower Jurassic) extinction event. In: Ryder G, Fastovsky D, Gartner S (eds) The Cretaceous-Tertiary Event and Other Catastrophes in Earth History. Geological Society of America Special Paper 307: pp 505-512

Masaitis VL, Mashchak MS, Naumov MV (1996) The Puchezh-Katunki astrobleme: A structural model of a giant impact crater. Solar System Research 30: 3-10

Masaitis VL, Pevzner LA (1999) Deep Drilling in the Puchezh-Katunki Impact Structure (in Russian). VSEGEI Press, Saint-Petersburg, 392 pp

Mossman DJ, Grantham RG, Langenhorst F (1998) A search for shocked quartz at the Triassic – Jurassic boundary in Fundy and Newark basins of the Newark Supergroup. Canadian Journal of Earth Sciences 35: 101 - 109

Olsen PE, Shubin NH, Anders MH (1987) New Early Jurassic tetrapod assemblages constrain Triassic-Jurassic tetrapod extinction event. Science 237: 1025-1029

Olsen PE, Kent DV, Sues H-D, Koeberl C, Huber H, Montanari A, Rainforth EC, Fowell SJ, Szajna MJ, Hartline BW (2002a) Ascent of dinosaurs linked to an iridium anomaly at the Triassic-Jurassic boundary. Science 296: 1305-1307

Olsen PE, Koeberl C, Huber H, Montanari A, Fowell SJ, Et-Touhami M, Kent DV (2002b) The continental Triassic-Jurassic boundary in central Pangea: recent progress and preliminary report of an Ir anomaly. In: Koeberl C, MacLeod KG (eds) Catastrophic Events and Mass Extinctions: Impacts and Beyond. Geological Society of America Special Paper 356, pp 505-522

Pálfy J (in press) Volcanism of the Central Atlantic Magmatic Province as a potential driving force in the end-Triassic mass extinction. In: Hames W, McHone G, Renne P, Ruppel C (eds) The Central Atlantic Magmatic Province. American Geophysical Union, Washington, DC

Pálfy J, Smith PL, Mortensen JK (2000) A U-Pb and $^{40}Ar/^{39}Ar$ time scale for the Jurassic. Canadian Journal of Earth Sciences 37: 923-944

Pálfy J, Demény A, Haas J, Hetényi M, Orchard M, Vetö I (2001) Carbon isotope anomaly and other geochemical changes at the Triassic-Jurassic boundary from a marine section in Hungary. Geology 29: 1047-1050

Pálfy J, Smith PL, Mortensen JK (2002) Dating the end-Triassic and Early Jurassic mass extinctions, correlative large igneous provinces, and isotopic events. In: Koeberl C, MacLeod KG (eds) Catastrophic Events and Mass Extinctions: Impacts and Beyond. Geological Society of America Special Paper 356, pp 523-532

Plotnick RE, Sepkoski JJ Jr (2001) A multiplicative multifractal model for originations and extinctions. Paleobiology 27: 126–139

Poag CW (1997) Roadblocks on the kill curve: Testing the Raup hypothesis. Palaios 12: 582-590

Rampino MR, Haggerty BM (1996) Impact crises and mass extinction: A working hypothesis. In: Ryder G, Fastovsky D, Gartner S (eds) The Cretaceous-Tertiary Event and Other Catastrophes in Earth History. Geological Society of America Special Paper 307, pp 11-30

Raup DM (1992) Large-body impact and extinction in the Phanerozoic. Paleobiology 18: 80-88

Sepkoski JJ Jr. (1996) Patterns of Phanerozoic extinction: a perspective from global data bases. In: Walliser OH (ed) Global Events and Event Stratigraphy in the Phanerozoic. Springer, Berlin, pp 35-51

Swisher CC, Grajales-Nishimura JM, Montanari A, Margolis SV, Claeys P, Alvarez W, Renne P, Cedillo-Pardo E, Maurasse F, Curtis GH, Smit J, McWilliams MO (1992) Coeval $^{40}Ar/^{39}Ar$ ages of 65.0 million years ago from Chicxulub crater melt rock and Cretaceous-Tertiary boundary tektites. Science 257: 954-958

Ward PD, Haggart JW, Carter ES, Wilbur D, Tipper HW, Evans T (2001) Sudden productivity collapse associated with the Triassic-Jurassic boundary mass extinction. Science 292: 1148-1151

Ziegler PA (1990) Geological Atlas of Western and Central Europe. Shell Internationale Petroleum, The Hague, 130 pp

Geochemistry of a Langhian Pelagic Marly Limestone Sequence of the Cònero Riviera, Ancona (Italy) and the Search for a Ries Impact Signature: A Progress Report

Dieter Mader[1], Christian Koeberl[1,2] and Alessandro Montanari[2]

[1] Institute of Geochemistry, University of Vienna, Althanstrasse 14, A-1090 Vienna, Austria (christian.koeberl@univie.ac.at);
[2] Osservatorio Geologico di Coldigioco, I-62020 Frontale di Apiro, Italy

Abstract. Samples from the mid-Miocene marine La Vedova section at the Cònero Riviera, Umbria-Marche sequence, Italy, were analyzed for their chemical composition to provide data for impact- and cyclostratigraphic studies. The rocks from the La Vedova section provide new data for the Mediterranean Langhian. Major and trace elemental compositions demonstrate a clayey and (phyllo)silicatic terrigenous input. The rare earth elements show typical upper continental crustal composition and do not indicate seawater-fractionated patterns. Thus, despite of carbonate dilution, the siliciclastic input in the pelagic carbonates may reflect the composition and provenance of the source areas by using trace element ratios and rare earth elements. The Middle and Upper Miocene in the Alpine-Apennine region is mainly represented by synorogenic reworked siliciclastic deposits. The pelagic carbonate sequence at the Cònero Riviera is nearly continuous, and one of the aspects of this work was a search for ejecta (and regional effects) of the medium-sized Ries impact crater event, about 600 km from the Cònero Riviera. No clear chemical signal of an extraterrestrial component, from the Ries or any other impactor, was obvious in the present data set. However, the Ries impactor has previously been proposed to have been of achondritic composition, in which case only a detailed search for shocked minerals and microspherules could help to recognize a possible Ries-derived distal ejecta layer.

1 Introduction

For impact-stratigraphic studies, as for any other geologic task, a continuous sedimentary sequence is an ideal and desirable condition. With the known age of a certain impact structure, its distal ejecta may be found in such time-related successions. Impact-related ejecta layers discovered in lithostratigraphic units several hundred or thousand kilometers away allow to investigate the effects of this particular impact event to the geological and biological environment (Koeberl 2001). First, however, a supposed ejecta layer has to be confirmed as impact

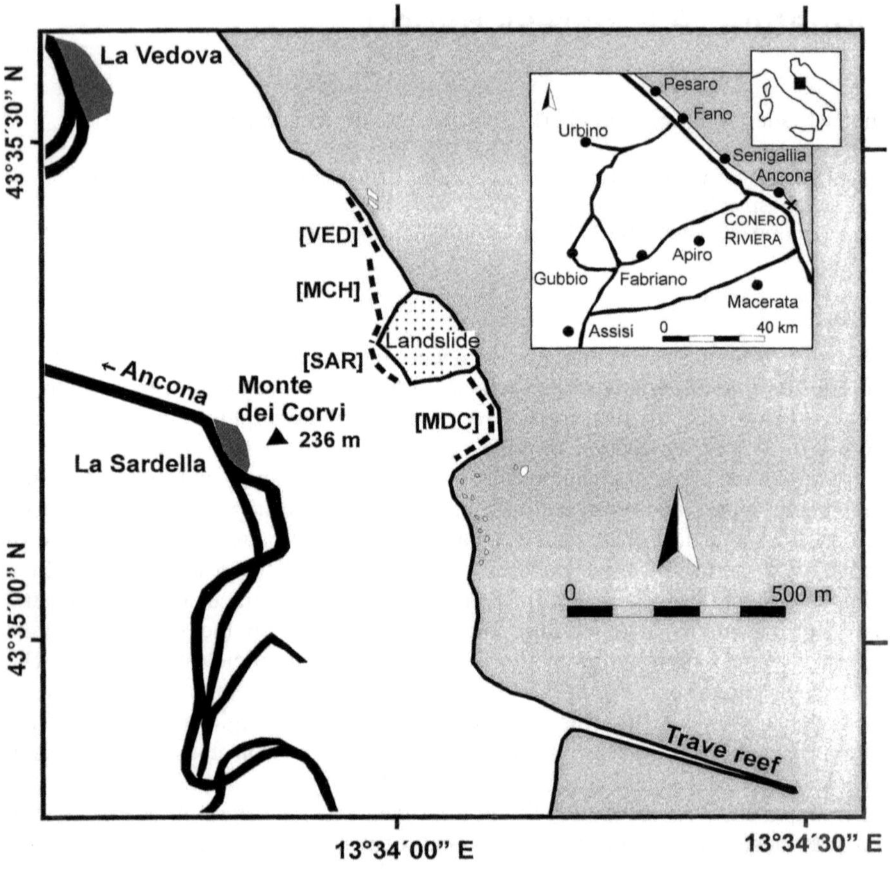

Fig. 1. Simplified map of the location of the study area (after Montanari et al. 1997)

induced. Thus, meteoritic components within the sedimentary record have to be identified in order to assign an impact origin to a suspicious layer (e.g., Koeberl 1998; Montanari and Koeberl 2000). In the Umbria-Marche (U-M) region in the northeastern Apennines, Italy, a nearly complete Late Triassic to Pleistocene sequence of marine sedimentary rocks exists, which provides an excellent opportunity to search for possible time-related impact events (Montanari and Koeberl 2000; and references therein). In this area of the Apennines the early-mid Miocene represents the beginning of the Alpine-Himalayan orogenesis. The onset of folding and thrusting was preceded by the formation of NW-SE elongated synorogenic siliciclastic foredeep troughs, which, as the orogenic front advanced eastward, were filled by thick sandy and marly flysch deposits. In the eastern part of the U-M basin, however, pelagic limestones and marls continued to accumulate until the Late Miocene. As a result in this easternmost part of the U-M basin, which is extensively exposed on the coastal cliffs of the Cònero Riviera, just south of the port city of Ancona (Fig. 1), a nearly continuous and undisturbed carbonate sequence is present. This provides the fortunate opportunity for detailed and integrated stratigraphic investigations of the Miocene Epoch (Montanari et al. 1997a; Montanari and Koeberl 2000).

Several volcaniclastic clay layers, in the form of ashfall deposits of distant volcanic eruptions, are interbedded and well preserved in the upper Paleogene to lower Neogene deep water carbonate sequence. Their provenance is still unknown, although a Sardinian source was proposed for an Early Miocene volcaniclastic layer in the Bisciaro Formation (Assorgia et al. 1994; Montanari et al. 1994). Such volcaniclastic layers within the carbonate sequences provide an excellent opportunity for calibrating the Paleogene and Neogene time scale by radioisotopic dating. Several detailed geochronologic studies were done on Eocene - Oligocene biotite-rich clay layers in Italy (Montanari et al. 1985; Odin 1985; Mattias et al. 1987; Montanari 1988; Montanari et al. 1988a; Odin et al. 1988; Odin et al. 1991). Some biotite-rich levels have also been dated with the $^{40}Ar/^{39}Ar$ method in the Miocene formations of the U-M sequence (Montanari et al. 1988b, Montanari et al. 1997a). Recently, a volcaniclastic biotite-rich layer within this section, named the Aldo Level, has been dated by $^{40}Ar/^{39}Ar$ laser fusion and yielded an apparent age of 14.9 ± 0.2 Ma (Mader et al. 2001). Besides being a contribution to constrain the Miocene geochronology, this age also has specific implications for impact-stratigraphic studies (Montanari and Koeberl 2000), as the Cònero Riviera is located within 600 km of the Ries impact structure in southern Germany, which has been dated at 15.1 Ma (Gentner et al. 1961; Staudacher et al. 1982), and 14.82 ± 0.32 Ma (Storzer et al. 1995). New Ar-Ar dating of moldavites show ages of 14.35 ± 0.05 Ma (Laurenzi et al. 2001) and 14.50 ± 0.16 Ma (Schwarz and Lippolt 2002), which imply a younger age for the Ries impact event than previously assumed.

In this study a high spatial resolution analysis for major and trace elements were done in the mid-Miocene La Vedova section at the Cònero Riviera (Fig. 1) in order to provide data for impact-stratigraphic, and also cyclostratigraphic, studies. As the age of the Aldo Level coincides (within error) with that of the Ries crater event, a detailed search for possibly occurring impact signatures related to the Ries event, such as shocked quartz or geochemical anomalies, or unusual bioevents, may be possible. Considering the mean sedimentation rate of 38.7 m/Ma in this area and taking the errors on the apparent $^{40}Ar/^{39}Ar$ ages into account, it is conceivable that probably present Ries ejecta are within close stratigraphic proximity of the dated Aldo Level. The Cònero Riviera area is at a distance of about 25 crater diameters from the Ries crater, which is in the distal ejecta region (Fig. 2). A 10 cm-thick layer of Ries ejecta was found in Switzerland, at a distance of about 200 km from the crater (e.g., Hofmann and Hofmann 1992), so it is reasonable to suspect that distal ejecta might also be present in the U-M basin.

In this paper we present of the complete geochemical data set and some preliminary general discussion, and provide suggestions regarding further studies.

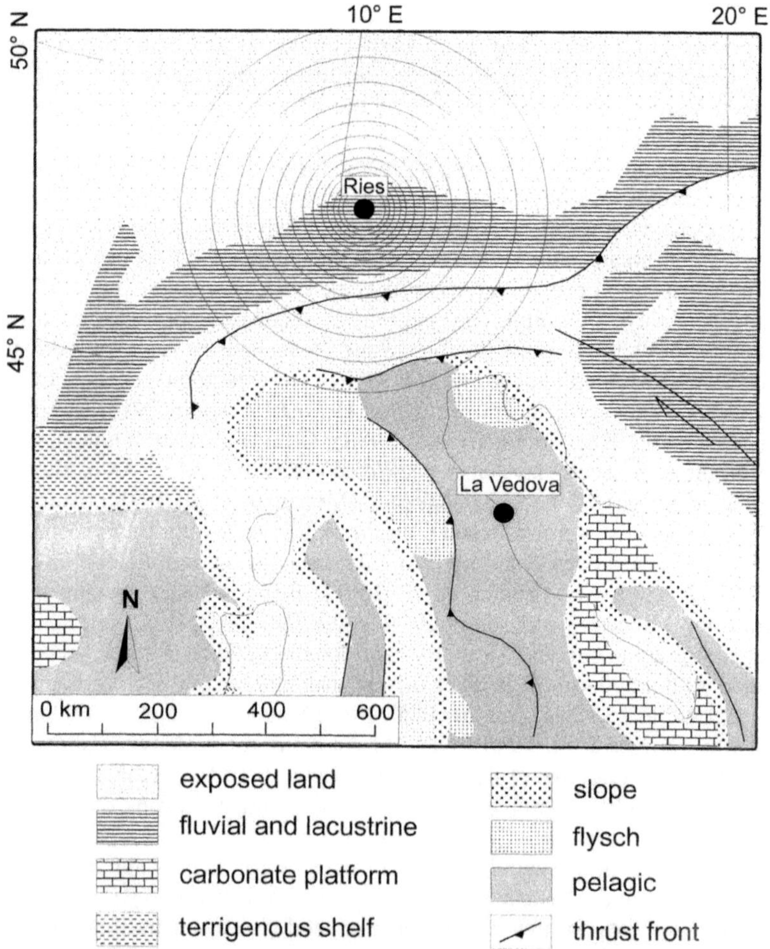

Fig. 2. Paleogeographic map of central Europe (Late Langhian), showing the distance of the Ries impact structure to the study area at the Cónero Riviera, Ancona (modified after Meulenkamp et al. 2000).

2
Geological Setting and Lithostratigraphy

The marly limestones and marls of the Cònero Riviera, south of the port city of Ancona, represent a 300-m-thick pelagic sequence covering the entire Miocene Epoch. The Middle to Upper Miocene portion of this sequence (upper Langhian to the lower Messinian) is continuously exposed along the coastal cliffs of Monte dei Corvi, northwest of the Trave Reef (Fig. 1). It contains the upper part of the Schlier Formation (Serravallian and Tortonian), and the Euxinic Shale unit at the base of the Gessoso-Solfifera Formation (Messinian), which represents the inception of the Messinian salinity crisis. The Schlier Formation is a rhythmic sequence of hemipelagic and pelagic marls and marly limestones interbedded with frequent black shales, and containing rare, distal volcanic ash layers.

The composite sedimentary sequence of the Monte de Corvi was derived by Montanari et al. (1997a) from three sections. From S to N, these sections are: (a) the Monte de Corvi (MDC) along the beach, (b) the combined sections of La Sardella and the higher cliffs of the Monte dei Corvi (SAR-MCH), and (c) the La Vedova (VED), which extends along the beach below the homonym locality (Fig. 1). With the exception of a 10-cm-thick turbidite called the Cavolo marker, which is located within the mid-Serravallian portion of the Schlier, the whole Langhian to Messinian sequence of Monte dei Corvi is virtually devoid of siliciclastic or detrital beds (Montanari et al. 1997a).

In contrast to the western areas of the U-M basin, where the Schlier Formation displays reduced thickness and is overlain by siliciclastic flysch deposits since the Langhian, the eastern part of the basin received by the siliciclastic turbidites not before the late Messinian. The Schlier Formation of the Cònero Riviera is subdivided into three lithostratigraphic subunits from base to top: (a) the Massive Member, (b) the Calcareous Member, and (c) the Marly Member. The Massive Member is a 35 m thick sequence of seven massive beds of hemipelagic marls (Sandroni 1985). Above them follows the Calcareous Member, with a rhythmic sequence of planktonic foraminiferal marly limestones and strongly bioturbated marls. Within the upper part of this member, frequent thin beds of black shale are interbedded between marls and marly limestone couplets. The Calcareous Member is overlain by the Marly Member, which consists of marls and several intercalated marly limestones and black shales.

The La Vedova section (Fig. 3), which was investigated for this study, consists of a 40-m-thick, continuous sequence of bioturbated marine marls and marly limestones at the foot of the Monte dei Corvi cliffs (upper Massive Member and lower Calcareous Member). A biotite-rich volcaniclastic layer, the Aldo Level, occurs within this section at 24.1 m (Montanari et al. 1997b, Mader et al. 2001).

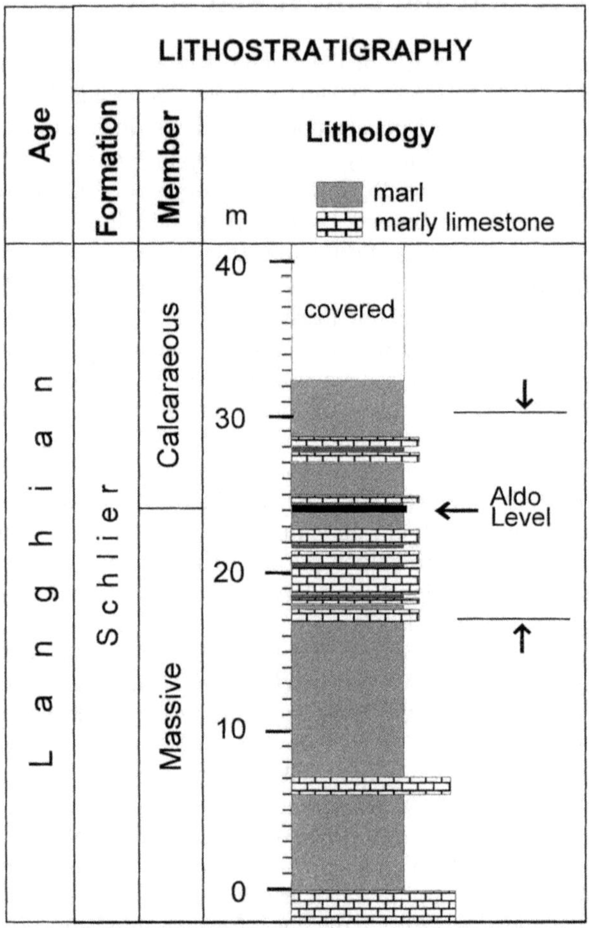

Fig. 3. Lithostratigraphy of the La Vedova section. The sampled interval of the section is indicated by arrows.

3 Methodology

From the La Vedova section, representative samples were collected every 10 cm from meter levels 17.0 to 30.0. From each sample, between meter levels 18.0 to 30.0, about 150-200 g were powdered in situ using a masonry drill. Determination

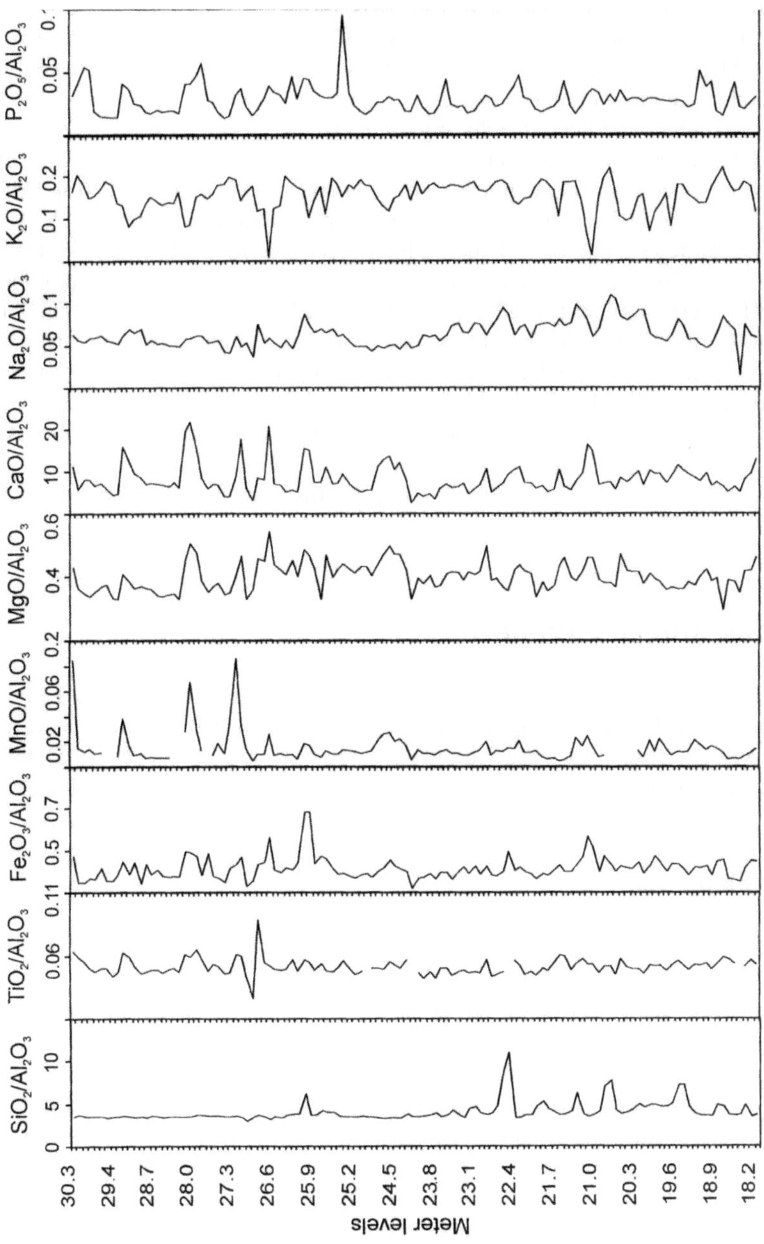

Fig. 4. Major element oxides normalized to Al_2O_3 for comparison of the individual variations within the La Vedova section (meters levels after Montanari et al. 1997a).

of the contents of major element oxides and some trace elements (V, Cr, Co, Ni, Cu, Zn, Rb, Sr, Y, Zr, Nb and Ba) was done by XRF spectrometry at the Department of Geology, University of the Witwatersrand, Johannesburg (South Africa). Details of the analytical method, as well as accuracy and precision, are described in Reimold et al. (1994).

All other trace elements (including again Cr, Co, Ni, Zn, Rb, Sr, Zr, and Ba) were determined by instrumental neutron activation analysis (INAA) at the Institute of Geochemistry, University of Vienna (Austria). About 140 mg of each sample and about 65 mg of the international rock standards granite ACE (CRPG-Nancy), granite G-2 (U.S.G.S.) and carbonaceous chondrite Allende (Smithsonian Institution, Washington) were sealed in polyethylene capsules. Samples and standards were irradiated together in the 250 kW Triga reactor of the Atomic Institute of the Austrian Universities for 8 hours at a neutron flux of 2.10^{12} n cm^{-2} s^{-1}. More details on standards, instrumentation, accuracy and precision of our method are described by Koeberl (1993).

Due to the low detection limit for iridium by INAA, seventeen samples have been remeasured for iridium contents by γ-γ coincidence spectrometry with the iridium coincidence spectrometer (ICS) at the Institute of Geochemistry (Koeberl and Huber 2000).

4 Results and Discussion

Analytical results of the major and trace element contents are listed in Table 1 (in the Appendix). For discussing the general geochemical pattern of the investigated stratigraphic section, only a few diagrams are presented in this report. Details on, e.g., cyclostratigraphic aspects of the data will be discussed in a separate paper.

4.1 Major Elements

All of the Ca can be attributed to $CaCO_3$, as indicated by separate $CaCO_3$ measurements on these samples (Cleaveland 2001). As the varying carbonate content does not allow a direct comparison of the terrigenous supply, these dilution effects are removed by normalizing the contents to Al_2O_3. In the Al-normalized data of the major elements, no trends are visible (Fig. 4). Only the SiO_2/Al_2O_3 and Na_2O/Al_2O_3 ratios display a slight decrease upwards in the stratigraphic section. In the upper part of the La Vedova section (interval ~ 23.5-30.0 m), the Si_2O/Al_2O_3 ratio shows little variation, which is not indicated by the other elemental ratios. In this upper part of the section, the CaO/Al_2O_3 ratios show larger variations than below. Similar patterns of higher variability are shown by the Fe_2O_3/Al_2O_3, MnO/Al_2O_3, P_2O_5/Al_2O_3 ratios.

A remarkable fact is that the biotite-rich Aldo level is not clearly depicted by the major elemental ratios. Only the lowest CaO/Al_2O_3 and Fe_2O_3/Al_2O_3 ratios, and the highest MgO/CaO ratio, point to a more significant terrigenous input.

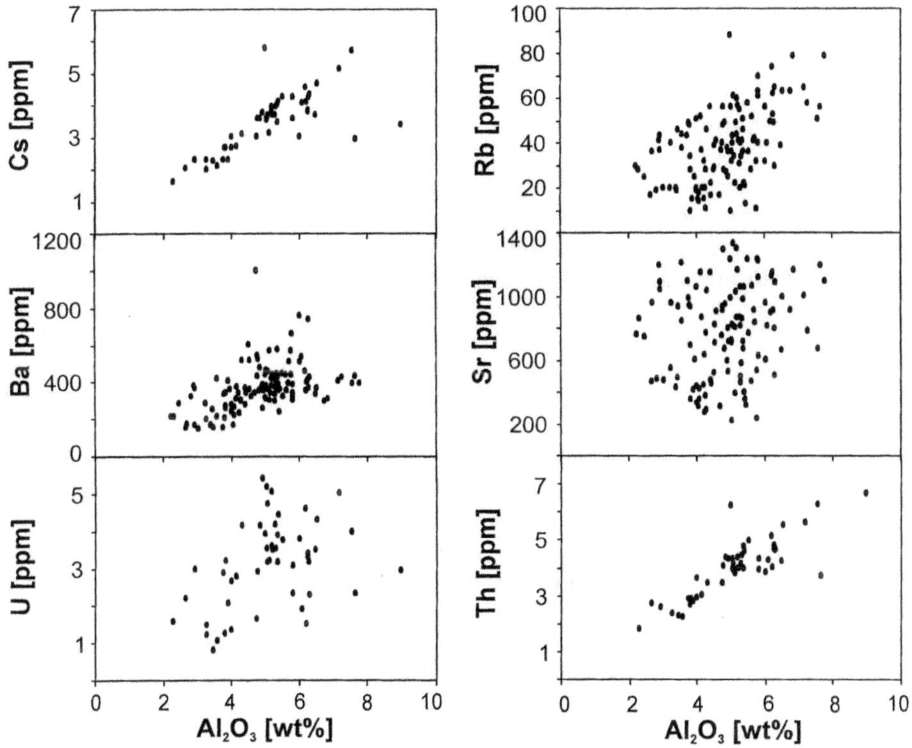

Fig. 5. Binary diagrams of the abundances of trace elements Rb, Cs, Ba, Sr, Th and U (in ppm) against Al_2O_3 content (in wt%), showing the degree of correlation of these elements to (phyllo)silicatic phases in the samples.

Whether the varying carbonate content indicates productivity, dilution or dissolution cycles (e.g., Einsele and Ricken 1991), or a mixture of some or all of them, is not yet understood. From preliminary $^{13}C/^{12}C$ isotope information by Montanari et al. (1997a), productivity cycles are suggested as the main cause for the rhythmic bedding in the Monte dei Corvi section (Goese 1999). The data in this study seem to confirm this suggestion, as the Si_2O/Al_2O_3 ratios – which indicate terrigenous quartz supply - show a nearly constant pattern, at least in the upper half of the La Vedova section, whereas the CaO/Al_2O_3 ratios are showing higher variations. This may indicate a variable $CaCO_3$ production during a more or less constant terrigenous supply. The few higher Si_2O/Al_2O_3 ratios in the lower part of the La Vedova section reflect the denser cyclicity between marls and marly limestone composition, which might depict a rapid and combined change in productivity and dilution cycles. Other terrigenous elemental ratio indicators, however, do not show this pattern. The ratios of Ti_2O/Al_2O_3, K_2O/Al_2O_3 and Fe_2O_3/Al_2O_3 are approximately constant through the section.

4.2 Trace Elements

With trace elemental compositions the evaluation of the terrigenous trace element carriers may be possible. The strong positive correlation of Cs and Th with Al_2O_3, as well as the moderate correlation of Rb, Ba, and U with Al_2O_3 (Fig. 5) indicate the phyllosilicatic control of these elements (e.g., Bauluz et al. 2000). A moderate positive correlation of Rb and K_2O may also point to illite phases as carrier for Rb. As K and Th are mainly linked to detrital phases (Plank and Langmuir 1998) a strong positive correlation between Th and K contents ($r = 0.75$; Fig. 6) most likely points to the presence of detrital aluminosilicate mineral phases (Jones and Manning 1994). Strontium in marine rocks is commonly thought to be associated with carbonate minerals (Plank and Langmuir 1998). In the La Vedova section Sr correlates neither with CaO nor with Al_2O_3 (Fig. 6). This possibly indicates a balanced mixture of calcite, feldspars or biotite as host phases for Sr.

According to Dypvik and Harris (2001), Zr/Rb ratios may reflect grain size variations in fine-grained siliciclastic sediments, as Zr is enriched in heavy minerals of coarser material and Rb is associated with the clayey and pyllosilicatic fraction. The Zr/Rb ratios in the La Vedova section shows higher values in the lower part, which is consistent with the Si_2O/Al_2O_3 ratio (Fig. 7). This points to coarser grained, quartz-rich siliciclastic component in the marls of the lower La Vedova section.

The average composition of all samples was normalized to deep sea carbonate composition (Turekian and Wedepohl 1961) in order to demonstrate the deviation of the marly samples from common marine limestones (Fig. 8a). With the exception of Ca, all considered elements differ from the average deep-sea carbonate composition. The elements Mn, Cu, Sr, and Y are depleted, whereas all others (above all Si, Mg, K, P, Sc, V, Cr, Rb, Cs) are more or less enriched, which indicates a marly composition of the samples, as these elements are mainly hosted in heavy minerals, clays and silicate phases of mafic and sialic terrigenous components.

Normalization of the average of all samples to PAAS (Post-Archean average Australian shale; McLennan 1981; 1989) shows the depletion of most elements and oxides in relation to the shale composite (Fig. 8b). Only CaO and Sr are enriched, which indicates a carbonate-rich composition. Thus, despite the fact that the contents of Sr and CaO do not correlate, some of the Sr may be although associated with calcite. In this case the variations in the Sr contents, and, in some cases, their low concentrations through the La Vedova section (216-1332 ppm), may not only indicate fluctuations in the detrital input, but may also point to some diagenetic mobilization of Sr (e.g., Baker et al. 1982).

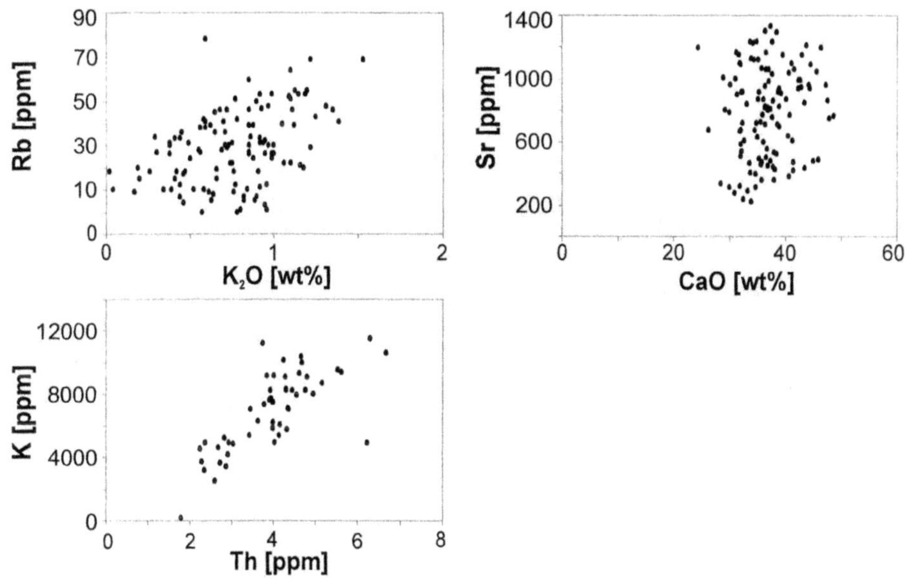

Fig. 6. Binary diagrams of the contents of Rb - K_2O and K - Th, indicating a clayey phyllosilicatic component of the samples, and of Sr - CaO, demonstrating that strontium is not mainly associated with the carbonaceous phase, as would be expected for limestones.

4.3
Rare Earth Elements (REE)

Chondrite-normalized REE distribution patterns do not show any significant variations within the measured samples of the La Vedova section. Only the absolute abundances of the REEs vary by about a factor of four. The chondrite-normalized (Taylor and McLennan 1985) average REE contents of the samples show the pattern that is common for post-Archean upper continental crust rocks, with enriched LREE, flat HREE and a negative Eu anomaly (Fig. 9a). The Eu anomaly compared to the neighboring REEs (Sm, Gd) was calculated from the relation

$$Eu/Eu^* = Eu_N/(Sm_N * Gd_N)^{0.5} \qquad (4.1)$$

(Taylor and McLennan 1985), and ranges from 0.58 to 0.91.

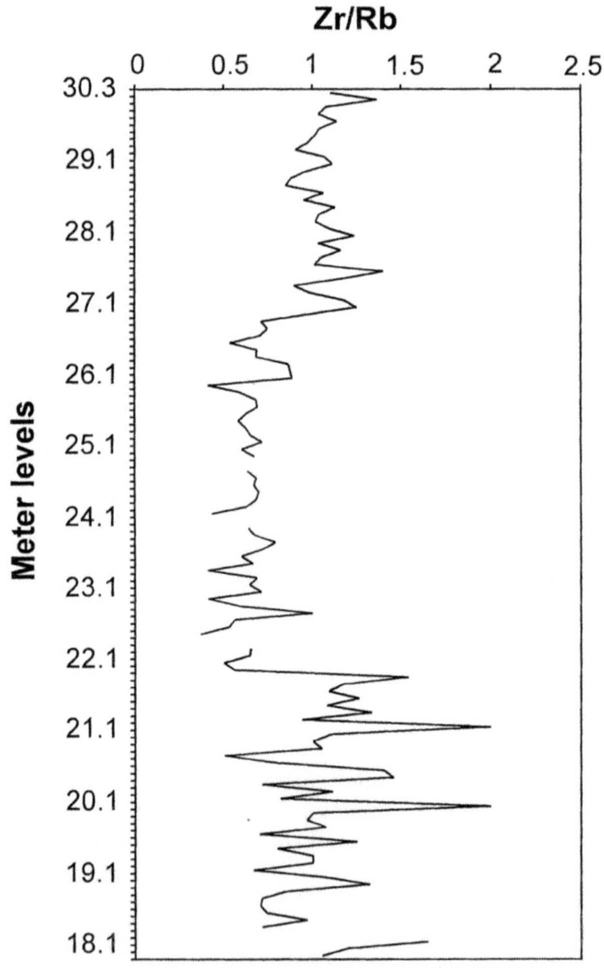

Fig. 7 Diagram showing the variation in the Zr/Rb ratio within the rocks of the La Vedova section (meter levels after Montanari et al 1997a), which is thought to demonstrate the grain size variation in fine-grained siliciclastic rocks (Dypvik and Harris 2001).

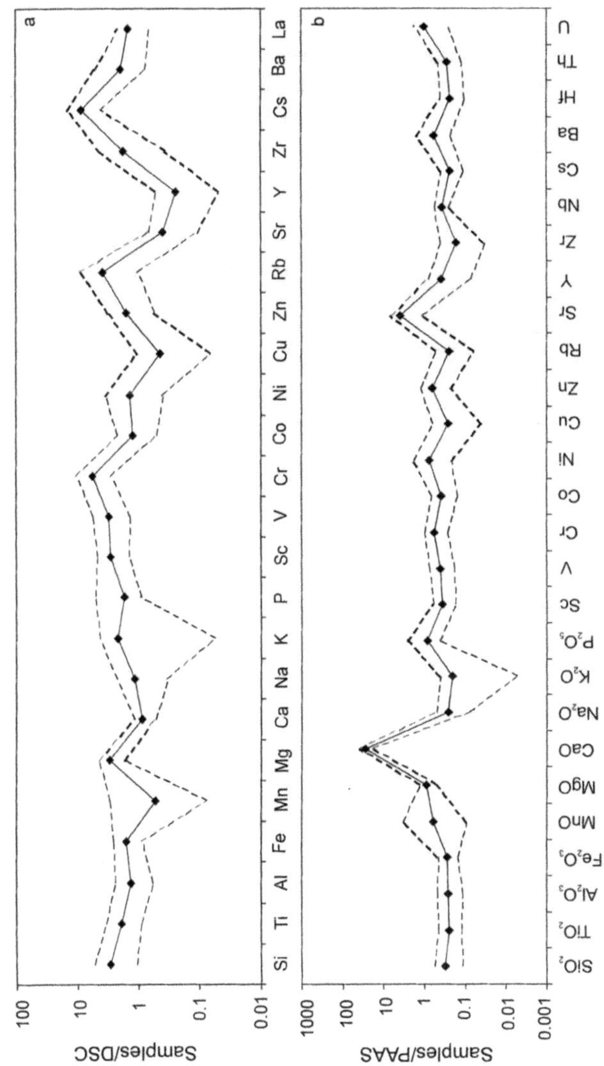

Fig. 8. a. Average sample major and trace element contents (in ppm) normalized to average deep sea carbonate compositions (Turekian and Wedepohl 1961), indicating the enrichment of elements that are mainly associated with detrital terrigenous material. **b** Average sample major and trace element contents (in ppm), normalized to post-Archean average Australian shale (McLennan 1981; 1989). Dashed lines show the maximum and minimum range of the the element contents.

Typical fine-grained siliciclastic rocks (shales) have absolute abundances of approximately 100 times chondritic for La and approximately 10-15 times chondritic for Yb (McLennan 1989). Due to carbonate dilution the absolute abundances are about a factor of six lower than those of shales. The abundance variations within all samples most probably indicate the dilution by variable carbonate contents, and also the different amounts of terrigenous siliciclastic input and its grain size fraction.

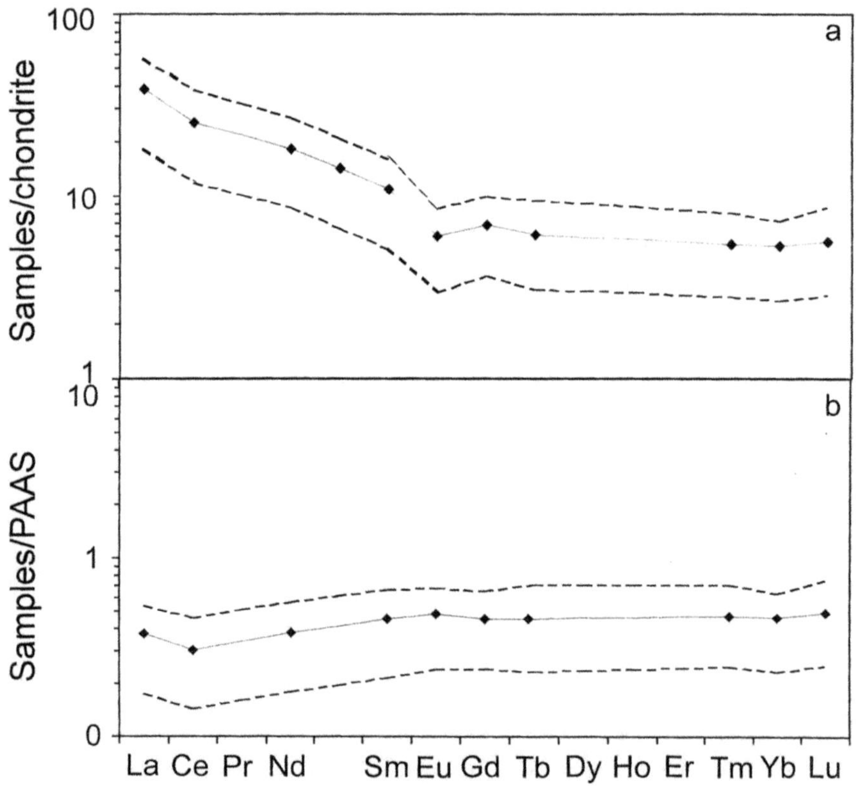

Fig. 9. a Chondrite-normalized average rare earth element contents of all analyzed samples (normalization values used are from Taylor and McLennan 1985). **b** Post-Archean average Australian shale-normalized average rare earth element contents of all analyzed samples. Dashed lines show the maximum and minimum REE contents of the samples.

The averaged samples, normalized to PAAS, show a lower REE abundance by a factor of five and a slight depletion of the LREE compared to PAAS (Fig. 9b). This pattern does not resemble the typical seawater REE signature, as there is no enriched trend of the heavy REE and no marked Ce depletion (e.g., Elderfield and Greaves 1982; Liu et al. 1988; Piepgras and Jacobsen 1992) and, therefore, support the dominance of siliciclastic detritus. However, the PAAS-normalized pattern shows a hint of a Ce-anomaly, which does not show up in the chondrite-normalized pattern. The Ce/Ce* values (= Ce anomaly) range from 0.75 to 0.94 (calculation of Ce/Ce* was done with the formula:

$$Ce/Ce^* = Ce_N/[(La_N)^{0.6667} * (Nd_N)^{0.3333}]) \quad (4.2)$$

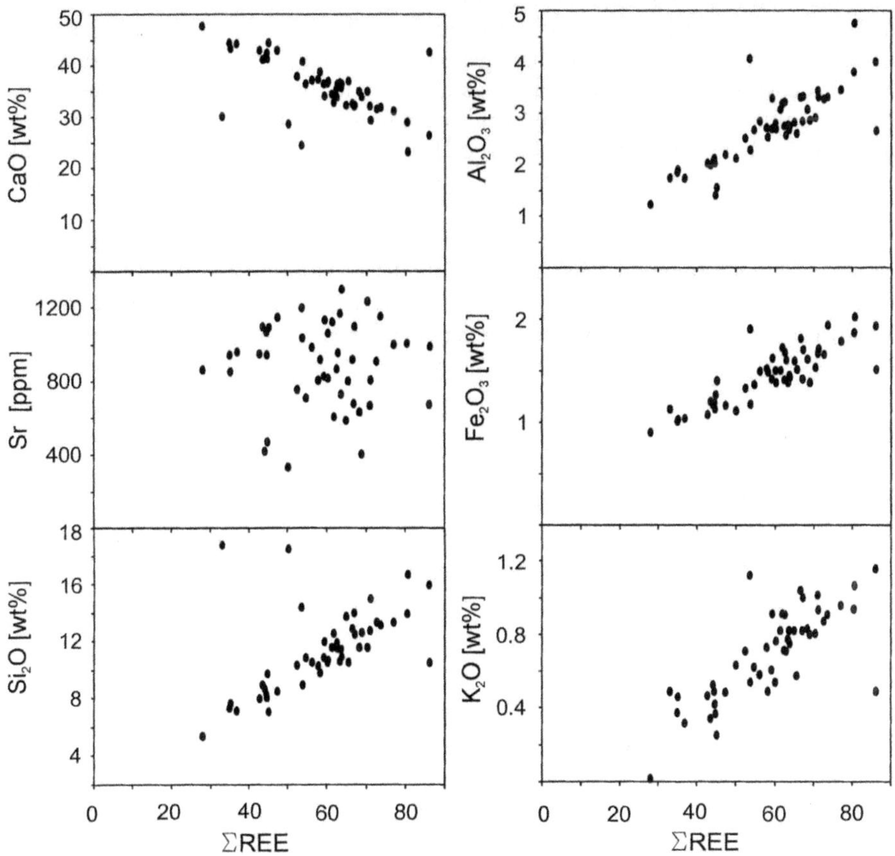

Fig. 10. Binary diagrams showing the degree of correlation of the abundances of the rare earth elements with those of Sr, P, Si, Al, Fe, K and CaO.

Commonly, seawater is regarded as the main source for REEs in marine sedimentary rocks (e.g., Liu et al. 1988; Bellanca et al. 1997). According to, e.g., Piper (1974), Toyoda et al. (1990), Ilyin (1998), and Plank and Langmuir (1998), the REEs are mostly associated with phosphates (e.g., bone detritus), and, therefore, their contents correlate with that of phosphorus. The sum of the rare earth elements (ΣREEs) of the La Vedova section does not display any correlation with Sr (r=0.01), or P (r=0.21), and a moderate negative correlation to CaO (r=–0.65), but show moderate to strong positive correlations with the contents of SiO_2 (r=0.55), Al_2O_3 (r=0.83), Fe_2O_3 (r=0.84) and K_2O (r=0.77) (Fig. 10). This points to clays and phyllosilicates as carrier phases of the REE and indicates terrigenous weathering products as their main source (e.g., Piper 1974; McLennan 1989; Holser 1997). Also the Eu/Eu* vs. $(Gd/Yb)_N$ diagram (McLennan 1989; McLennan et al. 1993) indicates a pattern similar to PAAS, but, due to carbonate dilution, with lower HREE concentrations compared to shales (Fig. 11). As shown in Fig. 12, the HREEs do not correlate with Zr (r=0.29), Y (r=0.35) and P (r=0.24), which suggests that the terrigenously supplied REE are not controlled by heavy accessory minerals, such as zircon, monazite, or apatite (e.g., Zhang et al. 1998; Bauluz et al. 2000). Ce/Ce* values ranging from 0.75 to 0.94 are closer to the average shale value of 1 (Haskin and Haskin 1966; Murray et al. 1991) than to seawater values of about 0.1 (Elderfield and Greaves 1982; Piepgras and Jacobsen 1992).

The generally uniform siliciclastic rock-like REE patterns is interpreted to demonstrate the original detrital source of the REE. Dilution by calcium carbonate seems to be the cause of the variations in the REE abundances within the La Vedova section, which is supported by the negative correlation of ΣREE and CaO.

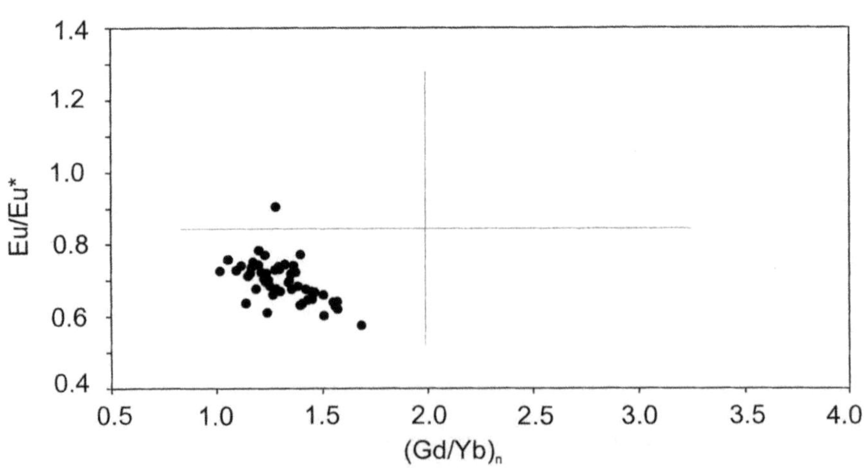

Fig. 11. Diagram of Eu/Eu* versus $(Gd/Yb)_N$, clearly showing ratios typical for post-Archean siliciclastic sediments.

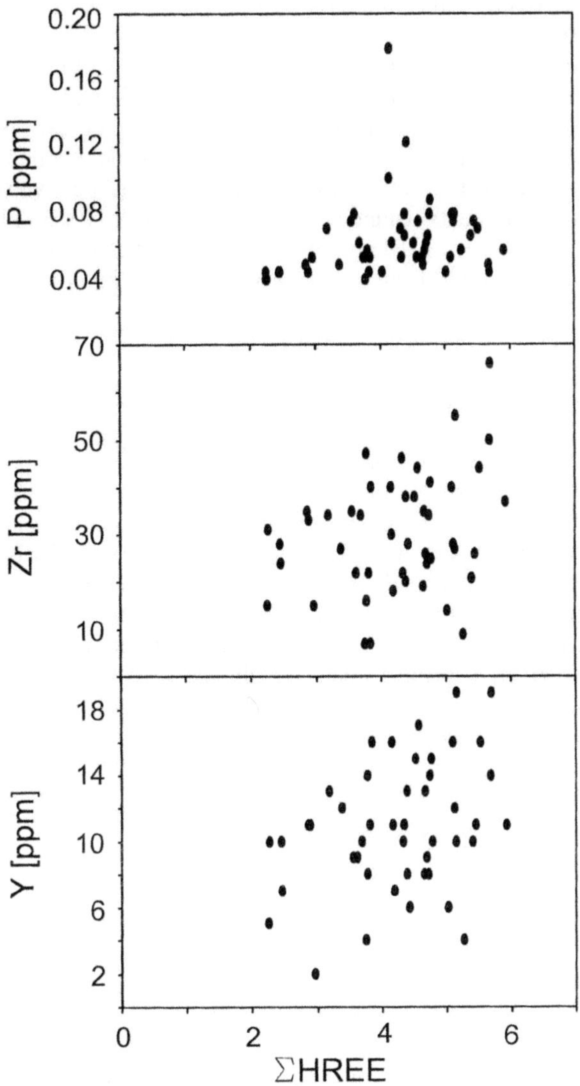

Fig. 12. Binary diagrams showing the degree of correlations between the contents of the heavy rare earth elements and those of Zr, Y, and P.

As the absolute abundances of the trace elements are diluted by carbonate, the values per se cannot be used to evaluate the provenance of the siliciclastic material. However, REE ratios might still be able to provide some information regarding source area composition. Although REE patterns are partly affected by source area weathering and diagenesis (e.g., Nesbitt 1979; Nesbitt et al. 1990; Zhang et al. 1998), they are generally considered to show source area composition (e.g., Taylor and McLennan 1985; McLennan 1989; McLennan et al. 1993; Nesbitt and Markovics 1997). Eu/Eu* values between 0.58 and 0.91 are typical for active continental margin sediments with varying degrees of intracrustal differentiation (McLennan 1989), mostly pointing to felsic composition. Average La/Yb, $(La/Yb)_N$ and Eu/Eu* ratios (10.9, 7.38, and 0.7, respectively) show values similar to those of continental island arc settings (La/Yb = 11 ± 3.6, $La_N/Yb_N = 7.5 \pm 2.5$, Eu/Eu* = 0.79 ± 0.13; Bhatia 1985). These patterns may reflect the early input of siliciclastic material from the synorogenic Apenninic flysch basins.

4.4 Impact-stratigraphic Markers?

The Ries impact event, which resulted in a crater with a diameter of 24 km, has been large enough to distribute ejecta of impact-produced material over hundreds of kilometers. Distal ejecta of the Ries event have been found about 250 km to the East (Central European strewn field; e.g., Montanari and Koeberl 2000, and references therein) and about 160 km to the South-West (Hofmann and Hoffmann 1992). The Cónero Riviera is located at a distance of about 600 km from the Ries crater. At such a distance, some distal ejecta material (shocked minerals, microspherules, Ni-rich spinels, etc.) might be expected and preserved in undisturbed deep sea sediments.

From empirical calculations of proximal ejecta at nuclear test sites and comparisons with lunar ejecta, McGetchin et al. (1973) formulated an equation that can be used to calculate the ejecta thickness depending on their distance from the source crater and the crater radius (in meters):

$$t = kR^m(r/R)^n, \qquad (4.3)$$

where t is the ejecta thickness, R is the radius of the transient cavity and r the distance of the ejecta from the crater rim/center. The empirical constants are $k = 0.14$, $m = 0.74$, and $n = -3$. A more recent relation, derived from distribution studies on Australasian tektites, was proposed by Glass and Pizzuto (1994). They used the following constants: $k = 0.02$, $m = 0$, and $n = -4.4 \pm 0.3$.

Using a value of about 7.5 km for the radius of the transient cavity, the distance to the Cónero Riviera, and the constants of McGetchin et al. (1973) and Glass and Pizzuto (1994), thicknesses of 0.63 µm and 0.2 mm, respectively, are derived for a possible impactoclastic layer within the Miocene sedimentary sequence. However, the direct application of these equations to terrestrial distal ejecta deposits may not yield reliable values. Impact melt derived ejecta, such as microtektites and spherules, have a different ballistic and depositional behavior in the atmosphere than solid ejecta, such as shocked quartz (e.g., Montanari 1991). Sedimentary processes (e.g., compaction, resedimentation), as well as ocean currents in the marine environment, may contribute to the thickness variation with distance (Smit 1999). This needs to be considered in the pelagic sediments investigated here.

Iridium contents determined with "normal" neutron activation analysis are mainly below the detection limit and are not useable for any interpretation. However, some values appeared relatively high. Thus, some samples were re-measured by iridium coincidence spectrometry (see Koeberl and Huber 2000) in order to verify that these values do not comprise measurement artefacts. According to these high-precision measurements, no Ir enrichment is present in the samples (Ir concentrations are between 0.02-0.12 ppb).

Also the contents of the siderophile elements Co, Cr, and Ni do not show any enrichment exceeding those of common sedimentary composition (Fig. 8).

The marls and marly limestones are somewhat enriched in detrital components compared to the average deep sea carbonate composition, but, in general, are similar to PAAS composition. Major and trace elemental data of samples several meters above and beneath the Aldo Level do not indicate any geochemical signatures pointing to impactogenic distal ejecta, which would make it easier to search for petrographic impact signatures, such as shocked quartz or microspherules. However, it has to be considered that the investigated interval is only within the error of the radiometric age of the Aldo Level (Mader et al. 2001) and does not comprise the whole interval of the age error of the Ries event (unfortunately, the La Vedova section is partly covered by landslide material and vegetation). Therefore, the distal ejecta layer may occur several centimeters, or even meters, above or below the investigated part of the La Vedova section. Moreover, the absence of a distinct geochemical signal is not unexpected, because of the absence of an unambiguous meteoritic component in the Ries structure itself and even in proximal Ries ejecta. This led to the suggestion that the Ries impactor might have been a siderophile element-poor achondrite (e.g., Morgan et al. 1979; Hörz 1982; Pernicka et al. 1987; Schmidt and Pernicka 1994). Achondrites are known to be relatively poor in platinum group elements and other siderophile elements, such as Ni and Co, but do have significant Cr contents (on the order of 0.2 wt%; e.g., Mason 1979). Thus, elevated Cr contents compared to average crustal values (83 ppm, McLennan 2001) could also indicate an achondritic contribution, but in absence of Cr-isotopic values, which may distinguish between terrestrial and extraterrestrial Cr sources, the present data (Table 1) do not show significant enough anomalies to make any suggestions. Thus, within the sampled interval the only opportunity to detect ejecta from the Ries event will be to search for a layer containing petrographic impact markers, such as shocked minerals, spherules, Ni-rich spinels etc. Finally, it should be noted that, as samples were collected at an interval of 10 cm, an ejecta layer of less than 1 mm thickness could have simply been missed during the sampling process. In order to find a distal ejecta layer, more samples with a higher spatial resolution have to be collected and studied, possibly with the help of drill cores along the Cónero Riviera. Another way would be the more accurate correlation of the age of the stratigraphic section with the Ries event. This can be done by tuning of cyclostratigraphic geochemical proxies to the insolation curve, which depends on the Earth's orbital parameters, such as precession (e.g., Hilgen et al. 1999); this study is currently in progress.

5 Conclusions

The geochemical whole rock data of the La Vedova section provide new information for detailed studies of the Mediterranean Langhian. Major and trace elemental compositions demonstrate a clayey and (phyllo)silicatic terrigenous input. The rare earth elements show typical upper continental crustal compositions and do not indicate seawater-fractionated patterns. Thus, in spite of carbonate dilution, the siliciclastic input from trace element ratios and rare earth elements in the pelagic carbonates, may reflect the composition and provenance of the source areas.

As the Middle and Upper Miocene in the Alpine-Apennine region is mainly represented by synorogenic reworked siliciclastic deposits, the nearly continuous pelagic carbonate sequence at the Cònero Riviera will allow future studies on the regional effects of the medium-sized Ries crater event. Because the Ries impactor did not seem to have contributed a large enough meteoritic signature at this distance, a detailed high-resolution search for shocked minerals and microspherules may be the only way to recognize a probably existing distal impact ejecta layer of the Ries crater event.

Acknowledgments

This study has been supported by the Austrian Science Foundation (FWF), project Y58 - GEO (to C.K.), the "Jubilaeumsfonds" of the Austrian National Bank (project 7915, to C.K.), the Austrian – Italian Scientific and Technical Exchange Program (ÖAD), project no. 14 (to C.K. and A.M.), and the University of Vienna (International Relations Office) (to D.M.). We are grateful to P. Claeys and an anonymous reviewer for helpful comments on an earlier version of this manuscript.

References

Assorgia A, Chan LS, Deino A, Garbarino C, Montanari A, Rizzo R, Tocco S (1994) Volcanigenic and paleomagnetic studies on the Cenozoic calc-alkalic eruptive sequence of Monte Furru (Bosa, mid-western Sardinia). In: Coccioni R, Montanari A, Odin GS (eds) Miocene Stratigraphy of Italy and Adjacent Regions, Giornale di Geologia 56: 17-29

Baker PA, Gieskes JM, Elderfield H (1982) Diagenesis of carbonates in deep-sea sediments - evidence from Sr/Ca ratios and interstitial dissolved Sr^{2+} data. Journal of Sedimentary Petrology 52: 71-82

Bauluz B, Mayayo MJ, Fernandez-Nieto C, Lopez JMG (2000) Geochemistry of Precambrian and Paleozoic siliciclastic rocks from the Iberian Range (NE Spain): implications for source-area weathering, sorting, provenance, and tectonic setting. Chemical Geology 168: 135-150

Bellanca A, Masetti D, Neri R (1997) Rare earth elements in limestone/marlstone couplets from the Albian-Cenomanian Cismon section (Venetian region, northern Italy): assessing REE sensitivity to environmental changes. Chemical Geology 141: 141-152

Bhatia M (1985) Rare earth element geochemistry of Australian Paleozoic graywackes and mudrocks: provenance and tectonic control. Sedimentary Geology 45: 97-113

Cleaveland LC (2001) Calcium carbonate and magnetic susceptibility analysis at Monte dei Corvi, Italy: Trends in the Mediterranean climate proxy record during Middle Miocene ice sheet expansion. Senior Integrative Exercise, Carleton College, Minnesota, USA, 34 pp

Dypvik H, Harris NB (2001) Geochemical facies analysis of fine-grained siliciclastics using Th/U, Zr/Rb and (Zr+Rb)/Sr ratios. Chemical Geology 181: 131-146

Einsele G, Ricken W (1991) Limestone-marl alternation - an overview. In: Einsele G, Ricken W, Seilacher A (eds) Cycles and Events in Stratigraphy, Springer Verlag, Berlin Heidelberg, pp 23-47

Elderfield H, Greaves MJ (1982) The rare earth elements in seawater. Nature 296: 214-219

Gentner W, Lippolt HJ, Schaeffer OA (1961) Das Kalium-Argon-Alter einer Glasprobe vom Nördlinger Ries. Zeitschrift für Naturforschung 16A: 1240

Glass BP, Pizzuto JE (1994) Geographic variation in Australasian microtektite concentrations: Implications concerning the location and size of the source crater. Journal of Geophysical Research 99: 19075-19081

Goese S (1999) Direct dating of Milankovitch cycles: A study of the rhythmic limestone-marl sequence at Monte dei Corvi, Italy. Senior Integrative Exercise, Carleton College, Minnesota, USA, 42 pp

Haskin MA, Haskin LA (1966) Rare earths in European shales, a redetermination. Science 154: 507-509

Hilgen FJ, Abdul Aziz H, Krijgsman W, Langereis CC, Lourens LJ, Meulenkamp E, Raffi I, Steenbrink J, Turco E, van Vugt N (1999) Present status of the astronomical (polarity) time-scale for the Mediterranean Late Neogene. Philosophical Transactions of the Royal Society London 357: 1931-1947

Hofmann B, Hofmann F (1992) An impactite horizon in the Upper Freshwater Molasse in Eastern Switzerland: Distal Ries ejecta? Eclogae Geologicae Helveticae 85: 788-789

Holser WT (1997) Evaluation of the application of rare-earth elements to paleoceanography. Palaeogeography, Palaeoclimatology, Palaeoecology 132: 309-323

Hörz F (1982) Ejecta of the Ries crater, Germany. In: Silver LT, Schultz PH (eds) Geological Implications of Impacts of Large Asteroids and Comets on the Earth. Geological Society of America, Special Paper 190: 39-55

Ilyin AV (1998) Rare-earth geochemistry of "old" phosphorites and probability of syngenetic precipitation and accumulation of phosphate. Chemical Geology 144: 243-256

Jones B, Manning DAC (1994) Comparison of geochemical indices used for the interpretation of paleoredox conditions in ancient mudstones. Chemical Geology 111: 111-129

Koeberl C (1993) Instrumental neutron activation analysis of geochemical and cosmochemical samples: a fast and reliable method for small sample analysis. Journal of Radioanalytical and Nuclear Chemistry 168: 47-60

Koeberl C (1998) Identification of meteoritic components in impactites. In: Grady MM, Hutchinson R, McCall GJH, Rothery DA (eds) Meteorites: Flux with Time and Impact Effects. Geological Society, London, Special Publication 140: 133-152

Koeberl C (2001) The sedimentary record of impact events. In: Peucker-Ehrenbrink B, Schmitz B (eds), Accretion of Extraterrestrial Matter throughout Earth's History, Kluwer Academic/Plenum Publishers, pp 333-378

Koeberl C, Huber H (2000) Optimization of the multiparameter γ-γ coincidence spectrometry for the determination of iridium in geological materials. Journal of Radioanalytical and Nuclear Chemistry 244: 655-660

Laurenzi MA, Bigazzi G, Balestrieri ML (2001) $^{40}Ar/^{39}Ar$ chronology of Central Europe tektites (moldavites). [abs.] Meteoritics and Planetary Science 36: A 109

Liu Y-G, Miah MRU, Schmitt RA (1988) Cerium: a chemical tracer for paleo-oceanic redox conditions. Geochimica et Cosmochimica Acta 52: 1361-1371

Mader D, Montanari A, Gattacceca J, Koeberl C, Handler R, Coccioni R (2001) $^{40}Ar/^{39}Ar$ dating of a Langhian biotite-rich clay layer in the pelagic sequence of the Conero Riviera, Ancona, Italy. Earth and Planetary Science Letters 194: 111-126

Mason B (1979) Meteorites. In: Data of geochemistry. Chapter B, Part 1, US Geological Survey Professional Paper 440-B-1, 132 pp

Mattias P, Mariottini M, De Casa G (1987) I minerali silicatici e gli altri minerali compresi nella sequenza eocenica-oligocenica della Valle della Contessa presso Gubbio (Appennino Centrale). Mineralogica et Petrographica Acta 30: 113-139

McGetchin TR, Settle M, Head JW (1973) Radial thickness variation in impact crater ejecta: implications for lunar deposits. Earth and Planetary Science Letters 20: 226-236

McLennan SM (1981) Trace element geochemistry of sedimentary rocks: Implications for the composition and evolution of the continental crust. PhD dissertation, Australian National University, Canberra, 609 pp

McLennan SM (1989) Rare earth elements in sedimentary rocks: Influence of provenance and sedimentary processes. In: Lipin BR, McKay GA (eds) Geochemistry and Mineralogy of Rare Earth Elements. Reviews of Mineralogy 21, Mineralogical Society of America, Washington, D.C., 169-200 pp

McLennan SM (2001) Relationships between the trace element composition of sedimentary rocks and upper continental crust. Geochemistry Geophysics Geosystems 2, Paper number 2000GC000109, 24 pp

McLennan SM, Hemming S, McDaniel DK, Hanson GN (1993) Geochemical approaches to sedimentation, provenance, and tectonics. In: Johnsson MJ, Basu A (eds) Processes Controlling the Composition of Clastic Sediments. Geological Society of America, Special Paper 284: 21-40

Meulenkamp JE et al. (23 co-authors) (2000) Early Langhian (16.4 – 15.5 Ma). In: Dercourt J, Gaetoni B, Vrielynck B, Barrier E, Biju-Duval B, Brunet MF, Cadet JP, Crasquin S, Sandulescu M (eds) Atlas Peri-Tethys, Palaeogeographic maps, Commission de la Carte Géologique du Monde/Commission for the Geologic Map of the World, Paris, map 21

Montanari A (1988) Geochemical characterization of volcanic biotites from the Upper Eocene - Upper Miocene pelagic sequence of the northeastern Apennines. In: Premoli Silva I, Coccioni R, Montanari A (eds) The Eocene-Oligocene Boundary in the Marche-Umbria Basin (Italy), IUGS Special Publication, Aniballi Publishers, Ancona, pp 209-227

Montanari A (1991) Authigenesis of impact spheroids in the K/T boundary clay from Italy: new constraints for high-resolution stratigraphy of terminal Cretaceous events. Journal of Sedimentary Petrology 61: 315-339.

Montanari A, Koeberl C (2000) Impact Stratigraphy – The Italian Record. Lecture Notes in Earth Sciences, Vol. 93, Springer Verlag, Berlin-Heidelberg, 364 pp

Montanari A, Drake R, Bice MD, Alvarez W, Curtis GH, Turrin BD, DePaolo DJ (1985) Radiometric time scale for the upper Eocene and Oligocene based on K/Ar and Rb/Sr dating of volcanic biotites from the pelagic sequence of Gubbio, Italy. Geology 13: 596-599

Montanari A, Deino AL, Drake RE, Turrin BD, DePaolo DJ, Odin GS, Curtis GH, Alvarez W, Bice DM (1988a) Radioisotopic dating of the Eocene-Oligocene Boundary in the pelagic sequence of the northeastern Apennines. In: Premoli Silva I, Coccioni R, Montanari A (Eds), The Eocene-Oligocene Boundary in the Marche-Umbria Basin (Italy), IUGS Special Publication, Aniballi Publishers, Ancona, pp 195-208

Montanari A, Langenheim VE, Coccioni R (1988b) Stratigraphy and geochronologic potential of the pelagic and hemipelagic sequence of the northeastern Apennines: a research note. Bulletin of Liaison and Informations, Project 196, 7: 17-23

Montanari A, Carey S, Coccioni R, Deino A (1994) Early Miocene tephra in the Apennine pelagic sequence: An inferred Sardinian provenance and implications for western Mediterranean tectonics. Tectonics 13: 1120-1134

Montanari A, Beaudoin B, Chan LS, Coccioni R, Deino A, DePaolo DJ, Emmanuel L, Fornaciari E, Kruge M, Lundblad S, Mozzato C, Portier E, Renard M, Rio D, Sandroni P, Stankiewicz A (1997a) Integrated stratigraphy of the Middle to Upper Miocene pelagic sequence of the Conero Riviera (Marche Region, Italy). In: Montanari A, Odin GS, Coccioni, R (eds) Miocene Stratigraphy: An Integrated Approach. Elsevier, Amsterdam, pp 409-450

Montanari A, Coccioni R, Fornaciari E, Rio D (1997b) Potential integrated stratigraphy in the Langhian L`Annunziata section near Apiro (Marche Region, Italy). In: Montanari A, Odin G S, Coccioni R (eds) Miocene Stratigraphy: An Integrated Approach, Elsevier, Amsterdam, pp 343-349

Morgan JW, Janssens M-J, Hertogen J, Gros J, Takahashi H (1979) Ries impact crater, southern Germany: search for meteoritic material. Geochimica et Cosmochimica Acta 43: 803-815

Nesbitt HW (1979) Mobility and fractionation of rare earth elements during weathering of a granodiorite. Nature 279: 206-210

Nesbitt HW, Markovics G (1997) Weathering of granodioritic crust, long-term storage of elements in weathering profiles, and petrogenesis of siliciclastic sediments. Geochimica et Cosmochimica Acta 61: 1653-1670

Nesbitt HW, MacRae ND, Kronberg BI (1990) Amazon deep-sea fan muds: light REE enriched products of extreme chemical weathering. Earth and Planetary Science Letters 100: 118-123

Odin GS (1985) Niveaux à biotite del Apennines autour de la limite Eocène- Oligocène. Bulletin of Liaison and Informations, International Union of Geological Sciences - Subcommission on Geochronology, 5, Paris, pp 17-24

Odin GS, Guise P, Rex DC, Kreuzer H (1988) K-Ar and $^{39}Ar/^{40}Ar$ geochronology of late Eocene biotites from the northeastern Apennines. In: Premoli Silva I, Coccioni R, Montanari A (eds), The Eocene-Oligocene Boundary in the Marche-Umbria Basin (Italy), IUGS Special Publication, Aniballi Publishers, Ancona, pp 239-245

Odin GS, Montanari A, Deino A, Drake R, Guise PG, Kreuzer H, Rex DC (1991) Reliability of volcano-sedimentary biotite ages across the Eocene-Oligocene boundary (Apennines, Italy). Chemical Geology (Isotope Geoscience Section) 86: 203-224

Pernicka E, Horn P, Pohl J (1987) Chemical record of the projectile in the graded fall-back sedimentary unit from the Ries Crater, Germany. Earth and Planetary Science Letters 86: 113-121

Piepgras DJ, Jacobsen SB (1992) The behavior of rare earth elements in seawater: precise determination of variations in the North Pacific water column. Geochimica et Cosmochimica Acta 56: 1851-1862

Piper DZ (1974) Rare earth elements in the sedimentary cycle: a summary. Chemical Geology 14: 285-304

Plank T, Langmuir CH (1998) The chemical composition of subducting sediment and its consequences for the crust and mantle. Chemical Geology 145: 325-394

Reimold WU, Koeberl C, Bishop J (1994) Roter Kamm impact crater, Namibia: Geochemistry of basement rocks and breccias. Geochimica et Cosmochimica Acta 58: 2689-2710

Schmidt G, Pernicka E (1994) The determination of platinum group elements (PGE) in target rocks and fall-back material of the Nördlinger Ries impact crater, Germany. Geochimica et Cosmochimica Acta 58: 5083-5090

Schwarz WH, Lippolt HJ (2002) Coeval Argon-40/Argon-39 ages of moldavites from the Bohemian and Lusatian strewn fields. Meteoritics and Planetary Science 37: 1757-1763

Sandroni P (1985) Rilevamento geologico al 1:10.000 e litostratigrafia di alcune sezioni dello Schlier nel bacino marchigiano esterno e studio mineralogico e petrografico di una sezione ricostruita nell'anticlinale del Cònero, Thesis, Univ. of Urbino, Italy, 188 pp

Smit J (1999) The global stratigraphy of the Cretaceous Tertiary boundary impact ejecta. Annual Reviews of Earth and Planetary Sciences 27: 75-91

Staudacher T, Jessberger EK, Dominik B, Kirsten T, Schaeffer OA (1982) ^{40}Ar-^{39}Ar ages of rocks and glasses from the Nördlinger Ries crater and the temperature history of impact breccias. Journal of Geophysics 51: 1-11

Storzer D, Jessberger EK, Kunz J, Lange J-M (1995) Synopsis von Spaltspuren- und Kalium-Argon-Datierungen an Ries-Impaktgläsern und Moldaviten. 4. Jahrestagung der Gesellschaft für Geowissenschaften, Nördlingen, Exkursionsführer und Veröffentlichungen der Gesellschaft für Geowissenschaften 195: 79-80

Taylor SR, McLennan SM (1985) The Continental Crust: Its Composition and Evolution Blackwell Scientific Publications, Oxford, 312 pp

Toyoda K, Nakamura Y, Masuda A (1990) Rare earth elements of pacific pelagic sediments. Geochimica et Cosmochimica Acta 54: 1093-1103

Turekian KK, Wedepohl KH (1961) Distribution of the elements in some major units of the Earth's crust. Bulletin Geological Society of America 72: 175-192

Zhang L, Sun M, Wang S, Yu X (1998) The composition of shales from the Ordos basin, China: effects of source weathering and diagenesis. Sedimentary Geology 116: 129-141

Appendix

Table 1. Major and trace elemental composition of rocks from the La Vedova section.

	v30.3	v30.0	v29.9	v29.8	v29.7	v29.6	v29.5	v29.4	v29.3	v29.2	v29.1
SiO_2	13.09	21.72	16.71	16.61	19.04	18.24	19.80	23.64	23.38	10.48	12.63
TiO_2	0.24	0.34	0.26	0.24	0.26	0.26	0.29	0.30	0.32	0.18	0.21
Al_2O_3	3.78	5.80	4.77	4.80	5.53	5.18	5.81	6.84	6.75	2.89	3.57
Fe_2O_3	1.79	2.01	1.65	1.78	2.02	2.16	2.08	2.43	2.64	1.29	1.38
MnO	0.32	0.09	0.06	0.07	0.06	0.06	0.02	0.02	0.06	0.11	0.06
MgO	1.62	2.11	1.65	1.62	1.94	1.91	2.18	2.26	2.21	1.18	1.37
CaO	42.36	33.85	38.87	38.38	35.95	37.62	35.2	31.27	32.2	46.33	43.71
Na_2O	0.24	0.33	0.26	0.28	0.33	0.32	0.33	0.37	0.35	0.18	0.25
K_2O	0.62	1.19	0.85	0.72	0.85	0.88	1.10	1.22	0.92	0.38	0.29
P_2O_5	0.12	0.24	0.26	0.25	0.11	0.08	0.09	0.10	0.10	0.12	0.13
LOI	34.86	31.40	33.73	34.16	32.98	32.92	32.1	30.36	30.32	36.48	35.57
Total	99.04	99.08	99.07	98.91	99.07	99.63	99.00	98.81	99.25	99.62	99.17
Sc											
V	48	70	54	45	68	68	77	80	82	47	56
Cr											
Co											
Ni	74	54	40	76	45	55	72	81	76	38	40
Cu	13	23	6	31	4	15	19	30	18	12	15
Zn	44	62	55	90	70	79	85	109	89	40	54
As											
Rb	49	63	40	56	36	49	61	79	63	41	44
Sr	988	1230	937	1300	769	1030	1120	1160	914	1200	1210
Y	10	18	12	18	12	16	18	22	15	8	12
Zr	54	86	43	58	41	51	62	77	57	44	49
Nb	7	10	6	7	7	7	7	10	8	5	7
Sb											
Cs											
Ba	335	661	544	523	325	347	388	310	298	380	423
La											
Ce											
Nd											
Sm											
Eu											
Gd											
Tb											
Tm											
Yb											
Lu											
Hf											
Ta											
Ir											
Au											
Th											
U											

Combined analytical results from INAA and XRF analyses. Major element data in wt%, trace element data in ppm, except Ir and Au, which are in ppb. Total iron as Fe_2O_3. Blank spaces: not determined, data below detection limit are marked b.d.l. Sample numbers refer to the meter level (see chapter 3).

Table 1. (cont.)

	v29.0	v28.9	v28.8	v28.7	v28.6	v28.5	v28.4	v28.3	v28.2	v28.1	v28.0
SiO_2	14.62	15.87	18.14	18.1	18.74	18.16	19.53	17.34	19.97	8.48	7.66
TiO_2	0.22	0.21	0.25	0.25	0.26	0.25	0.28	0.25	0.27	0.15	0.13
Al_2O_3	4.28	4.59	5.30	5.10	5.36	5.39	5.63	5.02	5.68	2.45	2.22
Fe_2O_3	1.89	1.57	2.29	1.97	2.19	2.03	2.10	1.89	2.12	1.21	1.09
MnO	0.04	0.05	0.04	0.04	0.04	0.04	0.04	b.d.l.	b.d.l	0.07	0.15
MgO	1.55	1.69	1.92	1.83	1.83	1.81	1.93	1.74	1.87	1.11	1.12
CaO	40.71	39.28	36.73	37.29	37.85	36.99	35.85	37.65	35.30	47.84	48.63
Na_2O	0.28	0.32	0.27	0.29	0.28	0.29	0.28	0.25	0.28	0.14	0.13
K_2O	0.42	0.49	0.71	0.77	0.75	0.71	0.78	0.68	0.92	0.20	0.19
P_2O_5	0.11	0.11	0.10	0.09	0.11	0.10	0.11	0.10	0.10	0.10	0.09
LOI	34.87	34.77	33.1	33.63	33.09	33.29	32.86	33.98	32.46	37.15	37.52
Total	98.99	98.95	98.85	99.36	100.50	99.06	99.39	98.90	98.97	98.90	98.93
Sc											
V	55	60	64	61	70	75	68	64	63	29	27
Cr											
Co											
Ni	38	47	54	76	47	67	75	71	57	21	18
Cu	<2	<2	<2	30	6	14	20	19	8	<2	<2
Zn	60	64	74	96	68	87	90	92	71	23	26
As											
Rb	28	41	40	61	41	46	52	56	41	25	30
Sr	771	909	814	1330	859	1060	1070	1230	917	748	761
Y	8	10	12	19	12	16	16	16	11	7	6
Zr	27	36	34	65	39	52	54	57	45	31	31
Nb	7	5	6	8	8	9	8	8	7	5	6
Sb											
Cs											
Ba	299	323	337	362	450	358	356	320	397	285	215
La											
Ce											
Nd											
Sm											
Eu											
Gd											
Tb											
Tm											
Yb											
Lu											
Hf											
Ta											
Ir											
Au											
Th											
U											

Table 1. (cont.)

	v27.9	v27.8	v27.7	v27.6	v27.5	v27.4	v27.3	v27.2	v27.1	v27.0
SiO_2	10.87	16.75	21.16	18.88	18.91	25.39	27.13	16.00	8.94	17.97
TiO_2	0.19	0.26	0.28	0.28	0.27	0.33	0.36	0.27	0.16	0.26
Al_2O_3	2.93	4.55	5.82	5.30	5.26	7.26	7.77	4.42	2.68	6.06
Fe_2O_3	1.39	1.74	2.82	2.00	1.95	2.51	3.25	1.91	1.25	1.99
MnO	0.09	0.06	b.d.l.	0.05	0.10	0.08	0.26	0.38	0.09	0.09
MgO	1.39	1.75	2.05	1.95	2.00	2.48	2.70	1.81	1.25	1.99
CaO	45.56	38.98	34.21	36.66	36.12	30.07	31.90	39.56	47.26	36.40
Na_2O	0.18	0.28	0.31	0.29	0.30	0.31	0.32	0.27	0.13	0.32
K_2O	0.44	0.72	0.85	0.84	0.94	1.31	1.53	0.85	0.38	0.97
P_2O_5	0.14	0.26	0.16	0.14	0.10	0.10	0.12	0.14	0.10	0.14
LOI	36.30	33.62	31.40	32.59	32.88	29.60	23.88	33.50	36.95	32.82
Total	99.48	98.97	99.06	98.98	98.83	99.44	99.22	99.11	99.19	99.01
Sc										
V	45	71	63	62	68	73	84	58	40	58
Cr										
Co										
Ni	38	33	71	33	44	52	76	45	26	43
Cu	11	3	29	<2	6	8	31	17	9	<2
Zn	45	42	98	40	63	69	97	49	31	47
As										
Rb	43	39	70	20	41	58	79	56	36	40
Sr	1040	825	1230	501	867	790	1090	1150	962	816
Y	8	12	22	7	14	12	21	16	11	10
Zr	50	41	71	28	48	52	78	66	45	39
Nb	6	7	8	6	8	8	10	9	6	7
Sb										
Cs										
Ba	371	520	345	414	434	427	398	367	173	536
La										
Ce										
Nd										
Sm										
Eu										
Gd										
Tb										
Tm										
Yb										
Lu										
Hf										
Ta										
Ir										
Au										
Th										
U										

Table 1. (cont.)

	v26.9	v26.8	v26.7	v26.6	v26.5	v26.4	v26.3	v26.2	v26.1	v26.0
SiO_2	26.50	18.25	16.72	7.18	18.19	18.21	23.23	22.55	23.76	16.52
TiO_2	0.20	0.44	0.26	0.12	0.26	0.26	0.31	0.34	0.30	0.15
Al_2O_3	7.65	5.00	4.79	2.29	5.29	5.37	6.25	6.01	6.24	2.65
Fe_2O_3	2.72	2.16	2.12	1.29	2.15	2.13	2.59	2.46	2.77	1.81
MnO	0.04	0.05	0.05	0.06	0.05	0.06	0.06	0.06	0.04	0.05
MgO	2.76	2.28	2.15	1.24	2.32	2.25	2.54	2.71	2.49	1.29
CaO	24.44	42.58	38.70	47.58	36.61	37.11	32.47	32.73	31.77	41.34
Na_2O	0.28	0.38	0.26	0.13	0.28	0.26	0.35	0.28	0.38	0.23
K_2O	1.35	0.59	0.59	0.02	0.65	0.70	1.25	1.10	1.09	0.44
P_2O_5	0.12	0.10	0.14	0.09	0.18	0.17	0.16	0.28	0.18	0.12
LOI	34.16	27.99	33.27	38.78	33.03	33.21	30.69	31.04	30.92	34.32
Total	100.22	99.82	99.05	98.78	99.01	99.73	99.90	99.56	99.94	98.92
Sc	5.01	9.35	5.89	2.75	5.87	6.72	6.69	5.57	6.84	3.52
V	44	115	63	27	73	70	74	65	60	38
Cr	64.1	112	66.6	29.9	65.5	70.5	73.9	57.6	73.4	44.4
Co	6.09	15.5	9.09	4.16	15.2	9.93	10.1	7.67	7.68	5.69
Ni	45	100	47	20	83	58	49	26	54	15
Cu	5	23	<15	<15	4	<15	<15	<15	5	<15
Zn	72	91	46	25	61	58	65	38	82	26
As	0.50	1.27	0.84	0.34	0.74	0.77	0.87	0.48	1.11	1.18
Rb	56	88	48	28	55	51	53	32	62	17
Sr	1200	988	915	861	1060	983	914	606	1150	466
Y	16	19	10	5	13	9	10	6	19	4
Zr	40	66	34	15	38	35	46	28	55	7
Nb	7	11	9	5	8	7	9	7	10	5
Sb	0.25	0.56	0.24	0.10	0.34	0.30	0.32	0.21	0.24	0.21
Cs	2.95	5.79	3.60	1.62	3.69	3.49	3.85	3.01	3.83	2.04
Ba	435	440	434	216	298	362	416	761	404	155
La	13.5	20.9	14.3	6.69	14.3	14.0	16.1	14.9	18.3	11.4
Ce	21.8	35.7	24.5	11.5	25.3	24.3	28.2	25.9	30.3	17.5
Nd	11.7	19.5	12.8	6.13	13.1	11.6	14.6	13.5	16.1	9.69
Sm	2.32	3.71	2.52	1.19	2.65	2.34	2.81	2.64	3.14	2.06
Eu	0.48	0.72	0.51	0.26	0.53	0.49	0.59	0.53	0.67	0.45
Gd	1.91	2.84	1.68	1.19	2.40	1.79	2.17	2.36	2.55	1.81
Tb	0.30	0.50	0.31	0.19	0.36	0.32	0.37	0.35	0.41	0.30
Tm	0.18	0.25	0.19	0.10	0.17	0.16	0.19	0.18	0.22	0.17
Yb	1.26	1.80	1.29	0.66	1.24	1.12	1.38	1.32	1.71	1.28
Lu	0.20	0.30	0.22	0.12	0.22	0.17	0.21	0.21	0.27	0.19
Hf	1.23	1.74	1.33	0.53	1.13	1.04	1.39	1.36	1.50	0.70
Ta	0.52	0.72	0.48	0.27	0.48	0.5	0.48	0.49	0.61	0.23
Ir	<0.9	<1	<0.8	0.1	<1	<0.4	<1	<1	0.2	<1
Au	0.5	0.6	0.3	0.1	0.5	0.8	0.2	0.3	0.6	0.5
Th	3.74	6.22	4.05	1.80	4.14	3.99	4.66	3.86	4.81	2.73
U	2.35	3.93	2.91	1.60	4.20	3.17	3.30	3.80	3.40	2.19

Table 1. (cont.)

	v25.9	v25.8	v25.7	v25.6	v25.5	v25.4	v25.3	v25.2	v25.1	v25.0	v24.9
SiO_2	10.75	18.16	20.28	14.92	20.14	18.51	14.89	18.29	20.50	22.41	21.10
TiO_2	0.16	0.24	0.26	0.18	0.24	0.26	0.24	0.26	0.26	0.30	n.a.
Al_2O_3	2.92	4.93	4.86	3.74	5.05	5.08	4.31	5.23	5.82	6.30	6.09
Fe_2O_3	2.00	2.16	2.29	1.72	2.08	1.97	1.68	1.98	2.15	2.43	2.39
MnO	0.05	0.05	0.04	0.05	0.05	0.05	0.06	0.07	0.07	0.07	0.07
MgO	1.36	2.09	1.59	1.75	2.01	2.15	1.90	2.21	2.39	2.72	2.62
CaO	44.42	36.96	36.37	41.01	35.62	36.89	40.66	36.56	34.44	32.13	33.80
Na_2O	0.22	0.32	0.34	0.24	0.35	0.31	0.27	0.28	0.28	0.31	0.30
K_2O	0.30	0.69	0.85	0.41	0.99	0.92	0.65	0.93	0.99	1.20	1.09
P_2O_5	0.13	0.17	0.15	0.11	0.15	0.17	0.41	0.18	0.14	0.12	0.10
LOI	36.54	33.16	32.07	35.28	32.2	32.69	34.43	32.88	32.13	30.93	31.26
Total	98.85	98.93	99.10	99.41	98.88	99.00	99.50	98.87	99.17	98.92	98.82
Sc	3.65	5.83	6.09	3.97	5.82	5.62	5.04	5.90	5.90	6.76	6.45
V	42	60	66	50	68	66	60	56	66	74	n.a.
Cr	44.1	68.9	70.5	47.6	74.2	65.5	55.8	62.4	58.4	65.9	65.3
Co	5.83	7.99	8.87	4.66	8.80	7.53	8.07	8.46	5.91	9.77	9.77
Ni	28	36	47	43	42	34	44	63	51	63	n.a.
Cu	<15	<15	<15	<15	25	<15	2	12	8	8	n.a.
Zn	45	57	78	64	81	54	53	79	78	89	n.a.
As	0.93	1.21	1.33	0.92	2.03	2.21	1.49	1.48	1.49	3.06	1.73
Rb	37	38	49	43	36	43	46	57	63	65	n.a.
Sr	1090	801	952	1100	727	816	1040	1170	1120	1090	n.a.
Y	11	11	14	12	10	10	11	15	15	17	n.a.
Zr	22	26	34	27	21	27	30	41	38	44	n.a.
Nb	7	7	7	7	7	8	7	9	8	9	n.a.
Sb	0.15	0.34	0.34	0.17	0.46	0.44	0.37	0.29	0.22	0.38	0.36
Cs	2.31	3.78	3.59	2.30	3.62	3.65	3.12	3.70	3.60	4.35	4.10
Ba	167	388	481	156	314	458	516	439	303	429	n.a.
La	11.6	16.1	15.3	11.4	15.7	14.8	13.0	15.6	15.3	16.0	15.6
Ce	17.3	25.9	25.4	17.1	24.6	24	22.2	25.8	25.4	28.3	26.8
Nd	10.1	14.4	14.0	9.36	14.4	13.1	11.5	14.0	13.1	15.0	13.1
Sm	2.03	3.15	2.85	2.04	3.02	2.82	2.51	2.68	2.49	2.71	2.53
Eu	0.40	0.58	0.57	0.42	0.58	0.54	0.49	0.55	0.61	0.60	0.55
Gd	1.81	2.67	2.37	1.53	2.86	2.91	2.25	2.52	2.34	2.37	1.98
Tb	0.34	0.47	0.43	0.26	0.48	0.43	0.37	0.39	0.38	0.39	0.32
Tm	0.17	0.26	0.23	0.16	0.24	0.21	0.17	0.22	0.23	0.19	0.18
Yb	1.29	1.75	1.48	1.22	1.54	1.40	1.17	1.43	1.36	1.40	1.37
Lu	0.20	0.30	0.23	0.21	0.28	0.19	0.20	0.21	0.21	0.22	0.19
Hf	0.69	1.17	1.03	0.75	1.03	1.34	1.07	1.22	1.10	1.24	1.21
Ta	0.31	0.52	0.46	0.33	0.60	0.39	0.38	0.38	0.35	0.45	0.41
Ir	<1	<1	<0.4	<1	<1	1.4	<1	<0.5	<1	0.04	<0.8
Au	0.5	0.5	0.5	0.3	0.4	0.3	0.1	0.3	0.6	1.4	0.5
Th	2.60	4.33	4.37	2.88	4.31	3.92	3.44	3.96	3.94	4.68	4.29
U	2.99	5.42	4.15	2.90	5.18	4.73	4.14	3.49	2.34	2.30	1.92

Table 1. (cont.)

	v24.8	v24.7	v24.6	v24.5	v24.4	v24.3	v24.2	v24.1	v24.0	v23.9	v23.8
SiO_2	21.33	12.70	11.33	10.90	13.47	12.02	17.86	31.39	22.91	25.56	24.17
TiO_2	0.31	0.19	0.17	0.18	0.21	0.18	0.27	n.a.	0.30	0.30	0.31
Al_2O_3	6.21	3.81	3.46	3.25	4.00	3.57	4.75	8.98	6.48	7.17	6.52
Fe_2O_3	2.32	1.54	1.44	1.48	1.70	1.47	1.90	2.88	2.38	2.67	2.55
MnO	0.08	0.08	0.09	0.09	0.08	0.08	0.08	0.05	0.09	0.08	0.07
MgO	2.50	1.68	1.62	1.62	1.89	1.68	2.00	2.95	2.56	2.70	2.62
CaO	33.95	42.84	44.3	44.19	41.53	43.36	37.73	23.18	32.00	28.97	31.16
Na_2O	0.27	0.19	0.16	0.16	0.20	0.16	0.26	0.42	0.32	0.44	0.39
K_2O	1.10	0.56	0.45	0.38	0.59	0.55	0.85	1.28	1.22	1.13	1.15
P_2O_5	0.12	0.10	0.09	0.10	0.11	0.10	0.09	0.17	0.20	0.16	0.11
LOI	31.20	35.26	35.97	36.35	35.15	35.77	33.24	27.57	30.88	29.77	30.11
Total	99.39	98.95	99.08	98.70	98.93	98.94	99.03	98.87	99.34	98.95	99.16
Sc	6.42	4.49	3.72	3.71	4.52	3.53	4.99	6.36	6.18	7.91	7.24
V	71	46	41	45	54	44	54	n.a.	69	76	73
Cr	66.5	45.7	39.2	40.5	45.8	37.7	49.0	66.3	67.3	80.9	73.4
Co	14.3	6.57	5.33	5.59	8.15	3.52	6.34	10.0	9.40	12.8	10.4
Ni	104	43	39	39	64	27	29	n.a.	39	70	62
Cu	15	2	2	<15	7	<2	<2	n.a.	<2	8	5
Zn	70	41	37	34	47	36	43	n.a.	52	104	92
As	2.54	1.08	0.57	0.86	0.98	0.65	1.09	1.59	1.54	2.83	2.19
Rb	74	48	46	40	51	38	37	n.a.	39	65	63
Sr	1130	949	940	958	1060	851	754	n.a.	668	1010	998
Y	14	11	10	10	11	7	8	n.a.	10	16	14
Zr	47	33	31	28	35	24	16	n.a.	25	44	50
Nb	9	7	8	7	8	6	5	n.a.	8	8	8
Sb	0.33	0.19	0.12	0.20	0.19	0.15	0.24	0.31	0.42	0.76	0.57
Cs	4.12	2.70	2.28	2.32	2.68	2.12	3.01	3.40	3.71	5.13	4.69
Ba	359	206	155	198	276	215	996	n.a.	337	409	372
La	14.0	10.6	8.44	8.65	10.8	8.54	12.8	19.2	17.7	18.7	17.9
Ce	25.9	17.9	15.2	15.7	19.1	14.8	22.0	35.3	29.5	33.9	32.2
Nd	12.9	9.42	7.45	8.26	9.61	7.69	11.4	17.8	15.7	18.1	17.3
Sm	2.35	1.63	1.35	1.54	1.84	1.43	2.07	3.12	2.74	3.55	3.32
Eu	0.52	0.36	0.31	0.33	0.41	0.32	0.47	0.62	0.58	0.70	0.65
Gd	1.94	1.43	1.12	1.21	1.39	1.21	1.93	2.35	2.33	2.82	3.08
Tb	0.33	0.25	0.18	0.23	0.22	0.21	0.31	0.41	0.42	0.45	0.48
Tm	0.16	0.14	0.12	0.12	0.17	0.12	0.19	0.20	0.26	0.23	0.27
Yb	1.19	0.94	0.74	0.76	0.94	0.82	1.16	1.42	1.52	1.69	1.59
Lu	0.16	0.14	0.11	0.13	0.15	0.11	0.18	0.21	0.25	0.33	0.26
Hf	1.24	0.92	0.77	0.86	0.99	0.82	1.00	1.98	1.27	1.65	1.47
Ta	0.42	0.28	0.2	0.26	0.3	0.22	0.28	0.69	0.39	0.64	0.57
Ir	<0.4	<1	0.2	<1	<1	0.6	0.6	<1	<1	<1	<1
Au	1.7	0.4	0.2	0.4	0.6	0.5	0.1	0.3	1.1	0.8	1.1
Th	4.03	2.68	2.29	2.36	2.94	2.25	3.45	6.67	4.24	5.63	5.54
U	1.52	1.27	0.80	1.22	1.37	1.06	1.67	2.95	3.50	5.02	4.30

Table 1. (cont.)

	v23.7	v23.6	v23.5	v23.4	v23.3	v23.2	v23.1	v23.0	v22.9	v22.8	v22.7
SiO_2	29.83	20.42	19.01	22.65	20.49	17.60	27.70	25.01	20.40	14.19	24.23
TiO_2	0.32	0.29	0.26	0.24	0.26	0.25	0.29	0.25	0.24	0.22	0.27
Al_2O_3	7.56	5.81	5.20	5.40	5.51	5.13	6.27	5.33	5.20	3.89	6.17
Fe_2O_3	2.76	2.30	2.05	1.97	2.19	2.17	2.45	2.27	2.02	1.66	2.36
MnO	0.07	0.07	0.07	0.06	0.06	0.05	0.06	0.06	0.07	0.08	0.06
MgO	2.76	2.15	2.13	2.23	2.27	2.00	2.60	2.16	2.17	1.93	2.40
CaO	26.42	34.99	36.42	33.75	34.93	37.33	29.27	32.11	35.19	41.32	31.48
Na_2O	0.47	0.32	0.32	0.39	0.42	0.34	0.40	0.40	0.39	0.24	0.44
K_2O	1.39	1.00	0.90	0.96	0.97	0.88	1.12	0.99	0.86	0.63	1.05
P_2O_5	0.13	0.14	0.23	0.13	0.12	0.12	0.11	0.10	0.13	0.12	0.18
LOI	27.65	31.62	32.56	31.53	31.81	32.91	29.22	30.46	32.25	34.80	30.45
Total	99.36	99.11	99.15	99.31	99.03	98.78	99.49	99.14	98.92	99.08	99.09
Sc	8.85	6.88	6.36	6.47	6.96	5.22	6.63	6.17	6.13	4.24	7.24
V	83	76	67	58	58	54	68	53	58	55	69
Cr	93.3	76.6	79.8	64.6	67.8	51.0	71.9	65.7	73.4	45.7	79.9
Co	13.7	11.7	12.5	9.14	8.71	7.12	8.91	9.00	8.87	6.56	11.0
Ni	50	45	84	23	66	37	52	35	42	15	58
Cu	<2	<2	18	<2	10	<2	<2	<2	<2	<2	<2
Zn	68	40	92	41	106	69	73	55	62	22	90
As	2.76	1.77	1.83	1.96	3.03	1.93	3.32	3.26	1.55	0.56	1.78
Rb	51	40	60	22	58	34	49	34	44	15	50
Sr	670	629	1300	403	1230	806	804	584	868	419	904
Y	11	8	16	4	16	11	13	6	9	2	12
Zr	37	24	40	9	40	22	35	14	26	15	28
Nb	8	7	9	6	7	6	8	6	7	6	7
Sb	0.56	0.33	0.50	0.44	0.58	0.36	0.38	0.41	0.41	0.16	0.48
Cs	5.73	4.29	3.96	4.14	4.26	3.15	4.26	3.97	3.84	2.30	4.56
Ba	394	325	420	387	450	573	371	414	381	406	461
La	20.1	16.2	15.0	16.6	16.6	14.1	17.8	15.9	15.6	11.1	18.2
Ce	37.1	28.8	27.0	28.5	29.0	23.6	29.8	26.5	25.3	18.3	29.5
Nd	18.5	15.3	14.3	15.0	15.9	12.8	15.3	14.1	13.6	9.74	15.9
Sm	3.71	2.78	2.76	2.99	3.15	2.58	2.95	2.89	2.79	1.83	3.34
Eu	0.74	0.58	0.71	0.61	0.65	0.54	0.62	0.60	0.54	0.38	0.64
Gd	3.07	2.47	2.08	2.83	2.59	2.25	2.24	2.67	2.44	1.39	2.50
Tb	0.55	0.41	0.36	0.47	0.43	0.37	0.40	0.43	0.38	0.24	0.42
Tm	0.29	0.21	0.18	0.23	0.26	0.19	0.23	0.23	0.20	0.15	0.22
Yb	1.72	1.41	1.32	1.48	1.57	1.32	1.56	1.44	1.42	1.03	1.71
Lu	0.29	0.22	0.21	0.25	0.24	0.21	0.24	0.25	0.25	0.16	0.27
Hf	1.85	1.24	1.23	1.25	1.41	1.15	1.26	1.25	1.24	0.96	1.3
Ta	0.66	0.45	0.42	0.92	0.44	0.35	0.49	0.44	0.53	0.40	0.65
Ir	<1	<1	<1	0.4	<1	<1	<1	<1	<0.7	<1	<1
Au	0.8	0.6	0.7	0.2	0.8	0.8	0.9	1.5	0.3	0.5	0.8
Th	6.29	4.31	4.00	4.57	4.96	3.8	4.62	4.46	4.35	2.83	5.16
U	3.98	3.08	5.06	4.44	3.76	3.21	3.17	3.53	3.60	2.08	4.59

Table 1. (cont.)

	v22.6	v22.5	v22.4	v22.3	v22.2	v22.1	v22.0	v21.9	v21.8	v21.7
SiO_2	25.70	35.31	35.84	13.92	12.92	18.95	18.87	25.84	26.13	26.80
TiO_2	0.24	0.19	n.a.	0.23	0.20	0.23	0.24	0.24	0.25	0.29
Al_2O_3	5.37	4.00	3.25	4.13	3.82	5.07	5.05	5.46	4.95	6.02
Fe_2O_3	2.03	1.59	1.61	1.66	1.61	2.03	1.95	2.00	1.94	2.30
MnO	0.07	0.05	0.05	0.06	0.08	0.06	0.06	0.07	0.04	0.04
MgO	2.12	1.46	1.15	1.74	1.67	2.12	2.07	1.82	1.89	2.14
CaO	32.07	28.60	30.03	42.99	42.35	36.38	36.39	31.99	32.40	30.16
Na_2O	0.43	0.38	0.28	0.26	0.27	0.38	0.30	0.40	0.37	0.46
K_2O	0.99	0.76	0.59	0.58	0.50	0.73	0.75	0.95	0.95	1.11
P_2O_5	0.12	0.10	0.10	0.16	0.18	0.15	0.14	0.11	0.09	0.12
LOI	30.27	26.91	26.02	34.72	35.32	33.17	33.24	30.09	30.09	29.44
Total	99.41	99.35	98.92	100.45	98.92	99.27	99.06	98.97	99.10	98.88
Sc	6.72	4.93	3.19	4.85	4.36	6.09	5.49			
V	55	44	n.a.	59	54	60	56	54	60	70
Cr	68.7	55.2	37.3	52.0	49.1	69.3	61.5			
Co	8.05	6.49	4.84	7.06	8.44	9.12	8.37			
Ni	32	15	n.a.	47	38	43	40	20	40	62
Cu	<2	<2	n.a.	6	<2	<2	<2	<2	<2	18
Zn	58	30	n.a.	50	45	71	64	37	67	95
As	1.27	1.02	1.22	0.91	0.86	1.56	1.34			
Rb	36	19	n.a.	52	34	40	32	13	36	56
Sr	676	332	n.a.	1150	941	827	710	318	721	962
Y	8	<3	n.a.	13	9	8	7	3	10	15
Zr	19	7	n.a.	34	22	20	18	20	42	61
Nb	6	6	n.a.	7	7	6	6	5	6	9
Sb	0.43	0.33	0.18	0.20	0.22	0.48	0.40			
Cs	4.06	3.01	2.02	2.73	2.70	3.69	3.54			
Ba	428	239	n.a.	309	341	366	382	364	358	511
La	16.5	10.6	8.25	12.0	10.9	15.0	13.5			
Ce	27.5	21.5	13.7	20.2	18.4	23.9	22.0			
Nd	14.8	11.5	7.16	9.72	9.41	12.9	12.0			
Sm	2.98	2.32	1.41	2.00	1.97	2.59	2.56			
Eu	0.59	0.48	0.31	0.42	0.40	0.55	0.50			
Gd	2.33	1.87	1.16	1.50	1.87	2.10	2.09			
Tb	0.37	0.33	0.19	0.27	0.30	0.37	0.34			
Tm	0.20	0.18	0.10	0.15	0.15	0.19	0.19			
Yb	1.51	1.24	0.69	1.09	1.08	1.47	1.34			
Lu	0.25	0.22	0.12	0.18	0.21	0.26	0.23			
Hf	1.23	0.94	0.62	0.89	0.98	1.07	1.00			
Ta	0.54	0.37	0.26	0.40	0.39	0.49	0.41			
Ir	<1	<1	<0.3	<1.5	0.3	<1	<0.9			
Au	0.5	0.6	0.7	0.6	0.4	1.0	0.8			
Th	4.77	3.65	2.37	3.04	2.91	4.17	3.99			
U	3.88	2.65	1.49	2.80	3.21	3.19	3.52			

Table 1. (cont.)

	v21.6	v21.5	v21.4	v21.3	v21.2	v21.1	v21.0	v20.9	v20.8	v20.7	v20.6
SiO_2	23.07	14.82	20.50	23.47	26.73	15.65	9.90	11.18	21.05	30.20	31.29
TiO_2	0.31	0.24	0.32	0.28	0.23	0.24	0.15	0.16	0.24	0.22	0.21
Al_2O_3	5.70	3.96	5.38	5.75	4.29	4.17	2.81	3.03	5.14	4.24	4.06
Fe_2O_3	2.33	1.70	2.14	2.29	1.85	1.96	1.60	1.56	2.05	2	1.77
MnO	0.04	0.02	0.03	0.05	0.1	0.07	0.07	0.05	0.04	0.04	b.d.l.
MgO	2.12	1.72	2.47	2.29	1.65	1.74	1.29	1.39	1.99	1.61	1.54
CaO	33.25	41.21	34.86	32.30	33.26	40.19	45.96	44.93	34.67	30.96	30.13
Na_2O	0.41	0.32	0.39	0.44	0.42	0.37	0.22	0.18	0.36	0.39	0.44
K_2O	0.95	0.41	1.00	1.06	0.80	0.56	0.17	0.04	0.77	0.82	0.89
P_2O_5	0.13	0.11	0.23	0.13	0.07	0.10	0.09	0.11	0.17	0.10	0.13
LOI	30.86	34.51	31.90	30.85	29.76	33.83	36.55	36.32	32.36	28.76	28.35
Total	99.17	99.02	99.22	98.91	99.16	98.88	98.81	98.95	98.84	99.34	98.81
Sc											
V	68	56	70	66	48	57	31	40	57	58	45
Cr											
Co											
Ni	50	20	32	35	17	38	14	17	25	14	17
Cu	7	<2	<2	<2	<2	<2	<2	<2	<2	<2	<2
Zn	66	32	45	54	25	50	19	25	42	29	32
As											
Rb	42	25	36	32	11	37	19	20	22	15	15
Sr	838	606	716	536	287	870	488	476	396	272	313
Y	12	6	7	8	3	11	3	5	5	<3	4
Zr	53	27	48	30	22	41	19	21	11	12	21
Nb	8	7	8	7	5	7	5	6	8	6	6
Sb											
Cs											
Ba	511	363	579	568	305	374	327	148	303	233	169
La											
Ce											
Nd											
Sm											
Eu											
Gd											
Tb											
Tm											
Yb											
Lu											
Hf											
Ta											
Ir											
Au											
Th											
U											

Table 1. (cont.)

	v20.5	v20.4	v20.3	v20.2	v20.1	v20.0	v19.9	v19.8	v19.7	v19.6
SiO_2	24.16	17.17	19.92	19.38	19.02	22.65	18.65	20.05	18.68	22.24
TiO_2	0.26	0.26	0.25	0.22	0.19	0.23	0.20	0.21	0.21	0.23
Al_2O_3	5.77	4.53	4.90	4.43	3.83	5.03	3.83	4.09	4.01	4.83
Fe_2O_3	2.30	1.93	2.04	1.82	1.70	1.96	1.64	1.94	1.78	1.93
MnO	b.d.l.	0.02	0.01	0.02	0.05	0.04	0.08	0.05	0.09	0.07
MgO	2.11	2.13	2.05	1.84	1.59	1.89	1.57	1.60	1.54	1.72
CaO	32.55	38.40	36.07	37.32	38.00	33.90	38.84	37.96	37.80	34.63
Na_2O	0.6	0.38	0.39	0.37	0.35	0.46	0.24	0.24	0.23	0.26
K_2O	0.96	0.47	0.46	0.44	0.57	0.78	0.26	0.46	0.52	0.76
P_2O_5	0.15	0.16	0.13	0.13	0.11	0.13	0.11	0.12	0.11	0.13
LOI	30.31	33.55	32.82	33.44	33.55	31.73	33.76	33.35	33.93	32.03
Total	99.17	99.00	99.04	99.41	98.96	98.80	99.18	100.07	98.90	98.83
Sc										
V	70	63	62	53	52	59	51	55	59	48
Cr										
Co										
Ni	19	24	33	25	18	15	30	18	20	35
Cu	<2	<2	<2	<2	<2	<2	<2	<2	<2	<2
Zn	30	34	54	40	35	28	50	33	36	60
As										
Rb	11	28	27	22	10	10	28	14	20	28
Sr	237	521	601	477	356	216	697	359	442	676
Y	3	5	7	6	4	<3	8	4	5	8
Zr	16	20	30	18	20	10	27	15	14	35
Nb	7	6	7	6	5	6	6	5	6	6
Sb										
Cs										
Ba	444	606	374	277	271	469	262	218	247	357
La										
Ce										
Nd										
Sm										
Eu										
Gd										
Tb										
Tm										
Yb										
Lu										
Hf										
Ta										
Ir										
Au										
Th										
U										

Table 1. (cont.)

	v19.5	v19.4	v19.3	v19.2	v19.1	v19.0	v18.9	v18.8	v18.7	v18.6	v18.5
SiO_2	21.11	23.26	24.79	18.81	17.32	17.57	15.09	19.68	22.99	24.81	23.44
TiO_2	0.22	0.18	0.18	0.20	0.24	0.26	0.23	0.27	0.25	0.31	0.36
Al_2O_3	4.20	3.23	3.43	4.09	4.54	4.94	4.26	5.43	4.70	5.29	6.29
Fe_2O_3	1.82	1.40	1.42	1.57	1.91	2.06	1.81	2.08	2.09	2.40	2.29
MnO	0.04	0.04	0.04	0.05	0.10	0.09	0.06	0.09	0.07	0.06	0.04
MgO	1.54	1.16	1.23	1.57	1.68	1.98	1.80	2.10	1.85	1.56	2.44
CaO	36.85	36.78	35.30	38.2	38.67	38.03	40.55	35.83	34.77	33.69	32.03
Na_2O	0.26	0.26	0.25	0.23	0.26	0.25	0.20	0.28	0.30	0.44	0.47
K_2O	0.34	0.58	0.61	0.64	0.66	0.66	0.58	0.92	0.90	1.16	1.18
P_2O_5	0.11	0.09	0.09	0.09	0.11	0.25	0.16	0.23	0.09	0.08	0.16
LOI	32.95	32.20	31.70	33.89	33.63	32.97	34.28	32.31	31.04	29.67	30.15
Total	99.44	99.18	99.04	99.34	99.12	99.06	99.02	99.22	99.05	99.47	98.85
Sc											
V	52	45	41	56	55	63	55	70	55	68	78
Cr											
Co											
Ni	21	20	21	21	30	31	17	22	16	27	20
Cu	<2	<2	<2	<2	<2	<2	<2	<2	<2	<2	<2
Zn	37	37	39	37	46	37	22	32	31	46	37
As											
Rb	20	20	19	18	29	25	20	21	17	31	30
Sr	447	553	492	427	714	533	376	356	309	458	507
Y	5	6	5	5	8	8	4	4	<3	5	6
Zr	16	20	19	12	31	33	17	15	12	23	29
Nb	6	6	5	6	7	6	6	7	6	7	7
Sb											
Cs											
Ba	341	284	252	269	364	259	308	240	347	373	740
La											
Ce											
Nd											
Sm											
Eu											
Gd											
Tb											
Tm											
Yb											
Lu											
Hf											
Ta											
Ir											
Au											
Th											
U											

Table 1. (cont.)

	v18.4	v18.3	v18.2	v18.1	v18.0
SiO_2	20.14	23.35	21.67	14.96	12.46
TiO_2	0.30	n.a.	0.23	0.24	0.18
Al_2O_3	5.61	6.35	4.45	4.25	3.39
Fe_2O_3	2.02	2.23	1.86	1.92	1.51
MnO	0.04	0.04	0.04	0.05	0.05
MgO	2.14	2.21	1.86	1.78	1.56
CaO	35.86	32.21	35.70	40.50	43.47
Na_2O	0.38	0.09	0.33	0.26	0.20
K_2O	0.91	1.06	0.82	0.74	0.39
P_2O_5	0.23	0.14	0.09	0.11	0.10
LOI	32.15	31.07	31.93	34.15	35.52
Total	99.78	98.75	98.98	98.96	98.83
Sc					
V	71		57	55	36
Cr					
Co					
Ni	21		24	34	12
Cu	<2		<2	<2	<2
Zn	36		44	37	21
As					
Rb	28		17	32	20
Sr	466		457	636	435
Y	6		5	6	5
Zr	20		28	38	21
Nb	7		5	6	6
Sb					
Cs					
Ba	440		342	235	171
La					
Ce					
Nd					
Sm					
Eu					
Gd					
Tb					
Tm					
Yb					
Lu					
Hf					
Ta					
Ir					
Au					
Th					
U					

Titan: A New World Covered in Submarine Craters?

Ralph D. Lorenz

Lunar and Planetary Laboratory, 1629 East University Boulevard, University of Arizona, Tucson AZ 85721-0092, USA.
(rlorenz@lpl.arizona.edu)

Abstract. Saturn's satellite Titan is a unique object that not only has a thick nitrogen atmosphere, but may also have widespread deposits of surface liquids. Its surface has not yet been studied closely, but will be investigated in detail by the NASA-ESA Cassini-Huygens mission in 2004-2008. If, like virtually everywhere else in the solar system, it has suffered bombardment by impactors, it will have a significant impact crater population. Many of these may have been formed in, or have filled with, the surface liquid deposits. This paper aims to explore some of the phenomena we may see, and to alert the impact community to the opportunities that Titan may offer in understanding impact processes.

1 Introduction

Titan (Lorenz and Mitton 2002) with a radius of 2575 km is the second-largest satellite in the solar system, fractionally smaller than Jupiter's Ganymede. It is larger than the Earth's moon and Mercury, but smaller than Mars. Like Ganymede, its bulk density is a little under 2000 kgm^{-3}, suggesting a composition dominated by water ice, with a rocky core. Unlike Ganymede, however, Titan has a thick atmosphere, with a surface pressure of 1.5 bar; this is the only significant nitrogen atmosphere in the solar system apart from that of the Earth. A comparison of the composition of the Earth and Titan is given in Table 1.

Titan is therefore an outstanding object in terms of its usefulness for comparative planetology. While essentially an icy giant satellite like Ganymede and Callisto, its thick atmosphere makes it a very different place. Also, while Titan's thermal history may have similarities with Ganymede and Callisto, Titan's much larger distance from its primary and from the sun profoundly affects impactor speed distributions and gravitational focusing.

Titan's impact crater population will tell us much about its history, and the cratering process itself (see, e.g., Chapman and McKinnon 1986; Lorenz 1997 for a detailed discussion, some of which is summarized below). In particular, it may offer a number of submarine impact craters for study, since Titan's surface may have significant exposed bodies of liquid hydrocarbons: many dark regions identified on Titan's surface (shown as bright areas in Figure 1) may be lakes or seas.

Table 1. Parameters for Titan and Earth (see Lorenz and Mitton 2002 for details).

Atmosphere		Titan	Earth
Composition		~95% N_2, ~5% CH_4	79% N_2,20% O_2
Surface Pressure	(mbar)	1440	1013
Surface Density	(kg m^{-3})	5.3	1.25
Temperature	(K)	94	288
Speed of Sound	(m s^{-1})	~200	~330
Scale Height	(km)	20	8
Gravity	(m s^{-2})	1.35	9.81
Ocean*			
Composition*		CH_4 (C_2H6)	H_2O
Density	(kg m^{-3})	450 (650)	1000
Viscosity	(Pa s)	$6x10^{-4}$?	10^{-3}
Speed of Sound	(m s^{-1})	1500 (2000)	~1500
Specific Heat Capacity	(J $kg^{-1}K^{-1}$)	~2000	4200
Latent Heat of Vaporization (J kg^{-1})		~$5x10^5$	$2.9x10^6$
Boiling Point at Surface Pressure (K)		115 (190)	373
Bedrock			
Composition		Ice (NH_3-rich ice)	Basalt etc
Subsurface Heat Flow	(mW m^{-2})	5	80
Thermal Conductivity	($Wm^{-1}K^{-1}$)	~4 (~1.5)	~2
Density	(kg m^{-3})	~900	~3000
Melting Point	(K)	273 (176)	~1500
Escape Velocity	(km s^{-1})	2.6	11
Typical Impact Velocity	(km s^{-1})	3-6 (Saturnocentric) 10-20 (Comets)	20-30

* Titan ocean composition is a mixture, predominantly CH_4 and C_2H_6, with some N_2 and traces of other compounds.

2
Crater Size Distribution

Previous studies (Plescia and Boyce 1985; Lissauer 1988; Lorenz 1997) suggested that if Titan's surface is old, then by analogy with other Saturnian there should be around 200 craters of greater than 20 km diameter every 10^6 km^2, and approximately

$$\mathrm{Log}_{10}N(>D) = -2.7\mathrm{Log}_{10}D \quad (1)$$

where D is in km and N in craters per km^2.

This gives, for Titan's surface area of 82 million km^2, a population of some 50 craters of 200 km diameter, and perhaps a few 500 km craters.

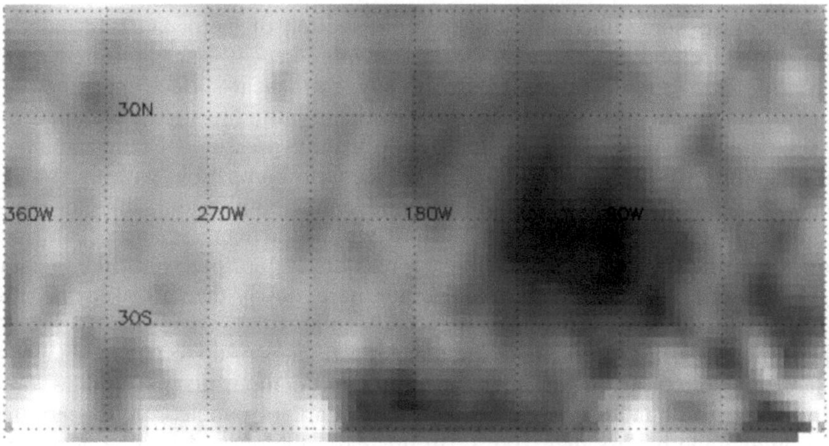

Fig. 1. Crude map of Titan from Hubble Space Telescope data (see Smith et al. 1996 for further details). Map is shown inverted – white regions above are actually black (and vice-versa) and may be crater lakes filled with liquid hydrocarbons.

These numbers are predicated on the assumption that Titan has not been significantly eroded or resurfaced by volcanic/tectonic processes; on energetic grounds alone (Lorenz 2002a) relative to the assumed cratering rate, Titan is probably more heavily tectonized but less heavily eroded than the Earth; see also Lorenz and Lunine (1996).

On the other hand, the irregular moon Hyperion, next out from Titan, may be the remnant of a once larger body that was disrupted by a giant impact. Orbital simulations (e.g., Farinella et al. 1990) indicate that most of the debris from such a collision would have impacted Titan in around a century - a period of very intense bombardment that would have essentially resurfaced (indeed, saturated) Titan with impact craters. Impacts may also be associated with the origin and evolution of Titan's atmosphere (Jones and Lewis 1987; Zahnle et al. 1992; Griffith and Zahnle 1995).

Modeling by Ivanov et al. (1997) - see also Engel et al. (1995) - shows that Titan's present atmosphere shields out smaller impactors such that a cumulative crater density plot flattens out at around 20 km diameter. Titan's atmosphere may have varied in thickness and composition in the past - it may be that studies of the crater size distribution at small sizes can constrain the atmospheric history.

3 Depth of Lakes and Seas

Photochemical models (e.g., Lara et al. 1996) predict of the order of 100-600 m of ethane produced over geological time. If photolysis has not occurred for the full 4.5 Gyr, then clearly the predicted ethane production is less.

There exists some uncertainty in the composition of the ocean, and indeed of the atmosphere above it. Voyager data and more recently near-infrared spectra taken from the Hubble Space Telescope suggest a near-surface methane abundance of a few per cent, consistent with methane humidities (i.e., obtained from the methane partial pressure divided by the vapour pressure of pure methane at that temperature) of around 40-60%. One scenario compatible with this value is that the ocean and atmosphere are in thermodynamic equilibrium, and thus the 40-60% humidity corresponds to equilibrium with a methane-ethane ocean where there is enough (involatile) ethane to suppress the vapour pressure (like the sugar in syrup forces a spill of syrup to evaporate more slowly than a spill of pure water) . In this case, the ocean depth may be 2-3 times the ethane depth, giving an upper limit of perhaps a couple of km (earlier estimates of up to 10 km were based on larger uncertainties in surface temperature and composition). The other possibility is that the humidity is controlled (as on Earth) by meteorology - humidity must on average be less than 100%. In this case, the ocean could be nearly pure methane, and the global average rather higher.

Note that because Titan's seas are composed of a fluid mixture, the thermodynamic behaviour can be complex - progressive heating to, say 150 K, might drive off all the methane that is present, while leaving almost all of the ethane. Thus the large-scale behaviour may depend on the exact assumed mixture (which would also include nitrogen). However, for small perturbations from the present conditions, assuming thermodynamic equilibrium between the present surface and atmosphere (Lorenz 2002b), shows that the vapour pressure curve has a slope of about 0.03 bar/K.

The brightest parts of Titan cover about 25% of the surface, and thus the typical depth of a lake (assuming all dark regions to be lakes) is a little higher than the global average equivalent.

Titan's volatile history is of course as poorly constrained as everything else about this body. If all the methane was delivered to the surface early on, the ocean would have begin deep, with a global average of perhaps a few kilometers. Over time, the progressive conversion of methane into more dense ethane (and the loss to other compounds) means the depth decreases monotonically with time, perhaps exposing formerly submerged impact structures.

4 Surface Morphology

4.1 Crater Morphology

On large icy satellites, large craters have upward-doming bottoms, presumably due to viscous relaxation of the crustal material. Schenk (1993) found that the depth h of craters on Ganymede could be related to their diameter D by

$$h=0.22*D^{0.44}, \qquad (2)$$

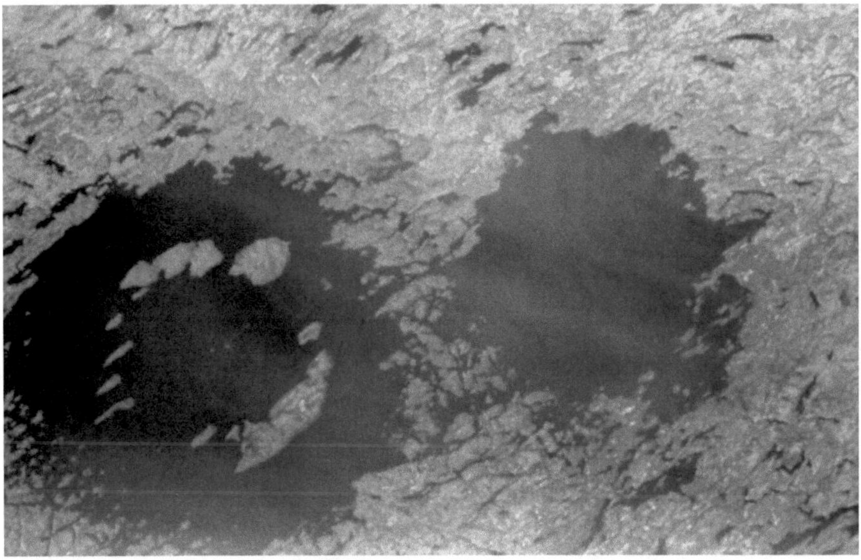

Fig. 2. A prototype of Titan's cratered, soaked surface ? Space shuttle picture of the Clearwater lakes in Canada - the left one is about 36 km across, the right 26 km. The right is thus the size of the smallest typical crater on Titan, smaller craters being depleted by atmospheric shielding. NASA Image 61A-35-86.

for craters below 40 km diameter width. Above 40 km diameter, craters had a near-constant depth (about 1 km) and feature central pits. Schenk (1993) (Fig.15 in that paper) shows that the transition diameter from central peak to central pit craters scales approximately with the reciprocal of gravity: thus a similar value for Titan might be expected as for Ganymede and Callisto (30-40 km).

It may be noted that these depths are considerably shallower than the lunar craters of the same diameter. Simple craters there have a depth/diameter ratio of about 0.2, whereas 0.1 is more common for the icy satellites. Also, the transition diameter from simple craters to those with central peaks is smaller – 10 km on the icy satellites, versus 20km on the moon. It is possible that Pit and Dome-style craters as found on Ganymede and Callisto (Schenk 1989) may be present on Titan, but their formation is not yet understood.

Multi-ring impact basins are formed by the crustal response to large impacts, especially if the lithosphere is thin. McKinnon and Melosh (1980) investigated multi-ring basins on Ganymede and Callisto. As Titan may have had a similar thermal history, it too may have a number of multi-ring basins. Note that if interior models of Titan pointing to the existence of large amounts of ammonia mixed with the ice are correct, then the water-ammonia liquid mantle may persist to the present day to depths of perhaps only 100 km. This depth is of course probed by 500 km-scale cratering events and thus multi-ring features are a distinct possibility.

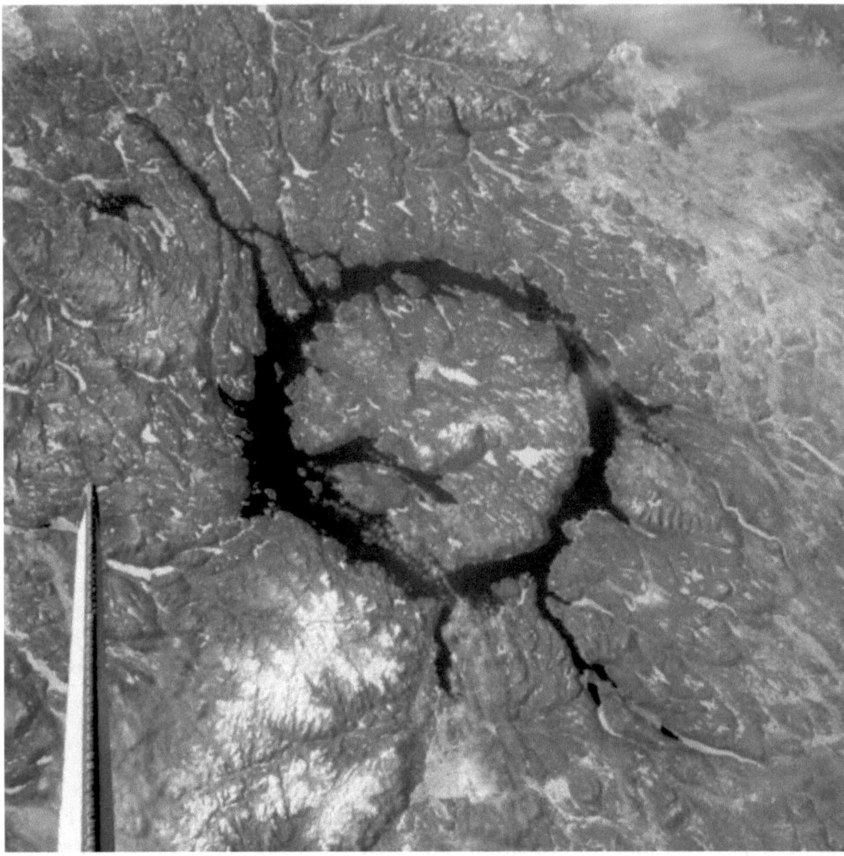

Fig. 3. Image of the Manicouagan crater in Canada (source NASA, STS-9). The fin of the space shuttle fin appears in the foreground. This crater is heavily eroded (hence the absence of a rim) although the hard floor caused by impact melt is resistant and now shows positive relief, much as Ashima in Figure 4 (although for a different reason) As on Titan, the presence of liquids bring stark contrast to the topography. It would be interesting to model (or observe!) the propagation of an impact tsunami arising from a subsequent impact in such an annular lake. Lake diameter is ≈70 km.

Note that the transient cavity of a 500 km diameter crater would be around 250-320 km across (Melosh 1989; see also Turtle and Pierazzo 2001 for recent modelling) and the cavity would probe to a depth of around 100 km.

4.2 Lake Formation

Lorenz (1994) noted that craters on Titan would slowly fill with photochemical products (mostly ethane) deposited from the atmosphere, forming circular, annular or multi-ring ('bullseye') lakes. The Manicouagan structure and the Clearwater lakes in Canada are excellent terrestrial examples of a ring and bullseye lakes - see Figures 2 and 3.

As most craters will have central peaks or updomed floors, the ring-lake type should be most common. By analogy with other icy satellites, most craters with updomed floors may have a central pit (which could also fill with liquid), "bullseye" lakes should not be uncommon.

Note that since liquid ethane and methane are rather less dense (approx. 500-600 kg m^{-3}) than the expected crustal material (water ice, 900 kg m^{-3}), there will be a restoring tendency to raise the centre of the crater by viscous relaxation, even if the crater is completely filled with liquid. Thus, as the crust bows upwards, liquid may be displaced, so crater lakes may be the *sources* of fluvial systems, rather than only their sinks.

If a multi-ring impact basin (which typically have inward-facing scarps) were partially filled with liquid, it would be a spectacular structure to observe.

4.3 Submarine Impact Craters

An impact into a sea with significant depth causes an impact feature ("hydrobleme") in the crust with substantially modified (subdued) appearance. On Earth, impactors into the ocean of less than ~0.2 km diameter will not affect the ocean floor, applying the rule of thumb that for water depth more than four times the crater depth, bottom effects are not seen in sand (Sonnett et al. 1987). On Titan, with a ~1 km ocean, of lower density, it may be expected that an impact which would produce a crater with a diameter of only a kilometer or two will generate a feature on the seabed. As this value is small compared with the diameter for impactors that are able to penetrate the atmosphere, it may be expected that 'hydroblemes' are as common (per unit area of sea) as 'normal' craters.

While no Cassini instrumentation offers the possibility to image the seabed, Titan may expose the bed to us. As a crater basin domes upwards by viscous relaxation (see Figure 4) , it will displace the liquid to the edges, forming a ring lake (Lorenz, 1994). The central region will form a circular island (Figure 3) - perhaps with an exposed 'hydrobleme' or two. There may be some evidence of such tsunamis by e.g., erosional or depositional features above lake shores. It might even be possible to determine the location of a still-submerged hydrobleme by the relative heights of tsunami features.

Because many lakes will be topographically constrained by crater rims, it is likely that 'slosh' from tsunami may be revealed by evidence of (possibly catastrophic) liquid flow just outside the crater rim. As an aside, it is interesting to speculate (and would doubtless be interesting to model) tsunami propagation in a ring-shaped lake.

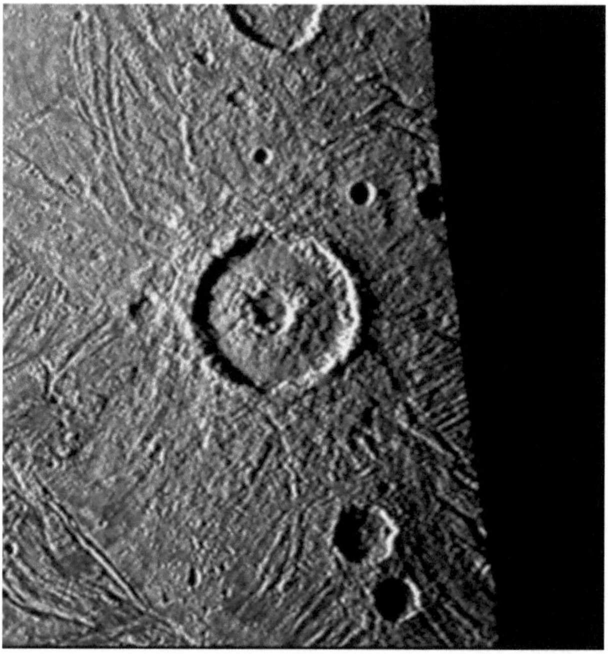

Fig. 4. Crater Ashima on the icy Galilean satellite, Ganymede (source NASA, Voyager image 20638.59). Relaxation pushes the floor of the crater into a dome. If a crater fills with liquid post-formation, the liquid will be pushed out into a ring, and the center will become shallow or dry, exposing any craters or hydroblemes that form subsequently. (The depression in this crater is a central pit, characteristic of Ganymede craters of this size. Rim diameter is 87 km.)

5 Cassini Prognosis

The Cassini orbiter is expected to map Titan's surface with a camera at resolutions of 100 m or better, with some smaller areas covered rather more closely. What exactly will be seen is open to question - HST imaging (Smith et al. 1996) has demonstrated that surface contrasts exist in the 0.94 and 1.07 μm atmospheric windows at scales of ~200 km, although how visible small-scale contrasts will be is not known. In addition to imaging at these wavelengths, the ISS (Imaging Science Subsystem) is also equipped with polarizers, which will reduce the light contribution of haze scattering that causes reduced contrast.

Surface albedo variations, over a rather wider (0.3-5.0 μm) wavelength range, will be measured down to resolutions of 500 m or so by the VIMS (Visual and Infrared Mapping Spectrometer) instrument, which can probe the atmosphere at window wavelengths of 1.28, 1.6 and 2 and 5 μm, as well as the 1.07 and 0.94 μm windows.

Additionally, a multimode RADAR instrument will be used on around 20 of the 44 flybys of Titan. Its synthetic aperture radar (SAR) mode will attain spatial resolutions of between 400 and 800 m, about 3 times poorer than Magellan, but nevertheless suitable for characterising large craters. Each swath is a few hundred km across, and over its ~3000 km length, covers about 1% of Titan's surface. A few small regions will also have detailed topographic measurements made by an altimeter mode, with approximately 80 m relative altitude resolution.

Titan's radar appearance, and the appearance of craters in particular, is not well-constrained. Although Titan's disk-integrated radar brightness is high, and spread across this disk, how the radar scattering area is distributed at small spatial scales is not known. The attenuation of radar waves on the terrestrial surface is mostly due to liquid water: where liquid water is scarce, in deserts or polar caps, the radar energy may propagate some distance into the ground. Thus terrestrial radar imaging has unveiled sub-Saharan river beds and the recently-discovered partially-buried crater structure at Aorounga, Chad (McHone et al. 2002). The Cassini radar may similarly penetrate a few metres, although it will not be able to sense the bed of lakes.

Finally, and perhaps most spectacularly, the ESA Huygens probe will descend for about 2.5 hours by parachute in Titan's atmosphere in January 2005, taking progressively higher-resolution pictures of the landscape as it does so.

6 Conclusions

Titan should have a rich impact crater population. Many impact events may have taken place in bodies of liquid. Hydroblemes and tsunami deposits may be common - any impactor large enough to penetrate the atmosphere is also likely to penetrate a 1 km ethane ocean. Crater lakes may be common, and may be sources, as well as sinks, of surface fluid flows. Erosive processes are fairly slow, so craters should have a long lifetime.

Acknowledgments

The author acknowledges the support of the Cassini mission. The organizers of the Svalbard meeting are thanked for orchestrating a very stimulating gathering.

References

Chapman CR, McKinnon WB (1986) Cratering of Planetary Satellites. In: Burns JA, Matthews MS (eds) Satellites. University of Arizona Press, Arizona, pp 492 - 580

Engel S, Lunine JI, Hartmann WK (1995) Cratering on Titan and implications for Titan's atmospheric history. Planetary and Space Science 43: 1059-1066

Farinella P, Paolocchi P, Strom RG, Kargel JS, Zappala V (1990). The fate of Hyperion's fragments. Icarus 83: 186-204

Griffith CA, Zahnle K (1995) Influx of cometary volatiles to planetary moons: The atmospheres of 1000 possible Titans. Journal of Geophysical Research 100: 16907-16922

Ivanov BA, Basilevsky AT, Neukum G (1997) Atmospheric entry of large meteoroids: implication to Titan. Planetary and Space Science 45: 993-1007

Jones TD, Lewis JS (1987) Estimated impact shock production of N_2 and organic compounds on early Titan. Icarus 72: 381-393

Lara LM, Lellouch E, Lopez-Morenz JJ, Rodrigo R (1996) Vertical distribution of Titan's atmospheric neutral constituents. Journal of Geophysical Research 101: 23261-23284

Lissauer J (1988) Bombardment history of the Saturn system. Journal of Geophysical Research 93: 13776-13804

Lorenz RD (1994) Crater lakes on Titan: Rings, horseshoes and bullseyes. Planetary and Space Science 42: 1-4

Lorenz RD (1997) Cratering on Titan: A pre-Cassini view. Planetary and Space Science 45: 1009-1019

Lorenz RD (2002a) Tectonic Titan: Landscape energetics and the thermodynamic efficiency of mantle convection [abs.]. Lunar and Planetary Science 32: 1165

Lorenz RD (2002b) Thermodynamics of geysers: Application to Titan. Icarus 156: 176-183

Lorenz RD, Lunine JI (1996) Erosion on Titan: Past and present. Icarus 122: 79-91

Lorenz R, Mitton J (2002) Lifting Titan's Veil. Cambridge University Press, Cambridge, 260 pp

McHone JF, Greeley R, Williams KK, Blumberg DG, Kuzmin RO (2002) Space shuttle observations of terrestrial impact structures using SIR-C and X-SAR radars. Meteoritics and Planetary Science 37: 407-420

McKinnon WB, Melosh HJ (1980) Evolution of planetary lithospheres: Evidence from multiringed structures on Ganymede and Callisto. Icarus 44: 454-471

Melosh HJ (1989) Impact cratering - A geologic process. Oxford University Press, Oxford, 245 pp

Plescia J, Boyce J (1985) Impact cratering history of the Saturnian satellites. Journal of Geophysical Research 85: 2029-2037

Schenk PM (1989) Crater formation and modification on the icy satellites of Uranus and Saturn : Depth/diameter and central peak occurrence. Journal of Geophysical Research 94: 3813-3832

Schenk PM (1993) Central pit and dome craters: Exposing the interiors of Ganymede and Callisto. Journal of Geophysical Research 98: 7475-7498

Smith PH, Lemmon MT, Lorenz RD, Sromovsky LA, Caldwell J, Allison MD (1996) Titan's surface, revealed by HST imaging. Icarus 119: 336-349

Sonnett CP, Pearce SJ, Gault DE (1987) The oceanic impact of large objects. Advances in Space Research 11(6): 77- 86

Turtle EP, Pierazzo E (2001) Thickness of a Europan ice shell from impact crater simulations. Science 294: 1326-1328

Zahnle K, Pollack JB, Grinspoon D, Dones L (1992) Impact-generated atmospheres over Titan, Ganymede and Callisto. Icarus 95: 1-23

Estimating Crater Size for Hypervelocity Impacts on Small Icy Bodies (e.g. Comet Nucleus)

Mark Burchell, Ellen Johnson and Ivan Grey

Centre for Astrophysics and Planetary Sciences, University of Kent, Canterbury, Kent CT2 7NR, United Kingdom.
(M.J.Burchell@kent.ac.uk)

Abstract. The morphology and size expected for impact craters on small icy bodies are presented. Such bodies are for example minor satellites of the outer planets (some of which are ice covered) or comet nuclei. The differences between the impact craters that result on such bodies, compared to those on more traditional ice targets (effectively large, well consolidated, semi-infinite ice surfaces) is discussed with particular reference to the impact on a comet nucleus expected in the Deep Impact space mission. Finally extrapolation of laboratory scale experiments is carried out to try and quantify crater size and shape for impacts on small, porous bodies. It is found that given our present knowledge, simple scaling with impact energy produces a result compatible to scaling via more sophisticated methods. The handling of the influence of the detailed composition of the ice target (porosity, volatile content, silicate content etc.) is less certain.

1 Introduction

Hypervelocity impacts on icy targets leave shallow craters (Croft 1981). Hypervelocity is usually taken to mean that the impact speed is at greater than a few kms^{-1}, i.e., the impact speed exceeds the speed of the resultant compression waves in both target and projectile material. The result is an impact, which usually vaporises the projectile and part of the target, with subsequent bulk movement of target material to form the classical impact crater shape. In addition there is formation of a vapour plume and high speed ejecta. This is standard wisdom obtained from impact experiments over many years (Melosh 1989). For impacts on Solar System scales this condition of hypervelocity impacts is usually met, since the driver of the impact speed is the combination of the gravitational attraction of the Sun and the mutual gravitational attraction of the two colliding bodies. Given that laboratory experimentation exists and the mechanical properties of ices are considered fairly well known, predicting the size and shape of craters that result from impacts on Solar System bodies should be straightforward.

Unfortunately this is not the case. There are several reasons for this. Simulations (either calculational or experimental) do not reliably reproduce realistic conditions in the Solar System. The problems with laboratory reproductions of impact cratering are well known; the impact speeds are usually not even close to hypervelocity, the sizes of projectiles are mm or cm instead of the required tens of metres or more, and the energy of the impacts is thus insufficient (by a factor of 10^{12} or more). However, some laboratory work is in the hypervelocity regime. Two stage light gas guns can achieve for example speeds in the range 5 to 10 km s^{-1}, but they cannot fire projectiles larger than the cm scale. This is a significant problem because at mm to cm sizes, the impacts in brittle materials produce craters which, in the late stages of growth, suffer significant effects (surface spallation of material) which do not occur on larger scales (where late stage growth is by bulk movement of material forming the classic high rim wall around a crater and is hence described as being in a gravity rather then strength dominated regime). However, experiments do yield impacts under controllable, reproducible conditions and as such can be used to probe hypervelocity impact cratering mechanics in general, albeit not at the scales of giant planetary impacts.

Computer simulations offer the opportunity to simulate impact events at the right scale. However, normally physics simulations are validated by comparison to known situations, but this is not possible with giant impacts. Furthermore, the codes include assumptions about the physics of an impact (e.g., how the energy and momentum of the projectile couple to the target) and as to how the materials respond to extreme shocks at such high stress rates (which can be tested by studying materials exposed to nuclear weapons testing). Also, assumptions are made for ease of calculation (e.g., turning off all target material strength until late in the crater formation process). Thus whilst such computer codes are well suited to studying the dependency of crater formation on different aspects of the impact conditions, they cannot be used with high confidence to predict actual crater sizes to better than for example an order of magnitude.

In addition, the condition of many types of surface in the Solar System are not themselves well constrained, e.g. composition of comet nuclei. In the case of comet nuclei many parameters are uncertain, composition of the ice (pure water ice is unlikely, but what mixture of volatiles can be expected ?), the presence of admixtures of silicates (in what proportion ?), and the bulk density can be affected by any porosity (i.e., compact object or fluffy dirty snowball ?). There is one attraction to considering impacts on such small bodies (or on natural satellites), since they are small their gravitational attraction is weak and craters may form which are on size scales where strength effects dominate as they do in the laboratory. Also, the lack of a "known" impact event against which to validate any predictions will be resolved in this case, as NASA is planning a mission (Deep Impact, see A'Hearn 1999) which involves dropping a 347 kg copper mass into a comet nucleus at a speed of some 10 km s^{-1}. The mission is due for launch in January 2004 with an encounter with comet Tempel 1 planned for July 4th 2005. The resultant impact event will be imaged by the main spacecraft which will fly past the comet. It therefore offers the chance to witness a real hypervelocity impact event on a body. However, the composition of the body will still be fairly uncercertain, indeed one of the aims of the mission is to use the impact to constrain the composition of the nucleus.

Finally, although not covered in this paper, it should be acknowledged that although the physical and mechanical properties of ice have been extensively studied in the laboratory, this has almost always been with a view to ice under conditions appropriate to those found on the Earth's surface. A review applicable to ice in the outer Solar System can be found in Durham and Stern (2001), where the point is made that more laboratory investigation of ice properties under Solar System conditions is still required.

It thus seems very appropriate to consider what the crater resulting from a hypervelocity impact on a small icy body might look like. The scale used is that of the Deep Impact mission. In the rest of this paper an extrapolation is made from laboratory experiments as a means of beginning to predict possible crater sizes and shapes. This is initially carried out for impacts on a well consolidated H_2O ice body. Then the results are elaborated to allow for a low temperature, porous icy body, with admixtures of other ices and silicates.

2 Method

The basic method used here was to take data from laboratory scale results for impacts on ice to extrapolate to scale of the Deep Impact mission. A typical impact crater in ice in the laboratory is obtained by firing a projectile at an ice target. Unfortunately, most experiments (e.g., Kawakami et al. 1983; Iijima et al. 1995; Kato et al. 1995) have used normal guns with consequent impact speeds of < 1 km s^{-1}. However, if a two stage light gas gun is used, impact speeds of up to 7 or 8 km s^{-1} have been obtained in ice cratering studies. This is comparable (i.e. only slightly slower) to the impact speed required for Deep Impact. A typical impact crater in ice is shown in Figure 1 and was obtained using the two-stage light gas gun of the University of Kent (Burchell et al. 1999). In principle, the parameters like crater depth, diameter and volume can all be measured. In this work we will focus on crater volume. The reason is that Koschny et al. (2002) demonstrated that for porous ice, crater shape differs markedly to that in solid ice. Koschny et al. (2002) make the point that the crater shape becomes more conical as the porosity increases. A typical depth to diameter ratio for a crater on solid ice is 0.2 with the shape of a shallow bowl (e.g., Shrine et al. 2000, 2002), but for porous ice Koschny et al. (2002) found this ratio to be ≈ 1, with, as stated, cone-shaped craters.

If data are taken from a variety of experimental sources a compilation can be built up and plotted vs. impact kinetic energy. In Figure 2 crater volume is shown

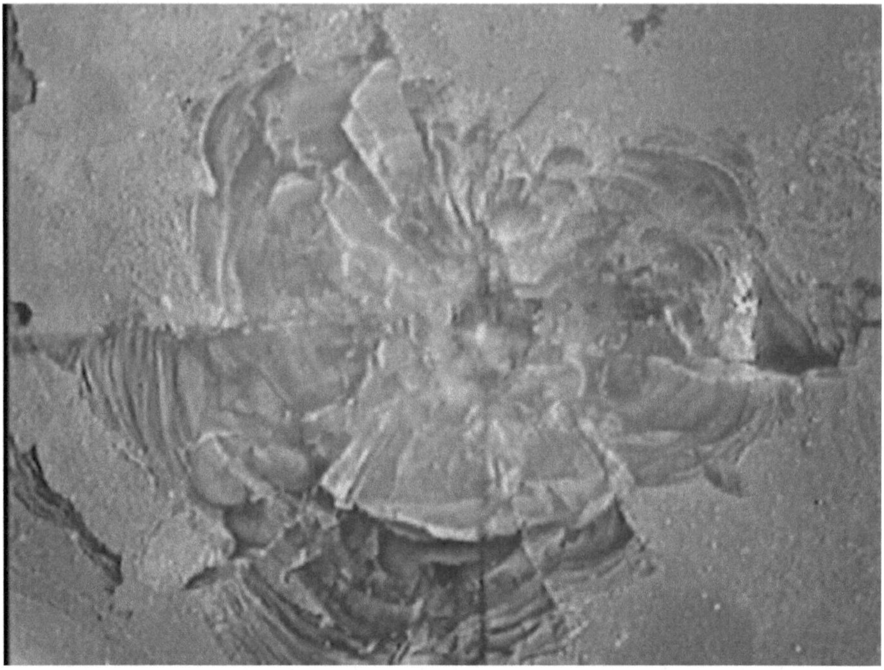

Fig. 1. Image of typical ice crater in the lab from impact of a 1 mm diameter aluminium sphere at 5.0 km s^{-1}. The average crater diameter is 49.6 mm. The overall image is 8 cm.

in this fashion, using data for impacts of 1 mm diameter aluminium projectiles at 1 to 7.3 km s^{-1} (Shrine et al. 2002) and 1 mm dia. steel projectiles at 5.2 km s^{-1} (Burchell et al. 1998). A fit to the data yielded

$$V = (0.278 \pm 0.05)\ KE^{\ 1.05 \pm 0.06} \qquad r = 0.98, \qquad (1)$$

where volume (V) was in cm^3 , kinetic energy (KE) in J and r was the regression coefficient of the fit. In principle one could now simply extrapolate this to the energy of Deep Impact (1.7×10^{10} J). However, this is some 7 or 8 orders of magnitude greater in energy than the data used to obtain the fit. This might be considered problematic, especially given that the range of energies covered by the fit was only some 2 to 3 orders of magnitude.

An improvement is to attempt to increase the range over which the fit was obtained. It is not possible to significantly increase the size of projectiles in hypervelocity work in the laboratory, but they can be decreased in size, lowering the impact energy. In figure 3 a plot of crater volume is shown for data spanning 10 orders of magnitude in energy (Burchell et al. 2001). The data at energies of some 10^{-7} J are from impact experiments using sub-micrometer sized dust projectiles at 1 to 50 km s^{-1} (Eichhorn and Grün 1993). The data at 10^{-2} J are from impacts on ice of

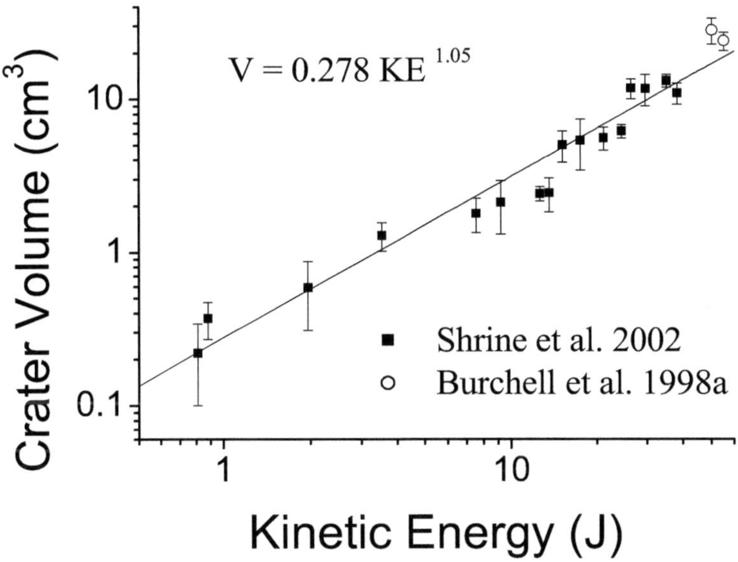

Fig. 2. Crater volume vs. kinetic energy

glass projectiles of size 10 to 100 micrometres at 1.8 to 9.6 km s^{-1} (Frisch 1993). The data at 10^1 J are an average from Figure 2 and is at 1 to 7.3 km s^{-1}, and that at 10^2 to 10^3 J are from Lange and Ahrens (1987), who used projectiles at 0.1 to 0.65 km s^{-1}. The fit shown gave

$$V = (0.087 \pm 0.04)\, KE^{\,1.06 \pm 0.07} \qquad r = 0.99, \tag{2}$$

again with units of cm^3 and J. The lower velocity data of Lange and Ahrens do not significantly distort the fit and extend the range of energies used by 2 orders of magnitude.

Extrapolation to the Deep Impact scale is still over many orders of magnitude, but now the range over which the fit was obtained stretches over a similar number of orders of magnitude. This doesn't of course remove the problem, but is the best that can be done with simple extrapolations.

Thus, at the Deep Impact scale (1.7×10^{10} J) we obtain from eqn. 1 that crater volume is 15,347 m^3, compared to the value of 6,078 m^3 predicted from eqn. 2. This difference (factor of 2.5) is itself comparable to the spread arising from the extrapolation of the energy over so many orders of magnitude when the exponent of the energy is estimated to ± 6 or 7%. This shows the problems of predictions made by extrapolating over so many orders of magnitude. Given that eqn. 2 was obtained over a wider range of impact energy it is taken as the more realistic formula and in the rest of this paper the predicted crater volume is thus taken as 6,078 m^3.

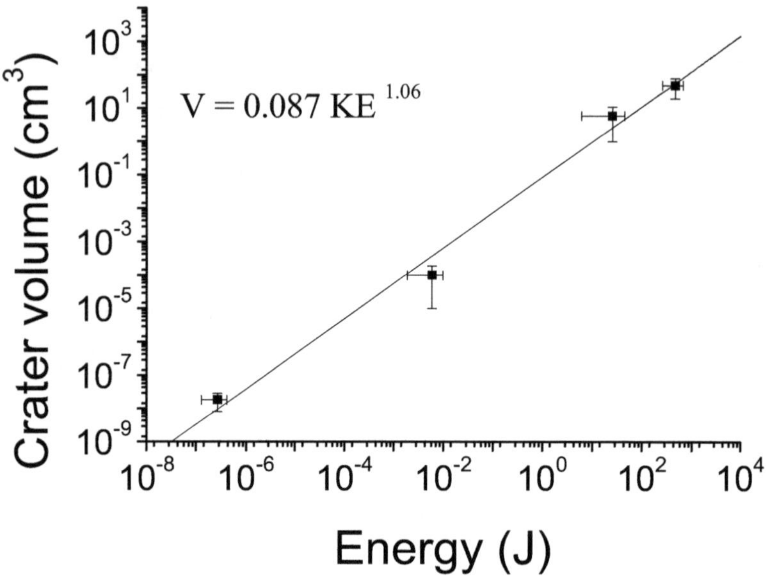

Fig. 3. Crater volume vs. kinetic energy (wide range of impact energies).

If the crater was in well consolidated ice, the shallow bowl like shape can be reasonably approximated by a disk. Given a typical depth diameter ratio of 0.2, we can thus predict that this crater would have a diameter of 34 m and a depth of 7 m.

An alternate method of scaling is to use Pi group scaling. This method (e.g., see Melosh 1989 for details) supposes that crater parameters can be expressed as dimensionless ratios of the relevant impact parameters. Data sets under different impact conditions can then be reduced to single parameterisations. Then, any impact which falls within the same range of the dimensionless constants, can be predicted to follow the same behaviour. This has been tested for hypervelocity impacts on water ice by Shrine et al. (2002).

The Pi groups for impact cratering are defined as:

$$\pi_v = V\rho \, / m \, , \tag{3}$$

$$\pi_r = r \, (\delta / m)^{1/3} \, , \tag{4}$$

$$\pi_2 = 2g(m / \delta)^{1/3} / v^2 \, , \tag{5}$$

$$\pi_3 = Y / \delta v^2 \, , \tag{6}$$

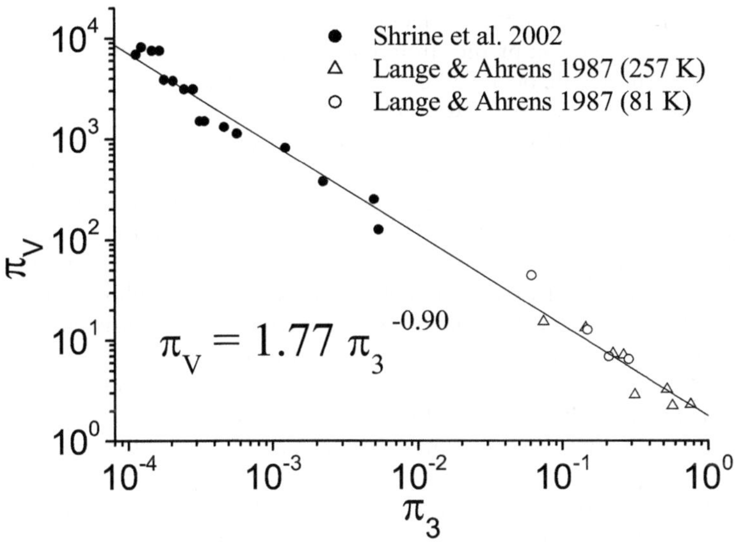

Fig. 4. Pi-scaled variables π_V vs. π_3.

where V and r are crater volume and radius, m and v are projectile mass and velocity, δ and ρ are projectile and target densities respectively, Y is target strength (taken as 17 MPa, following Lange and Ahrens 1987), and g is surface gravity. The data from Shrine et al. (2002) and Lange and Ahrens (1987) (at lower velocity) are shown in Figure 4 and can be seen to follow a common behaviour. A fit to the data (solid line) yielded:

$$\pi_V = (1.76 \pm 0.20)\ /\ \pi_3^{\ (0\ 90 \pm 0.02)}, \quad r = 0.99 \qquad (7)$$

The range over which the data are applicable can be expressed as ranges of impact speed (v) and the product ga (local gravity × projectile radius). The current data supports limits of:

$$0.2 < v < 18.6 \text{ km s}^{-1} \qquad (8)$$

and

$$2.6 \times 10^{-6} < ga < 1074 \text{ m}^2\text{s}^{-2}\,. \qquad (9)$$

Notice that these are outside the ranges of speeds etc. achieved in individual experiments, as they appear as combinations in the definitions in equations 3 - 6.

For an icy body like a comet nucleus, the local gravity will be small. If we assume a nucleus to be a solid sphere 10 km across, with density 1000 kg m^{-3}, then g

= 1.4 × 10^{-3} m s^{-2} . If we consider the Deep Impact mission's projectile has a diameter (*a*) of order 1 m, then the product *ga* will fall well within the range covered above. Accordingly we can predict from eqn 7 that for an impact at 10 km s^{-1} of the Deep Impact projectile on a well consolidated H_2O ice target, the result will be a crater of volume 11,982 m^3. If we again assume that the crater bowl is shaped and can be approximated by a disc with depth/diameter = 0.2, we obtain crater diameter = 42 m and crater depth = 8 m. These values are very similar to those obtained by simple energy scaling. The results can also be compared to the prediction by Belton and A'Hearn (1999). They used similar scaling (but based on less data) to predict that a crater of depth order 20 m and diameter order 100 m would result from the impact at 10 km s^{-1} of a 500 kg mass into ice. Although the projectile mass is slightly greater than that used here the results are consistent. It should be noted that this seems to be the favoured scenario for Deep Impact by the mission scientists, but they do acknowledge that if the cratering process is dominated by other mechanisms this result will change. For example, if the impact energy goes predominantly into compression of porous material the crater will be smaller but relatively deeper.

However, these predictions suffer from several deficiencies. The targets used in all the experiments were solid blocks of H_2O ice at approximately 255 K. In reality, the comet nucleus will have a lower temperature, will be porous, will not be pure H_2O ice but will have admixtures of other volatiles in the ice and probably silicate grains as well. The influence of all of these differences on the size of an impact crater has to be allowed for. At present, there is insufficient data to allow an extrapolation as in Figure 3 or 4 for any hypervelocity impact data set taken under conditions other than those shown here. However, preliminary individual studies of the effects of each of these influences have been carried out. In each case the resultant craters can be compared to those obtained in pure H_2O ice at 255 K. If we assume that the comparisons are independent of scale then we can predict that any estimated crater size should be modified by a series of coefficients that can be obtained from the individual studies. That is, we can say that the true value of any crater parameter (C_{true}) can be found from

$$C_{true} = C_{scaled}C_TC_VC_PC_S \qquad (10)$$

where C_{scaled} is the value predicted from energy scaling (e.g. eqn. 2), C_T is a coefficient that corrects for the actual temperature of the icy body being considered (rather than the value of 255 K used in the laboratory), C_V is a coefficient that corrects for the influence of the volatile content in the ice, C_P corrects for the porosity of the ice and C_S for the silicate content.

If we accept the validity of eqn. 10 the coefficients C_T, C_V, C_P and C_S all have to be estimated. The temperature effect has been studied by Grey *et al.*, 2001a. They looked at the variation in crater size for constant impact conditions as the temperature of the ice target was changed over the temperature range 150 to 260 K. Using their data it is possible to obtain:

$$C_T = C_{255} \times (-0.564 + 0.0061T)\,, \qquad (11)$$

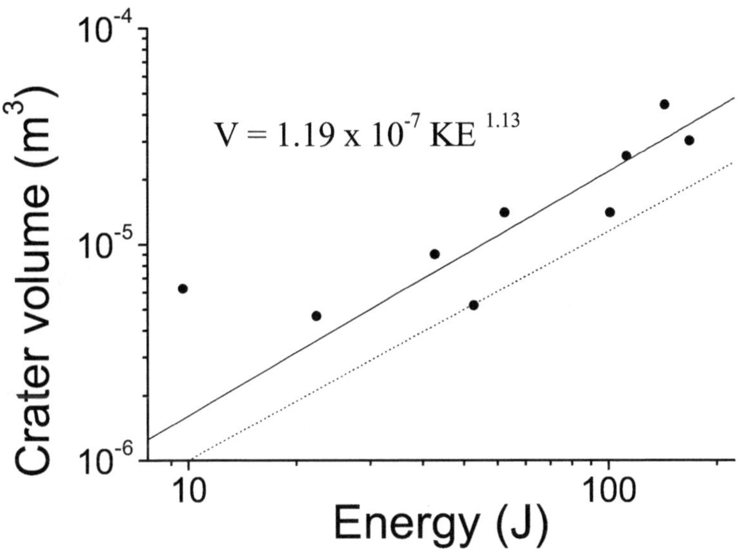

Fig. 5. Crater volume vs. impact kinetic energy for porous ice (Koschny et al. 2002). Solid line is fit to data, dashed line is fit to well consolidated ice (eqn. 2).

where T is the temperature in Kelvin, C_T is crater volume at any temperature T and C_{255} is the crater volume at 255 K.

The influence of the ice being other than H_2O is harder to estimate. Two studies exist. Burchell et al. (1998b) compared hypervelocity impacts in CO_2 and H_2O ices. They found that for crater volume, C_V (that is the ratio of crater volumes in CO_2 / H_2O) was 0.55. In Grey et al. (2001b), hypervelocity impacts in ammonia-hydroxide were compared to those in H_2O ice. They found that for ices with a 50% concentration of ammonia, C_V = 0.42. These values are only illustrations of possible ranges for C_V, as the value required for any particular application will depend on the particular composition of the ice target. In reality, it is unlikely that a body will be pure CO_2 ice or have an ammonia content approaching 50%. Thus the values given are probably maximal values and a more generally realistic choice would be to have C_V approximately equal to 0.75.

The influence of porosity can be judged from work by Koschny et al. (2002), where data were obtained, at up to 4 km s^{-1}, for impacts in ice with porosity of approximately 48%. Here a simple definition of porosity is used, namely the measured density of a body divided by the bulk density of solid, non-porous water ice. Koschny found that the excavated volume of a crater varied inversely with the ratio of bulk density of the target / density of ice. We thus obtain that C_P = density of non-porous ice / target density. The data of Koschny et al. are shown in Figure 5, along with a fit (solid line) and the fit given by eqn. 2 (dashed line). The average difference between the two fits is indeed approximately the ratio of two expected from use of ice with porosity of 48%.

For C_S there has been until recently, little concrete data allowing a comparison between hypervelocity craters in ice and ice-sand mixtures. Croft et al. (1980) did carry out some such studies, but did not report on crater volumes. At lower speeds Arakawa et al. (2000), studied impacts in ice-silicate mixtures, but used layered targets with a mixture of volatile contents and porosities as well as a silicate content. However, Koschny and Grün 2001 have recently reported on hypervelocity impacts into ice-silicate mixtures, where the targets were contained up to 20% silicate by mass. They found that crater volume decreased as the silicate content was increased. Although subject to much uncertainty, Koschny and Grün (2001) extract from their data a formula for the influence of silicate content (G, the fractional silicate content by mass) on crater yield (eqn. 4 in their paper). The dependence on G can be extracted in turn and an expression for C_S obtained:

$$C_S = 0.0149^G. \tag{12}$$

For a body which is, for example, 10% silicate by mass ($G = 0.1$), the value of C_S that is obtained from eqn. 11 is $C_S = 0.66$.

With formulae or values for these various terms in eqn. 10, the estimated crater size can be scaled to reflect a more realistic target. The next difficulty is to choose the composition of a typical comet nucleus. The dirty snowball model of Whipple from the 1950's is generally accepted. But details of the parameters such as density, dust content, temperature etc. are still only estimated for relatively few comets. Worse, even for a comet such as 1P-Halley, visited by 5 spacecraft during its last passage through the inner Solar System in 1986, estimates of a parameter as simple as bulk density range from 100 to 900 kg m^{-3}. Comet nucleus surface temperature is also not so well constrained. Beyond 4 AU temperatures can be modelled by the simple thermodynamics of Solar radiation on an airless rotating body with a typical albedo. This would give a typical surface temperature of about 150 K at 4 AU. Once closer than 3 AU, the temperature rises such that almost all the received Solar energy will go into sublimation of ices. This is a complicated process and temperature can no longer be predicted by a simple relation balancing incident and reflected Solar radiation. Instead, a comet's surface temperature will probably maintain itself between 180 and 200 K, with individual hot spots (active regions) at higher temperatures (up to 300 K) triggering outbursts of volatiles. The concentration of volatiles in the nucleus will depend on the region where the comet was formed and its history of passages through the inner Solar System. For example, whilst nitrogen is amongst the 5 most abundant elements, the chemical form it takes in a hydrogen rich gas is as N_2 or NH_3 depending on temperature and pressure. Thus the ammonia concentration in a comet will depend on where it formed in the Solar nebula. Worse, ammonia in comet nuclei is usually observed via its photodissociation products NH_2 and NH. These observations are then combined with models to predict the actual ammonia concentration in the comet nucleus itself. Ammonia was detected in the coma of Halley's comet, but again it was hard to obtain actual comet nucleus concentrations. Indeed, the first report of the ion composition of the coma of Halley's comet (Balsiger et al. 1986) explicitly states that ammonia itself was not observed, although confusingly the same paper is often cited as the first detection of ammonia at the Halley encounter. This discrepancy is explained by subsequent papers, e.g., Allen et al. (1987); Meier et al. (1994), which model the observed ion mass spectra in terms of the breakdown products of ammonia and deduce production rates of 1 – 2 % relative to water. The first probable observation of ammonia via radio observations was that of Altenhoff et al. (1983) who reported ammonia as comprising 6% of the gas subliming from the nucleus of comet IRAS-Araki-Alcock. However, it is only been recently that ammonia has been unambiguously observed in radio wavelength observations, in comets Hyakutake and Hale-Bopp (e.g., Bird et al. 1997). It should thus now be possible to obtain data for a statistically significant sample of comets, but again models are required to relate the observations to the concentration in the comet nucleus itself. Thus whilst concentrations in the nucleus are commonly predicted to be as low as 0.1% - 1%, different values are actually reported depending on the assumptions made in the models. The concentration of mineral grains is also hard to establish, suffering from similar problems. Accordingly, there is still a lot of leeway in choosing the composition of a "standard" comet nucleus.

Here we define the target as a comet nucleus at a temperature of 200 K (i.e., $C_T = 0.56$), with 20% porosity (i.e., $C_P = 1.25$), choosing a value of $C_V = 0.75$, (i.e., not the maximal values found for 50% ammonia or 100% CO_2 ice) and choose the silicate content as 10% (i.e., $G = 0.1$ in eqn. 12, and hence $C_S = 0.66$). The resulting product $C_TC_VC_PC_S = 0.35$. Thus the volume of the crater will be some 1/3rd of that predicted by scaling from impacts on well consolidated ice targets at 255 K. It is interesting that in the product $C_TC_VC_PC_S$ some terms are greater than one and some less, i.e. that some of the changes required by the individual corrections cancel each other. What should also be remembered is that the shape of the crater will not be a shallow bowl shape as seen of impacts on solid ice. Rather it will be cone shaped, leading to a crater with a smaller diameter but greater relative depth.

3 Conclusions

Using two different methods of scaling laboratory data it is possible to make simple estimates of crater volume for hypervelocity impacts on ice. It was shown that the predicted crater volumes are within a factor of approximately two of each other if either a simple extrapolation is made with impact energy, or a slightly more sophisticated scaling with dimensionless Pi-groups is used. In addition, it was also shown that the influence of the target not being pure, solid (i.e., non-porous) water ice can be crudely allowed for by use of coefficients to adjust the crater size dependent on each assumed variation, i.e., porosity, non-water ice content etc. For a hypothetical composition comet this correction was deduced to be a reduction in crater volume of some two-thirds. This is of similar magnitude to the spread in the two methods of extrapolation. If no allowance is made for the particular properties of the target comet nucleus, then at Deep Impact energy scales, all the methods used here predict a crater of volume 6 - 15×10^3 m^3. If, however, allowance is made for the particular composition of a comet nucleus in terms of its porosity, volatile/mineral content etc, then at this stage, the best estimate is that an impact crater from the Deep Impact mission will be of order 2 - 5. $\times 10^3$ m^3 in volume, and the entrance diameter and depth will depend strongly on the porosity of the ice.

There are several areas that are still poorly understood as regards cratering in comet nuclei. The influence of the target composition on the shape of the crater is of critical importance. The impacts may result not only in relatively deeper craters, but in craters with a narrow entrance hole and a wider interior. This in turn may cause much of the ejecta and vapour plume traditionally associated with hypervelocity impacts to be retained inside the crater. In the Deep Impact mission this would greatly affect what is observed by the cameras on the main spacecraft. In addition, something not considered at all here is the influence of the shape (spherical, conical, cylindrical, peanut shaped ?) and detailed composition of the projectile (inhomogeneous, with voids, etc ?) on the resultant crater. Nor have we considered what happens for impacts at non-normal incidence. Clearly, much more work (both experimental and via modelling) is still needed to better understand this problem of cratering in non-homogeneous ice targets.

Acknowledgements

We gratefully acknowledge financial support from PPARC (UK). We also thank Mr. M. Cole for operation of the light gas gun, which provided the data used in this work.

References

A'Hearn MF (1999) Deep Impact. Bulletin of the American Astronomical Society 31: 1114

Allen M, Delitsky M, Huntress W, Yung Y, Ip WH, Schwenn R, Rosenbauer H, Shelley E, Balsiger H, Geiss J (1987) Evidence for methane and ammonia in the coma of comet P/Halley. Astronomy and Astrophysics 187: 502-512

Altenhoff WJ, Batrla W, Huchtmeier WK, Schmidt J, Stumpff A, Walmsley M (1983) Radio observation of comet 1983D. Astronomy and Astrophysics 125: L19-L22

Arakawa M, Higa M, Leliwa-Kopystynski J, Maeno N (2000) Impact cratering of granular mixture target made of H_2O ice-CO_2 ice-pyrophylite. Planetary and Space Science 48: 1437-1446

Balsiger H, Altwegg K, Buhler F, Geiss J, Ghielmetti AG, Goldstein BE, Goldstein R, Huntress WT, Ip WH, Lazarus AJ, Meier A, Neugebauer M, Rettenmund U, Rosen-

Rosenbauer H, Schwenn R, Sharp RD, Shelley EG, Ungstrup E, Young DT (1986) Ion composition and dynamics at comet Halley. Nature 321: 330-334

Belton MJS, A'Hearn MF (1999) Deep sub-surface exploration of cometary nuclei. Advances in Space Research 24: 1167-1173

Bird MK, Huchtmeier WK, Gensheimer P, Wilson TL, Janardhan P, Lemme (1997) Radio detection of ammonia in comet Hale-Bopp. Astronomy and Astrophysics 325: L5-L8

Burchell MJ, Brooke-Thomas W, Leliwa-Kopystynski J, Zarnecki JC (1998a) Hypervelocity impact experiments on solid CO_2 targets. Icarus 131: 210-222

Burchell MJ, Leliwa-Kopystynki J, Vaughan B, Zarnecki, JC (1998b) Comparison of H_2O and CO_2 ices under hypervelocity impact. In: Schmidt SC, Dandekar DP, Forbes JW (eds) Shock Compression of Condensed Matter-1997. American Institute of Physics CP429, pp 949-952

Burchell MJ, Cole MJ, McDonnell JAM, Zarnecki JC (1999) Hypervelocity impact studies using the 2 MV Van de Graaff accelerator and two-stage light gas gun of the University of Kent at Canterbury. Measurement Science Technology 10: 41-50

Burchell MJ, Grey IDS, Shrine NRG (2001) Laboratory investigations of hypervelocity impact cratering in ice. Advances in Space Research 28: 1521-1526

Croft SK, Kieffer SW, Ahrens TJ (1979) Low velocity impact craters in ice and ice saturated sand with implications for Martian crater count ages. Journal of Geophysical Research 84 (B14): 8023-8032

Croft SK (1981) Hypervelocity impact craters in icy media [abs.]. Lunar and Planetary Science 12: 190-191

Durham WB, Stern LA (2001) Rheological properties of water ice – Applications to satellites of the outer planets. Annual Reviews of Earth and Planetary Sciences 29: 295-330

Eichhorn K, Grün E (1993) High velocity impacts of dust particles in low temperature water ice. Planetary and Space Science 41: 429-433

Frisch W (1993) Hypervelocity impact experiments with water ice targets. In: McDonnell JAM (ed) Hypervelocity Impacts in Space. University of Kent, Kent, pp 7 – 14

Grey IDS, Burchell MJ, Shrine NRG (2001a) Laboratory investigations of the temperature dependence of hypervelocity impact cratering in ice. Advances in Space Research 28: 1527-1532

Grey IDS, Burchell MJ, Johnson E (2001b) Hypervelocity impact craters in ammonia-water ice [abs.]. Lunar and Planetary Science 32: 1200 (CD-ROM)

Iijima Y, Kato M, Arakawa M, Maeno N, Fujimura A, Mizutani H (1995) Cratering experiments on ice; dependence of crater formation on projectile materials and scaling parameters. Geophysical Research Letters 22: 2005 – 2008

Kato M, Iijima Y, Arakawa M, Okimura Y, Fujimura A, Maeno N, Mizutani H (1995) Ice-on-ice impact experiments. Icarus 113: 423-441

Kawakami S, Mizutani H, Takagi Y, Kato M, Kumazawa M (1983) Impact experiments on ice. Journal of Geophysical Research 88 (B7): 5806-5814

Koschny D, Grün E (2001) Impacts into ice-silicate mixtures: Crater morphologies, volumes, depth-diameter ratios and yield. Icarus 154: 391-401

Koschny D, Kargl G, Rott M (2002) Experimental studies of cratering process in porous ice targets. Advances in Space Research, 28: 1533-1537

Lange MA, Ahrens TJ (1987) Impact experiments in low temperature ice. Icarus 69: 506-518

Maier R, Eberhardt P, Krankowsky D, Hodges RR (1994) Ammonia in comet P/Halley. Astronomy and Astrophysics 287: 268-278

Melosh HJ (1989) Impact Cratering. Oxford University Press, Oxford, 245 pp

Shrine NRGS, Burchell MJ, Grey IDS (2000) Velocity scaling of impact craters in water Ice with relevance to cratering on icy planetary surfaces [abs.]. Lunar and Planetary Science 31: 1696 (CD-ROM)

Shrine NRGS, Burchell MJ, Grey IDS (2002) Velocity scaling of impact craters in water ice over the range 1 to 7.3 km s^{-1}. Icarus 155: 475-485

Survivability of Bacteria in Hypervelocity Impacts on Ice

J.R. Mann[1], M.J. Burchell[1], P. Brandão[2], A.W. Bunch[2] and I.D.S. Grey[1]

[1] Centre for Astrophysics and Planetary Science, School of Physical Sciences, University of Kent Canterbury, Kent, CT2 7NR, United Kingdom. (M.J.Burchell@kent.ac.uk)
[2] BioSciences Laboratory, University of Kent at Canterbury, Canterbury, Kent, CT2 7NJ, United Kingdom.

Abstract. As part of the arrival stage of the Panspermia process, organisms must endure a hypervelocity impact either into the atmosphere or onto the surface of the destination planet. The impacts associated with this arrival stage are studied in this paper. To this end, the two-stage light gas gun at the University of Kent has been used to fire bacteria-laden projectiles, at velocities of approximately 5 km s^{-1}, onto semi-solid nutrient medium, and solid and porous ice targets, representing planetary oceans and icy surfaces. The targets were then analysed to investigate whether the bacteria survived the impacts. It was found that bacteria can survive hypervelocity impacts at 5 km s^{-1}, with a survival rate of 1 per 3.5 million using targets of nutrient gel. With ice targets no survival has been found yet with a limit on survival of less than 1 per 0.4 million.

1 Introduction

The relatively short time period between the end of the period of late heavy bombardment of the early Earth, and the advent of life on Earth, has led many scientists to suggest that the complex biomolecules required for life, or indeed primitive life itself, may have been delivered to Earth, ready-formed, inside comets, asteroids, or meteorites. This way of thinking has existed for many hundreds of years, dating back to circa 500 BC and the Greek philosopher Anaxagoras of Clazemonae. The most well-known proponent of this school of thought was the Swedish physicist Svante Arrhenius, who formulated the ideas into a coherent theory, called Panspermia, in 1908. Since then, the Panspermia theory was relatively neglected until 1996 – the time of publication of the analysis of purported Martian meteorite ALH84001 (McKay et al. 1996). Inside natural cracks in the rock, tiny carbonate grains were found, originally thought to imply liquid water. On close analysis of these grains, nanometre-sized structures were found clinging to the surface of the carbonate, which resembled terrestrial bacteria in their morphology. These structures were consequently described as possible fossilised Martian bacteria (McKay et al. 1996). Although these claims have been refuted many times (e.g., Grady et al. 1997), the idea of bacteria travelling through space inside meteorites has received a clear boost, leading to renewed interest in the theory of Panspermia.

There exist three variations of the Panspermia idea (Parsons 1996). The first is the original idea proposed by Arrhenius, and is known as Radiopanspermia (Arrhenius 1908). This form of the theory suggests that the organisms originate in space and are propelled through space by stellar radiation pressure. Secondly, there is Directed Panspermia, where the organisms are moved around deliberately via, for example, spacecraft. Although it may sound far-fetched, this form of Panspermia has already occurred with the advent of the lunar landings and the Mars probes accidentally delivering terrestrial microbes to other Solar System bodies. Finally, perhaps the most plausible form of the theory is Lithopanspermia, sometimes also known as rocky or icy satellite Panspermia (Melosh 1988). This describes a way in which the organisms can be transported through space whilst on, or inside, meteorites.

The Lithopanspermia variant of the theory can be described as a journey that can be split into three stages: leaving the planet of origin, travel through space, and arrival at a destination planet. Each of these three stages is fraught with danger for the organisms being transported. Firstly, they must survive a process providing enough energy to eject themselves, and a portion of the rock they live on, off the surface of the planet at a speed greater than local escape velocity (e.g., approximately 11 km s^{-1} for Earth, and approximately 5 km s^{-1} for Mars). This is usually considered to be associated with the hypervelocity impact of a large body onto the planet's surface causing ejection of material from the surface at high speed. A hypervelocity impact is one in which the speed of propagation of the shock waves caused by the impact exceeds the speed of sound in both the projectile and target materials. Transfer through space requires any organisms that survived ejection, having to survive the harsh vacuum and intense radiation environment of space, possibly for hundreds of thousands of years. Finally, upon arrival at the destination planet, any viable organisms need to survive another hypervelocity impact as the meteorite they are travelling on impacts either the atmosphere or surface of the destination planet. Both these types of impacts (i.e., surface and atmosphere) would expose the projectile to extreme temperature and pressure conditions. For example, with projectiles impacting the atmosphere and possibly being aerobraked (depending on the size of the impactor), airburst or explosion events (e.g., Tunguska) can occur, paying testament to the extreme temperature and pressure conditions experienced. On Earth objects arriving from interplanetary space generally have speeds on the order of 25 km s^{-1} (Hughes and Williams 2000).

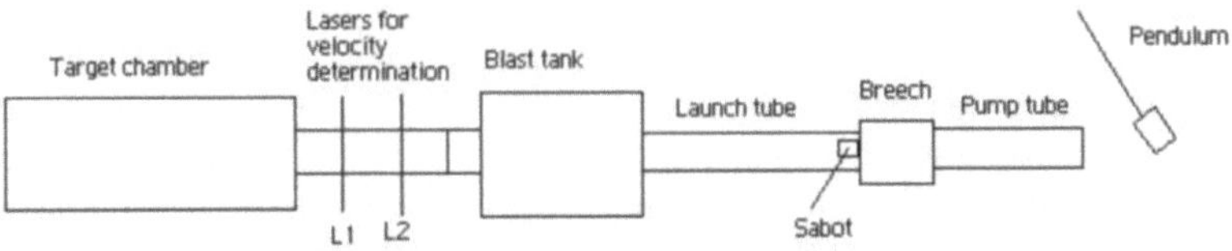

Fig. 1. Schematic of the light gas gun at the University of Kent at Canterbury. Full details regarding the operation of the gun can be found in Burchell et al. (1999).

It is clear that any life forms being transported through space in this way must survive all of these conditions. Experiments and calculations have previously been performed to test the survival of bacteria in the first two stages of the Panspermia journey (e.g., Mileikowsky et al. 2000a, Mileikowsky et al. 2000b; Horneck et al. 2001), which have shown that some bacteria survived the constraints associated with these first stages. However, data are lacking for the third stage describing the effect on bacteria of the impact involved with the arrival stage of the process. A preliminary investigation has been performed by firing bacteria-laden projectiles at targets including rock, glass, and aerogel (Burchell et al. 2000, 2001a, 2001b), in which viable bacteria were harvested from the ejecta produced by impact into rock, but further investigation is required. This additional investigation has since been performed, and is described in the rest of this paper.

2 Experimental Program

In order to simulate planetary impacts in the laboratory, the two-stage light gas gun at the University of Kent was used to fire bacteria-laden projectiles at various targets. The two-stage gun is capable of firing projectiles at speeds of up to approximately 7.5 km s^{-1}.

Figure 1 shows a schematic of the light gas gun. The gun is fired from right to left as shown in the diagram, with the projectile being loaded inside a sabot, and the target loaded in the target chamber. The gun's target chamber is held under vacuum during each shot, at around 20 - 100 Pa. Because the launch tube is rifled, during the shot the pieces of the sabot are spun off-axis and impact a stop-plate at the exit aperture of the blast tank. The projectile continues its flight along the central axis of the gun and impacts the target. On route, the projectile passes through two laser light curtains, separated by a known distance, and the timing of the disruption of these beams can be used to determine the velocity of the shot. A more detailed description of the light gas gun and its operation can be found in Burchell et al. (1999).

For the present work, the projectile material used was a porous ceramic. Depending on the target material, the ceramic was either fired as a single projectile approximately 1.7 mm in size, or as a cloud of smaller particles between 90 and 200 μm in size. The choice of ceramic as the carrier material was made because it contained internal pores with sizes large enough to house bacterial cells, and its capability to absorb liquids meant that the solution in which the bacteria were grown could be absorbed into these pores. This ensured that the bacteria were given some protection against the harsh conditions of the impact process, as they would have if they were trapped inside meteorites.

The bacteria used for this project were a specific strain of a species called *Rhodococcus erythropolis* DSM13002. It is a physically hardy, non-pathogenic species, with rod-shaped cells 1 μm in diameter and 2 μm in length. Rhodococci appear red in colour when grown on glucose yeast extract nutrient medium (a common biological nutrient medium), which made them easily distinguishable from common laboratory contaminants. Because these cells are not airborne, they do not exist naturally inside a research laboratory. Most species of *Rhodococcus* are known to be able to withstand high pressures, indeed some are indigenous to ocean sediments at depths of around 6000 m. Rhodococci are also identical, in structural features, to some prokaryotic organisms obtained in pure culture after cryopreservation in Antarctic cold desert soils (Soina et al. 1995), highlighting their ability to survive extreme conditions. Thus, *Rhodococcus erythropolis* is a suitable choice for impact related studies.

2.1
Preliminary Experiments

Intuitively, the chance of any living organism surviving a hypervelocity impact is low, given that in most such impacts the projectile is vaporised. So in the first instance, preliminary shots were performed in order to ascertain whether or not to embark on a full experimental program. For these initial shots, a target material was used that would increase the chance of survival as much as possible. The target material used was a standard biological nutrient medium (glucose yeast extract) set into a semi-solid state in a petri dish using agar. This target material was chosen as it supplied the bacteria with all the nutrients they needed for optimum growth immediately upon impact. This target also eliminated the need to harvest the bacteria from the impact crater post-shot. The velocity used for these initial shots was approximately 5 km s^{-1}.

A program of seven shots was performed, each of which was coupled with a control shot, in which projectiles that had not been contaminated with bacteria were fired under the same conditions. This provided a way to test that accidental contamination of the targets during handling did not occur, and also provided a measure of the existence of other species of bacteria within the laboratory. After the shot, the target plates were hermetically sealed and incubated at 30 °C for 7-10

Table 1. The results obtained in the preliminary shot program, firing bacteria into semi-solid nutrient medium.

Type of Shot	Velocity (km s^{-1})	Survival?
Control	5.24	No
Doped	5.76	No
Control	5.2 ± 0.5	No
Doped	5.2 ± 0.5	No
Control	5.2 ± 0.5	No
Doped	5.2 ± 0.5	Yes, 2 colonies
Control	5.0 ± 0.2	No
Doped	4.98	No
Control	5.01	No
Doped	5.30	No
Control	5.79	No
Doped	5.53	No
Control	4.47	No
Doped	5.271	No

days to allow optimum growth of any surviving cells.

The results from this preliminary program are shown in Table 1. All the control shots carried the negative result expected with respect to the presence of *Rhodococcus erythropolis*. Of the shots where the projectile was doped with bacteria, most also carried negative results. However, one of the target plates showed evidence of growth of *Rhodococcus erythropolis* in two separate colonies (Figure 2), indicating that at least two individual cells survived the impact process.

The knowledge that two cells survived the impact process can be expressed as a survival rate, i.e. as a fraction of how many cells were fired in each shot (and in the program as a whole). Considering the projectile consists of a cloud of particles of size 90-200 μm, an average size for a particle can be assumed to be 150 μm. By also assuming that 10% of the original projectile volume is bacteria, and assuming that ten pieces of projectile hit the target (based upon examination of previous nutrient medium targets), then the total volume of bacteria per shot is equal to the volume of one particle. If the particles are assumed to be spherical in shape, then the total volume of bacteria in one shot is approximately 10^{-12} m^3. Since each individual bacterium is cylindrical in shape, with diameter 1 μm and length 2 μm, the volume of an individual cell is approximately 10^{-18} m^3. This means that in each shot, there are approximately 10^6 bacteria fired. So for the seven shots performed, the rate of survival of the bacteria is of the order of 2/7,000,000, i.e. 1 per 3.5 million. That the cells recovered are of the same strain as those fired has been confirmed using a suite of standard biological analyses, including gram staining, growth on a selective nutrient medium, and pyrolysis mass spectrometry. These results are presented in more detail in Burchell et al. (2001b).

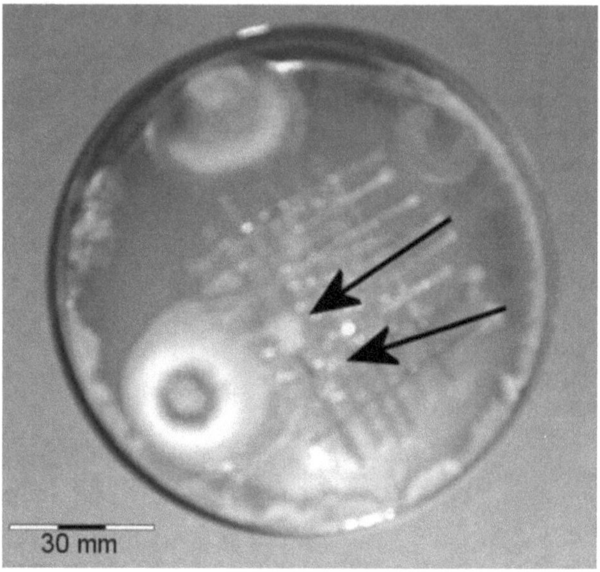

Fig. 2. Target plate from the shot in which *Rhodococcus erythropolis* survived. The arrows point to the two surviving colonies.

2.2
Ice Experiments

Having established that survival in general was possible, the program was continued using another relevant Solar System target material. Given that rock targets had already been tested with negative results, other planetary surface materials were considered as possible targets. On the basis of its abundance in the Solar System, ice was considered to be a suitable choice for a target material. Examples of ice in the Solar System include the icy satellites, comets and their nuclei, and in history, the Earth ice ages. The ice used in the experiments was made from sterilised, distilled water allowed to freeze at temperatures ranging from –18 to –150 °C, in order to investigate whether temperature affects survival in any way. To reduce the level of contamination, the ice was made in a sealed container, and at all times handled whilst wearing sterile surgical latex gloves, hairnets, and face-masks. As in the preliminary shot program, all of the ice shots were fired at a velocity of around 5 km s^{-1}, and the same strain of *Rhodococcus erythropolis* that survived the impacts into the nutrient medium targets was used. The same type of ceramic was also used as the projectile material, but instead of a cloud of particles, only one projectile piece was used in each shot, typical diameters of these pieces ranging from 1 mm to 1.7 mm. Four shots were performed with ice targets, each of which was again accompanied by a control shot containing no bacteria.

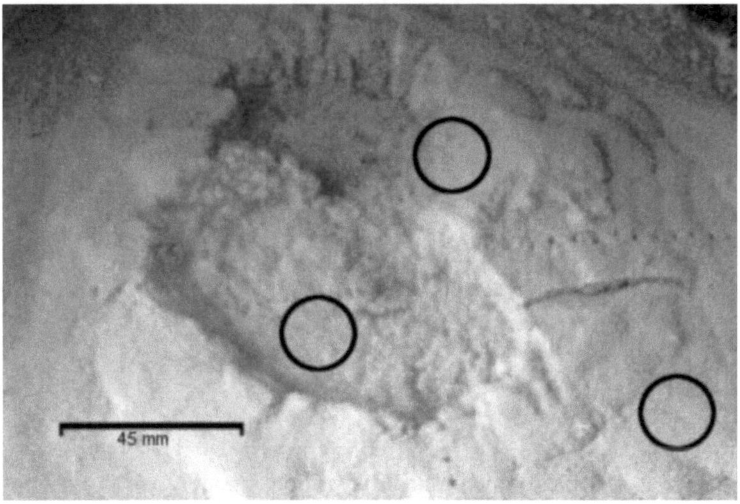

Fig. 3. A typical crater formed by a ceramic projectile impacting ice. The circles indicate the regions of the crater that were sampled for analysis - the centre of the crater, the edge of the crater, and the edge of the target outside the crater.

Figure 3 shows an example of a typical crater formed in the ice targets. In Figure 3 the double crater bowl indicates that the projectile fragmented in flight and split into two pieces, each of which successfully impacted the target. This type of fragmentation was fairly typical for the ceramic projectile pieces, since their highly porous nature meant they had low structural integrity. The arrows in the figure indicate the regions of the crater that were sampled for biological analysis, so it can be seen that although ice samples are taken from each of the main areas of the target, only approximately 10% of the total crater area is sampled at most. This difficulty with sampling the entire crater area effectively introduces a 90% inefficiency when attempting to harvest surviving bacteria.

The conditions under which each of the ice shots was fired, and the results from the shots are shown in Table 2. The temperatures given are accurate to within a few degrees, and the velocities are accurate to within 1%. From the summary given in Table 2 it is clear that as yet no cells have been detected surviving on impact into the ice targets.

Despite the fact that this result is currently a negative one, it can still be quantified. Assuming an average projectile diameter of 1.35 mm, and allowing for some fragmentation of the particle to occur during flight, it is assumed that only a 1 mm particle impacts the target. Therefore, the volume of the total projectile, again assuming a spherical shape, is approximately 10^{-10} m^3. Taking 10% of this value to estimate the total volume of bacteria in each shot, as before, gives 10^{-12} m^3. The volume of an individual cell, as in the preliminary experiment calculations, is estimated as 10^{-18} m^3, thus the total number of cells per shot is approximately 10^6, as before.

Table 2. The results obtained in the preliminary shot program, firing bacteria into semi-solid nutrient medium.

Type of Shot	Velocity (km s^{-1})	Temperature (K)	Survival?
Control	4.69	225	No
Doped	4.94	225	No
Control	4.93	150	No
Doped	4.87	150	No
Control	4.92	200	No
Doped	4.81	200	No
Control	4.82	250	No
Doped	5.16	250	No

However, as previously mentioned, only approximately 10% of the total crater area is sampled, which implies that only 10% of the cells that impact the target are cultured. So for the four shots fired into ice targets, the current rate of survival of the bacteria is less than 1 per 400000. A sample of ice from the edge of the target, away from the crater, is also cultured in order to investigate levels of contamination within the ice. All of these samples showed negative results so are not included in the 10% sampling area.

2.3 Porous Ice Experiments

In an effort to increase the chance of bacteria surviving an impact, porous ice has been considered as a target material. It has been shown by Koschny et al. (2001) that craters formed in hypervelocity impacts on porous ice are different in shape than those on solid, well-consolidated ice. In the latter, the craters are as shown in Figure 3, a relatively broad, i.e. shallow, crater. By contrast, in porous ice Koschny et al. showed that the craters are deeper, resembling carrot-shaped tracks. Indeed, unusually for hypervelocity impacts, Koschny et al. reported that at 4 km s^{-1} (the maximum speed in their study) fragments of the projectile were found at the bottom of the crater/track. Interestingly, the impacts into the nutrient medium in the first part of this study (Figure 4) were also long, thin carrot-shaped tracks, with an entry hole diameter of 200 micrometres, and length 1-2 mm. This gives a depth-to-diameter ratio of approximately 10:1.

Given this similarity between impacts in porous ice and nutrient gel, there may be good reason to consider porous ice as a target in such work. The porous ice targets were made by allowing crushed ice flakes, each typically 10-15 mm in size, to melt slightly then freezing them so that they bonded together into a single target. The resultant target has a bulk density approximately 55 % of that of normal solid ice, so although it is still in solid state, it provides a softer impact.

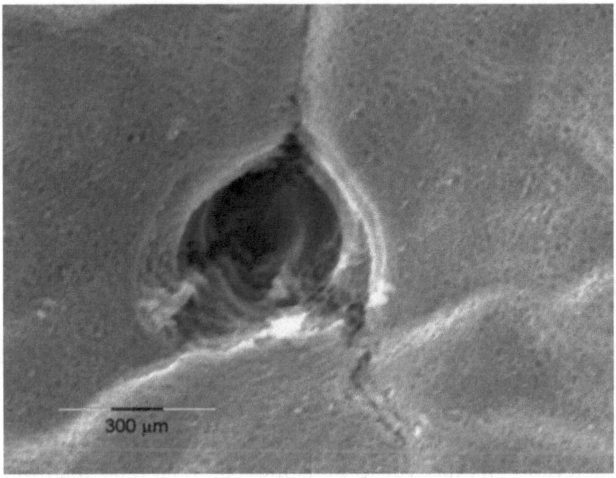

Fig. 4. A typical crater formed when a 100 - 200 micrometre-sized ceramic projectile impacts semi-solid nutrient medium.

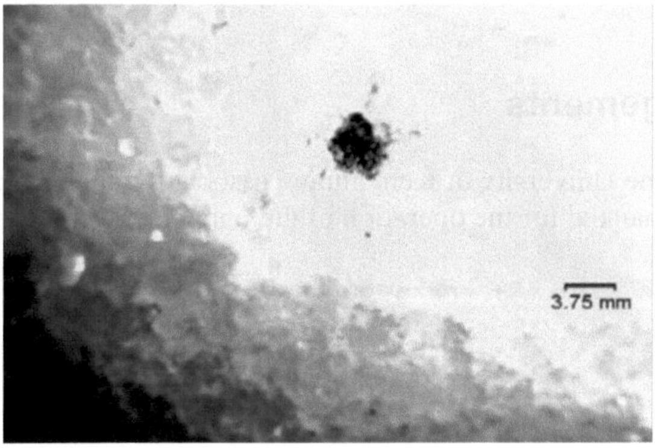

Fig. 5. A typical crater formed when a 1 - 1.7 mm ceramic projectile impacts porous ice.

This is shown in Figure 5, which is an image of a porous ice impact crater, produced by the same type of projectile (diameter 1 – 1.7 mm, velocity 5 km s^{-1}) used in the standard ice shots. The crater excavated in the porous ice is thin and deep, similar to the craters produced in the semi-solid nutrient medium. This suggests that the impact process involved in excavating this crater is more similar to that involved in the nutrient medium than to that for the solid ice, so survival may be more likely. Porous ice is also relevant to simulating Solar System impacts, as it can be found in abundance in comets and areas such as the deep permafrosts around the Martian polar caps. One shot firing bacteria into a porous ice target has been performed, firing a 1 - 1.7 mm ceramic projectile into the target at 5 km s^{-1}. As yet no surviving bacteria have been observed, but further shots into porous ice targets are in progress. Future experiments will also include sampling of the inside of the target chamber in order to investigate whether viable bacteria are being ejected from the crater.

3 Conclusions

A preliminary investigation that involved firing bacteria-laden projectiles into semi-solid nutrient medium targets confirmed the possibility of survival of the bacterial cells, with a survival rate of 1 per 3.5 million. Similar experiments using solid ice targets at various temperatures have also been performed, although survival of any bacterial cells has yet to be achieved. The total load of cells fired in the ice shots is currently only of the order of 400000 cells (i.e., survival of less than 1 per 0.4 million), so this program of shots is being continued to gain equivalent statistics to those of the nutrient medium program. In addition, the possibility of using porous ice targets has been shown to be feasible. Our preliminary results suggest that the arrival stage of the Lithopanspermia process is a viable concept.

Acknowledgements

J. Mann thanks the University of Kent alumni association for financial support. Mr. M. Cole is thanked for the operation of the light gas gun used in this work.

References

Arrhenius S (1908) Worlds in the making: the evolution of the Universe. Harper Bros., London New York, 230 pp

Burchell MJ, Cole MJ, McDonnell JAM, Zarnecki JC (1999) Hypervelocity impact studies using the 2 MV Van de Graaff accelerator and two-stage light gas gun of the University of Kent at Canterbury. Measurement Science and Technology 10: 41-50

Burchell MJ, Shrine NRG, Bunch AW, Zarnecki JC (2000) Exobiology: Laboratory tests of the impact related aspects of Panspermia. In: Gilmour I, Koeberl C (eds) Impacts and the early Earth, Springer, Heidelberg, pp1-26

Burchell MJ, Shrine NRG, Mann J, Bunch AW, Brandão PFB, Zarnecki JC, Galloway JA (2001a) Laboratory investigations of the survivability of bacteria in hypervelocity impacts. Advances in Space Research 28: 707-712

Burchell MJ, Mann J, Bunch AW, Brandão PFB (2001b) Survivability of bacteria in hypervelocity impact. Icarus 154: 545-547

Grady MM, Wright IP, Pillinger CT (1997) Microfossils from Mars: a question of faith? Astronomy and Geophysics 38: 26-29
Horneck G, Stöffler D, Eschweiler U, Hornemann U (2001) Bacterial spores survive simulated meteorite impact. Icarus 149: 285-290
Hughes DW, Williams IP (2000) The velocity distributions of periodic comets and stream meteoroids. Monthly Notices of the Royal Astronomical Society 315: 629-634.
Koschny D, Kargl G, Rott M (2001) Experimental studies of the cratering process in porous ice targets. Advances in Space Research 28: 1533-1537
McKay DS, Gibson EK, Thomas-Keprta KL, Vali H, Romanek CS, Clemett SJ, Chillier XDF, Maechling CR, Zare RN (1996) Search for past life on Mars: Possible relic biogenic activity in Martian meteorite ALH84001. Science 273: 924-930
Melosh HJ (1988) The rocky road to panspermia. Nature 332: 687-688
Mileikowsky C, Cucinotta FA, Wilson JW, Gladman B, Horneck G, Lindegren L, Melosh J, Rickman H, Valtonen M, Zheng JQ (2000a) Natural transfer of viable microbes in space: 1. From Mars to Earth and Earth to Mars. Icarus 145: 391-427
Mileikowsky C, Cucinotta FA, Wilson JW, Gladman B, Horneck G, Lindegren L, Melosh J, Rickman H, Valtonen M, Zheng JQ (2000b) Risks threatening viable transfer of microbes between bodies in our solar system. Planetary and Space Science 48: 1107-1115
Parsons P (1996) Dusting off panspermia. Nature 383: 221-222
Soina VS, Vorobiova EA, Zvyagintsev DG, Gilichinsky DA (1995) Preservation of cell structures in permafrost: A model for exobiology. Advances in Space Research 15: 237-242

Impact Cratering of Icy and Rocky Targets in Planetary Sciences and in the Laboratory

Jacek Leliwa-Kopystynski[1], Mark J. Burchell[2]

[1]Institute of Geophysics, University of Warsaw, ul. Pasteura 7, 02-093 Warszawa, Poland; and Space Research Center of the Polish Academy of Sciences, ul. Bartycka 18A, 00-716 Warszawa, Poland. E-mail: jkopyst@mimuw.edu.pl
[2]Centre for Astrophysics and Planetary Science, Physics Laboratory, University of Kent at Canterbury, Kent CT2 7NR, United Kingdom. E-mail: M.J.Burchell@ukc.ac.uk

Abstract. A review of impact phenomena past and present in the solar system is presented. The particular aim of this paper is to integrate impacts on icy surfaces alongside the traditional impacts on rocky surfaces. The data are given ordered in terms of the parent solar system bodies, thus are related to solar system bodies of different mass, density, gravity, orbital velocity, and surface composition. The results of some of the most interesting impact events and the impact structures recorded on the surfaces of various bodies are presented in tables. It is shown that a synthesis of impacts on ice and impacts on rocky surfaces is a natural thing to do. Crater structures and the consequences of impacts are common to both materials. There is thus no logical reason, other than past practice, to ignore one type of target material (ice) when discussing impact craters in the solar system.
In addition to structures on planetary scales, we survey laboratory experiments with icy targets. It is shown that in recent years this previously relatively neglected area of study has developed rapidly. There is now much data available, under a wide range of conditions, for hypervelocity impacts onto ice in the laboratory.
Finally, a new area of impact cratering can be identified, namely that into small solar system bodies such as asteroids and comet nuclei. In the latter case, impact cratering in porous, inhomogeneous ices still needs to be better understood.

1 Introduction

Impact and/or collision events have played and still play a crucial role in the evolution of the bodies belonging to the Solar System. This became recognised observationally from the time of the first manned landings on the Moon (1969), as well as when space probes began (in the 1970s) to collect the first detailed pictures of the surfaces of the planets and their satellites, as well as the asteroids. The role of impacts/collisions events have been strongly supported by modern modelling of the formation of the Solar System bodies. The fundamental analytical studies started from the work of Safronov (1969). A statistical and numerical approach for collision-generated sticking and growth of planetary-like bodies from a swarm of smaller grains began from the papers of Wetherill (for reviews of his work see Wetherill 1980 and 1998). The giant impact hypothesis was introduced by Hartmann and Davies (1975) and by Cameron and Ward (1976). Primarily, the hypothesis was used for modelling the origin of the Moon. Next, these authors and many others have shown that the bodies of our System were born during very long series of impact-generated coagulations and, in some cases, of abrupt impact disruptions.

Studies of impact craters are done by means of geological/geophysical methods on the Earth and by imaging of the surfaces of other bodies, as well as by laboratory or field impact and explosion experiments combined with appropriate scaling laws. Large craters fall into two categories, simple basin (or bowl) shaped craters and, for the largest craters, a more complex structure involving central peaks and extra ring structures in addition to the classic rim wall. Observed craters have been heavily modified since their creation. The crater initially formed by the impact itself is often referred to as the transient crater. This is immediately followed by a collapse or modification stage (see for example Melosh and Ivanov 1999) that causes partial in-filling of the crater (reducing its apparent depth and, for the larger craters adding the central peak etc.). This stage of crater growth is not usually modelled, nor is it reproduced in laboratory scale experiments. Finally, the resultant crater is then subject to long-term erosive effects and/or geological processes involving the planetary materials the crater is made of.

Numerical methods (hydrocodes) are frequently applied for studying impacts of arbitrary scale up to the end of the stage of the impact involving the formation of the transient crater. Ideally, they can be verified at small scales by the results of laboratory or field experiments. However, for large scale (for the scale of planetary events) numerical methods give results that stand as a hypothesis only. They could be verified by the final, present day, observed state of a body considered as being a target long time ago. For example, Cameron's hypothesis of the Moon's origin (see Cameron 1997, 2000; the paper of 1997 is illustrated by very impressive computer-generated pictures of consecutive stages of impact and accretion events) is considered verified by arguments contrasting the rocky Moon and rocky-iron Earth. These arguments show how a series of physical steps implicit in an impact event on the Earth lead to observed similarities and systematic differences (volatile depletion, systematic enrichment of more refractory materials etc.) between the resultant Moon and the Earth. The other scenarios of the Moon's origin and composition are much less probable or contain incompatibilities with the known properties of the Moon. However, such verifications as this only involve general end results of impact processes. More definable outcomes such as crater size, extent of ejecta blankets, degree of mixing of projectile remnant and target material in the bottom of a crater, etc., are not usually predicted by impact modelling and are thus not verifiable.

Scaling laws established by Holsapple and Schmidt (1980, 1982, 1987), Schmidt and Holsapple (1982) and Schmidt and Housen (1987) are intended for the extrapolation of experimentally obtained data to other scales. Such extrapolation must span many orders of magnitude and it remains rather doubtful if it is at all valid to do this. Indeed, a typical size ratio: (size of meteoroid producing a hectometre crater)/(size of impactor in the laboratory) = (10 m)/(1 mm) gives the mass ratio of the order of 10^{12}. Taking in mind that impact velocities in the laboratory are rather smaller than these in the Solar System, the ratio of kinetic energy of a 10 m sized meteoroid to kinetic energy of impactor in the laboratory could even be considerably larger than 10^{12}. It is rarely safe to assume that the underlying physics is similar over such a large change in energy scales.

What can be tested by hydrocode modelling and scaling laws is how the general appearance of impact craters are influenced by the physical processes behind crater development during an impact. In addition, experiments on laboratory scales allow controlled tests to see how crater size and shape depend upon the nature of the impacts (projectile size, density and speed, target density, etc.).

One common deficiency when discussing impact cratering is to assume all craters form in rocky targets. Whilst this is true for many Solar System bodies, there is a growing interest in impacts on icy surfaces (e.g., Moore et al. 1998 and 2001; Turtle and Pierazzo 2001, etc.). In the rest of this paper there is a discussion of impacts, which adds icy bodies and impacts on ices to the more general discussion of impact processes on rocky bodies normally encountered in the literature.

2 Classification of Impact/Collision Events

When considering a collision it is necessary to determine whether it is in fact an *impact* or a *collision* and to correctly label the impactor and target bodies. In addition there is the somewhat arbitrary definition of what constitutes a *giant impact.* All three of these terms (collision, impact, giant impact) are used somewhat loosely in the literature, it is therefore useful to have a definition of each. Let us consider two bodies with masses M (*target*) and m (*impactor*) $\leq M$. Let escape velocity from mass M be v_{esc}. Let masses M and m collide with each other with a relative velocity v. We can thus introduce a simplified, and even a little bit arbitrary, classification of the collision events.

If $m/M < 10^{-2}$ an event is classified as an *impact.* In many cases, in space as well as in laboratory, "the classic condition" $m/M = 0$ for an impact can be assumed.

The "classic result" of an impact is production of a flux of ejecta and, finally, crater formation.
If mass ratio m/M is of the range $10^{-2} - 1$ (approximately corresponding to the size ratio of the range 0.2 – 1) and if v_{esc} is small gravity forces can be neglected, then this event is called a *collision*.

Table 1. Classification of the surfaces of solar system bodies: Rocky planets and the Moon. Surfaces of the atmosphere-less bodies are covered by a regolith that is a layer of fragmentary debris produced by meteorite impacts on the surface. Source: Cole and Woolfson (2002) (either directly or derived from quantities presented therein).

Body	Mean radius R (km)	Mean density ρ ($g\ cm^{-3}$)	Mean orbital velocity ($km\ s^{-1}$)	Surface gravity g ($m\ s^{-2}$)	Escape velocity from the surface v_{esc}, ($km\ s^{-1}$)	Surface composition and atmospheric pressure on the surface (if relevant)
Mercury	2440	5.43	47.9	3.70	4.25	Rock (regolith)
Venus	6052	5.20	35.0	8.87	10.4	Rock; 10^7 Pa
Earth	6378	5.52	29.8	9.81	11.2	Surface covering: Approximately 28% rock, 70% water, 2% ice, 10^5 Pa
Moon	1738	3.34	1.0	1.62	2.37	Rock (regolith)
Mars	3393	3.93	24.1	3.69	5.03	Rock (regolith) Deep permafrost is possible; 10^3 Pa

If the ratio m/M is of the range $10^{-2} - 1$, and v_{esc} is of the order of at least 10^2 $m\ s^{-1}$ gravity becomes important and therefore the collision is called a *giant impact*. The last case corresponds to the radius of mass M being of the order of 10^2 km at least. The typical examples of the giant impacts are those which possibly led to: (i) Formation of the Earth/Moon system; (ii) Catastrophic break-up of "Proto-Miranda", a pristine Uranian moon; from its debris the present-day Uranian satellite Miranda was (possibly) formed. (iii) Unusual inclination of the rotation axis of Uranus. Numerical modeling plays an essential role for studies of the dynamic phenomena concomitant to giant impacts.

From the point of view of the larger body (the target) an impact/collision event should lead to one of the following results: mass decrease of the target (destruction), rebound of impactor (no mass change of the target), partial or full

Table 2. Classification of the surfaces of Solar System bodies: Gaseous planets and their main satellites. "Ices" mean water ice plus possibly another solidified volatiles plus possibly some minerals. Source: Various, including Burns and Matthews (1986), Bergstralh et al., (1991), Cruikshank (1996), Pater and Lissauer (2001), Cole and Woolfson (2002).

Body	Mean radius R (km)	Mean density ρ (g cm^{-3})	Mean orbital velocity (km s^{-1})	Surface gravity g, (m s^{-2})	Escape velocity from the surface v_{esc}, (km s^{-1})	Surface composition and atmospheric pressure on the surface (if relevant)
Jupiter	69800	1.34	13.1	26.0	60.3	Gaseous
Io	1815	3.55	17.3	1.80	2.57	Rock and volcanic deposits of lava, sulfur and SO_2.
Europa	1569	3.04	13.7	1.33	2.04	Ice (mostly pure water ice).
Ganymede	2631	1.93	10.9	1.42	2.73	Ices (dirty ice).
Callisto	2400	1.83	8.2	1.23	2.43	Ices (dirty ice).
Saturn	57800	0.70	9.6	11.3	36.2	Gaseous
Mimas	199	1.14	14.3	0.063	0.16	Ices
Enceladus	251	1.2	12.6	0.084	0.21	Ices
Tethys	530	1.21	11.3	0.18	0.44	Ices
Dione	560	1.43	10.0	0.22	0.50	Ices
Rhea	765	1.33	8.5	0.28	0.66	Ices
Titan	2575	1.88	5.6	1.35	2.64	Ices and/or liquid hydro-carbonates; 1.4×10^5 Pa
Iapetus	730	1.16	3.3	0.24	21.5	Ices (with leading hemisphere covered by dark deposits).
Uranus	25200	1.25	6.8	9.1	21.5	Gaseous
Miranda	236	1.15	6.7	0.76	0.19	Water ice remains the only spectrally identified constituent on these moons (Veverka et al. 1989).
Ariel	579	1.56	5.5	0.25	0.54	
Umbriel	585	1.52	4.7	0.25	0.54	
Titania	789	1.70	3.6	0.37	0.80	
Oberon	761	1.64	3.1	0.35	0.73	
Neptune	24600	1.63	5.4	11.2	23.5	Gaseous
Triton	1353	2.06	4.4	0.78	1.45	Ices, including N_2 ice.

accumulation of impactor by the target (accretion process). For an example of experimental mapping of the impact results versus impact velocity and versus impact angle see Leliwa-Kopystynski et al. (1984).

3 Classification of the Solar System Bodies

Post-impact effects recorded on a target surface depend on many factors. Roughly speaking, the most important of these are impactor energy and the physical properties of a target. The properties of an impactor are less important. Therefore, classification of the potential targets is crucial for studies of the result of an impact. Targets in the solar system are solid (rocky, icy/rocky, icy), or gaseous. Solid targets can be atmosphere-less and liquid-free bodies or they are covered by a fluid layer and therefore from the point of view of an impactor they could well be perceived as gaseous to a degree. In addition, targets built of solid materials could be well-consolidated bodies or they could be porous, rather weak structures. In Tables 1-3 a classification of potential targets belonging to the solar system is presented. It should immediately be noted that large rocky planets are not the only bodies in the Solar System that have undergone impact processes. Thus impact studies should be generalised to include the minor bodies and/or icy surfaces.

4 Collision Events in the Solar System

The best preserved post-impact features exist on atmosphere-free bodies: e.g. Mercury, the Moon, asteroids, and the satellites of the giant planets. The appropriate motto is "larger structure is rarer". For statistics of impact features on the solar system bodies see, e.g., Melosh (1989) or Neukum and Ivanov (1994), as well as the papers related to cratering of the particular bodies.

The size of an impact structure depends both on impactor mass and on its velocity. The latter is the superposition of the orbital velocity of a target and that of an impactor, enlarged due to mutual gravitational interaction between the two. Orbital velocities of a planet-target or of a satellite-target are equal to their velocities on a circum-solar or on a circum-planetary orbit, respectively. For numerical data see Tables 1 - 3. Approximately, for near-circular orbits these are:

$$v_{\text{planet}} = (2GM_{\text{Sun}}/a_{\text{planet}})^{1/2} \text{ and } v_{\text{satellite}} = (2GM_{\text{planet}}/a_{\text{satellite}})^{1/2} \qquad (1)$$

Here M is the mass of the central body (the Sun or a planet), a is the mean orbital radius (of a planet round the Sun, or of a satellite round a planet), and G =

6.672×10^{-11} m^3 kg^{-1} s^{-2} is the gravitational constant. The velocity of an impactor is equal to its orbital velocity on an appropriate circum-solar or circum-planetary orbit.

Mutual gravitational interaction between target-impactor means that the free-fall-impact-velocity is equal to the escape velocity v_{esc}. For a case when the target mass M is much larger than the projectile mass m there is:

$$v_{esc} = (2GM/R)^{1/2} \quad \text{or in another form} \quad v_{esc} = (\rho/1.789)^{1/2}R \qquad (2)$$

Here M, R, and ρ are the target mass, radius, and mean density. If ρ is in g cm^{-3} and R is in km, then v_{esc} is in m s^{-1}. The second form of Eq. (2) is very convenient for the icy satellites as well as for the giant planets. In these cases $(\rho/1.789)^{1/2} \approx 1$ and therefore escape velocity from a surface in metres per second is approximately equal to the satellite (or the giant planet) radius in kilometres (compare values in Tables 1 – 3). The velocity v_{esc} is equal to the lowest possible value of the impact velocity.

The satellites of the outer planets and the cometary nuclei are typically icy, icy/rocky or dirty icy targets. Impact craters on the various satellites are well documented. Those on cometary nuclei are not identified as yet. Indeed, craters on cometary nuclei moving on orbits with small perihelion distance can not survive for long intervals of time due to surface erosion driven by the radiation flux of energy from the Sun.

Several types of data relevant to impact craters on various bodies are gathered in Tables 4 – 6. They are related to some impact structures identified on the Earth (Table 4) and to some impact craters observed in the solar system (Table 5). Finally, a list of some impacts or potential results of impacts is presented (Table 6). Table 6 contains five examples of impacts of special importance for our knowledge about formation and evolution of the bodies belonging to the solar system.

5
Statistics of Craters and Impact Rates

The impact rate is different on different bodies of the solar system. Moreover, within the 4.6×10^9 year long history of the System, it was time dependent. During the post-accretion period, when the planetary building material was not exhausted yet, like it is now, there was a relatively short period of heavy bombardment. This probably lasted from 4.0 to 3.9 Ga and involved not just the Earth but also other bodies in the inner solar system, see Kring and Cohen (2002) for a discussion. Next, the impact rate significantly decreased with time.

As well as a steady state rate of bombardment, there exists a cometary showers hypothesis related to quasi-periodic amplifications of the impact rate on the Earth.

It is not well documented as yet. Nevertheless, Rampino and Haggerty (1994) suggested a period of 26-30 Myr (although this is doubtful, even according to the authors). The discovery was made on the basis of the Fourier analysis of 22 extinction of species events that happened within the last 515 Myr, as documented by palaeobiologists. Further, some time-correlations have been suggested between the now well documented impacts events on the Earth and the major mass extinction of species (see Table 4, after Rampino and Haggerty 1994). However, it should be noted that as yet there is only one well accepted impact event correlated with a mass extinction, namely the KT boundary event. All the other possible correlations are subject to uncertainties in either the timing of the impact event or the timing of the mass extinction.

Grieve and Shoemaker (1994) estimated that an average cratering rate on the Earth is $(5.6 \pm 2.8) \times 10^{-15}$ km^{-2} yr^{-1}, for craters with diameter ≥20 km. Assuming, that this figure is valid for the last billion years we can write it in more intuitive manner: (5.6 ± 2.8) $(10^6$ km$^2)^{-1}$ $(10^9$ yr$)^{-1}$. It means that on an area equal to France and Germany together, about 5 large impact craters were produced within the last billion of the years. However, due to the craters' degradation, their density on the surface of the Earth is considerably less. Time-dependent processes of erasing of the craters from the planetary surfaces are caused by four main processes. They are of exogenous or endogenous origin:

- Weathering, therefore atmosphere and water caused erosion (e.g.: observed on the Earth and on Mars).
- Volcanic activity (e.g., tectonic-driven on the Earth; tidally-driven on Io, the innermost galilean satellite of Jupiter).
- Tectonic effects leading to resurfacing (e.g., plate tectonic on the Earth; tidally-driven movements of the icy crust on the Jovian moon Europa).
- Impact activity leading to the overlapping of the older impact sites by the more recent ones (observed on the atmosphere-less bodies: the Moon, the icy satellites, and the asteroids).

These processes lead to gradual degradation of the impact craters' structures or to their complete disappearance. The best example of the latter is burial of a crater in a process of the lithosphere subduction, observed on the Earth. The areas of most dense distribution of the well recognised impact structures on the Earth are situated on the continental shields (for a map see Uchupi and Emery 1993). Since the shields are the oldest geological structures, they have therefore had the longest time interval to accumulate the impact craters.

The size distribution of the craters, i.e., the number N of the craters per unit of the surface of the target versus the crater diameter D (cumulative size-frequency distribution) can be approximated by a power-law function (e.g., Melosh 1989):

$$N \propto D^{-\alpha}, \quad \alpha = \alpha\,(g, D, \text{target body material}) \qquad (3)$$

It is usually presented in the form of a log-log plot. The exponent α is different for different target bodies (g-dependence and material-dependence). It depends as well on the craters' size D: it can be taken as a constant value within some ranges of D. According to data collected by Neukum and Ivanov (1994), for lunar craters

$$\alpha = -2.9 \quad \text{for} \quad 10\ \text{m} < D < 100\ \text{m},$$
$$\alpha = -(1.8 - 2.0) \quad \text{for} \quad 1\ \text{km} < D < 100\ \text{km}.$$

Table 3. Classification of the surfaces of Solar System bodies: Pluto and some examples of Kuiper Belt Objects and minor bodies. The figures in italics are rough estimates. Sources: Various, including NASA and Yoder et al. (1989) for the Saturnian satellites.

Body	Mean radius R (km)	Mean density ρ (g cm^{-3})	Mean Orbital velocity (km s^{-1})	Surface gravity G (m s^{-2})	Escape velocity from the surface, v_{esc} (km s^{-1})	Surface composition and atmospheric pressure on the surface (if relevant)
Kuiper Belt Objects (the largest known)						
Pluto	1145–1200	1.9-2.1	4.7	0.66	1.2	Ices
Charon (Pluto satellite)	600-650	1.5-1.8	0.22	0.29	0.6	Ices
"Average large"	*100*	*1–2*	*1 – 5*	*0.04*	*0.1*	Ices
Minor bodies (representative examples). Since they may differ considerably from regular spheres, gravity acceleration and escape velocity (in the following in m s^{-1}) are not constant values on their surfaces.						
2060 Chiron (centaur-class)	*100*	*1*	*8*	*0.04*	*100* m s^{-1}	Chiron is an active (comet-like) asteroid
243 Ida (asteroid)	15.7	2.6±0.5	18.3	*0.01*	*20*	Rock
433 Eros (asteroid)	8.4	2.67	25	*0.006*	*10*	Rock
Phobos (Mars satellite)	11	1.9	2.13	*0.006*	*10*	Rock; porous

For craters on the icy satellites, the typical value is α = -3 (Chapman and McKinnon, 1986; McKinnon et al., 1991). The size-frequency plot of the crater distribution can be presented in a relative form that is called R-plot. This is the ratio between actual crater distribution and a distribution (eqn. 3) with a fixed slope

Table 3. (cont.)

Body	Mean radius R (km)	Mean density ρ (g cm^{-3})	Mean Orbital velocity (km s^{-1})	Surface gravity G (m s^{-2})	Escape velocity from the surface, v_{esc} (km s^{-1})	Surface composition and atmospheric pressure on the surface (if relevant)
Minor bodies (cont.).						
Deimos (Mars satellite)	6.3	1.7 ± 0.5	1.35	*0.003*	*6*	Rock; porous
Janus (Saturn satellite)	90	0.67	15.9	*0.02*	*50*	Ices, porosity >0.3
Epimetheus (Saturn satellite)	64	0.64	15.9	*0.01*	*40*	Ices, porosity >0.3
Cometary nuclei	*1 - 10*	*0.1 – 1.0*	1 – 620 very variable along orbit	10^{-5}-10^{-3}	*0.5 – 5*	Ices; porosity 0.3 –0.8

(α = -2, see Melosh, 1989; α = -3, see McKinnon et al., 1991). Such presentation allows for better analysis of the particular populations of the craters.
As an alternative, instead of the simple formula (eqn. 3), Neukum and Ivanov (1994), derived an 11-power polynomial dependence of logN versus logD. They gave the appropriate numerical coefficients for crater distribution on the Moon and on Mars.

Present-day cumulative size-frequency distributions of craters can manifest a different density on a different area. Heterogeneity in density of craters' distribution on the surface of a particular body testifies the complex geological history of that body. For example, an existence of global or mantle-only convection within the icy satellites can explain heterogeneity of the craters' distribution on their surfaces: low density cratered terrains on some of the satellites (e.g., Enceladus, Dione, Ariel, Titania, ...) are certainly younger than the surrounding, high density cratered terrains. Therefore, these low density cratered terrain's are, possibly, located on the top parts of the connective up-streams (Czechowski and Leliwa-Kopystynski 2002a,b). In addition it should be noted that once a critical crater density has been reached, new impacts over-print the old ones. There is thus an upper limit on observed crater density on a planetary surface (Squyres et al. 1997). Apart from crater statistics, the individual features of the craters are also very interesting. One of them is the ratio of the crater diameter D to the radius R of the target body. The observed ratio D/R in some cases is even as large as about one (Table 5). The relatively largest craters on the icy satellites are these on Saturnian satellite Mimas and Neptunian satellite Protheus (Thomas 1999). For comparing crater morphology on globes with different gravity acceleration g, a scaling as g^{-1} can be used (Thomas 1999). The statistics of the craters gives information about the geologic age of cratered surface. For example, the most part of the surface of Mercury and Moon are saturated with impact craters and are old; the whole surfaces of Europa and Io, with only a small amount of craters are young. However, absolute ages cannot be obtained from saturated surfaces because of the overprinting of craters that occurs as described above. In addition, relative ages for surfaces in different parts of the solar system are difficult as the impact rate can vary substantially with location. For example, when measured, the rate impacting the inner planets will be seen to differ from that for the satellites of the outer planets.

Table 4. Selected impact structures identified on the Earth. The list is from the Earth Impact Database (University of New Brunswick) March 2003, available at http://www.unb.ca/passc/ImpactDatabase, and lists various typical craters. Present discovery rate of new terrestrial impact craters is ~3 to 5 per year (Grieve and Shoemaker 1994). Extinction data are from Rampino and Haggerty (1994).

Impact site	Diameter D (km)	Age (10^6 years)	Remarks
Vredefort, South Africa	≈300	2023±4	Oldest and largest impact structure known on Earth
Sudbury, Canada	250	1850±3	Region of extensive mineral mining
Chicxulub, Yucatan, Mexico	170	64.98±0.05	Extinction of 38.5% of genera (67% of species).
Manicouagan, Quebec, Canada	100	214±1	Extinction of 34.1% of genera (62% of species).
Popigai, Russia	100	35.7±0.2	Extinction of 11.4% of genera (25.5% of species).
Acraman, South Australia	90	≈590	Found after its ejecta blanket was recognised first
Chesapeake Bay, Atlantic shelf, close to the East USA	85	35.5±0.3	Impact happened on the sea. Extinction of some species.
Puchezh-Katunki, Russia	80	167±3	Extinction of 20.1% of genera (43% of species).
Kara, Russia	65	70.3±2.2	Twin structure of Kara and Ust-Kara.
Beaverhead, Montana, USA	60	~600	--
Mjølnir, Barents Sea	40	142.0±2.6	Impactor size 1 – 2 km. Impact happened in a 400 m deep sea.
Aorounga, Sahara, northern Chad	12.6	< 345	Impactor size 1 – 2 km.

Table 5. Impact structures in the solar system. Craters with $D/R \geq 1.3$ are the relatively largest known impact craters. Craters with smaller values D/R are some interesting examples chosen from a numerous population. Source: Thomas (1999), Head et al. (1993), Passey and Shoemaker (1982).

Target body	Target mean radius R (km)	Crater: name, location, remarks	Crater diameter D (km)	D/R	Target material	Target mean density, (g cm^{-3}).
4 Vesta (asteroid)	265	South pole crater	460	1.7	Rock	3.5 ± 0.4
Deimos, (satellite of Mars)	6.2		10	1.6	Rock	1.8 ± 0.3
Moon	1738	South Pole Aitken Basin (the largest known Lunar impact structure)	2500	1.4	Rock	3.34
Mathilde (asteroid)	26.5		33.4	1.3	Rock, porous.	1.3 ± 0.2
Amalthea (satellite of Jupiter)	83.5		88	1.1	Rock	
Proteus (satellite of Neptune)	209		210	1.0	Ice and rock.	0.7 – 2.0
Mimas (satellite of Saturn)	198	Herschel	145	0.7	Ice and Rock. Possibly porous.	1.14
Halley's comet nucleus	7		3.5	0.5	Ice and rock. Highly porous ?	0.2 – 1.0
Ganymede (satellite of Jupiter)	2631	Valhalla basin (palimpsest, a large, multiring structure).	800	0.3	Ice and rock.	1.94

Table 6. Some special impacts or the results of potential impacts in the Solar System.

Target body	Impact event	Comments
Jupiter	Several fragments of Comet Shoemaker-Levy 9 disrupted on an orbit around Jupiter. Event begun 18 July 1994 and spanned the next several days. Energy of fragments spanned within a range 10^{19} - 10^{22} J.	Series of about 20 fragments struck the Jovan atmosphere. Impact velocity was about 60 km s^{-1} The traces of the impacts were observable on Jupiter for several months.
Proto-Earth	Impactor with mass of an order of 0.1 mass of the Earth. Impact velocity around 20 km s^{-1}. Energy was of on order of 10^{32} J. The event was ca. 4.5 billion years ago	Impact happened after the Proto-Earth core formation. Much of the rocky material of the mantle and only a few of core material was dispersed round the Proto-Earth orbit. So, only minor amount of metal (iron) was incorporated into the Moon during its accumulation.
Proto-Miranda (satellite of Uranus)	Catastrophic (disruptive) impact that possibly broke down the pristine satellite	Unusual surface features of present-day Miranda suggest that it could be formed by means of gravitational gathering and bonding of separate chunks on the orbit.
Proto-Uranus	Was Uranus' high obliquity (82°05') caused by impact of a late-arriving planetesimal? Bergstralh and Miner (1991)	
Europa (satellite of Jupiter)	Only very few small diameter (D < 10 km) impact craters are observed on the surface.	Conclusion: Europan surface is geologically young. Resurfacing processes driven by internal heat production are still active.

6 Laboratory Experiments

6.1 Overview of Experiments

A list of the most important features of impact experiments in the laboratory is given in the following. The ranges of the key parameters are also given:

- Mass regime, typical: M from g to kg, m from μg to g.
- Target: flat ("infinite") or spherical; uniform or layered; nonporous or porous.
- Impactor: spherical or cylindrical (including flat coin-shaped cylinders).
- Impactor accelerating tube orientation: horizontal, vertical, or, more rarely, variable.
- Impact geometry: head-on (central) or oblique. (Studies of oblique impacts are much less typical even though the most probable angle of incidence for a planetary impact is 45°, see Pierazzo and Melosh, 2000).
- Impact velocity regime: from as low as mm s^{-1} (e.g. sticking experiments related to fluffy aggregates collisions at the very early stages of planets' formation) to as high as order of 10 km s^{-1} (hypervelocity impacts related to the early and the present-day impacts in the Solar System).
- Impact energy regime: up to 10^3 J, approximately.

As well as impact experiments, high energetic explosions (TNT or nuclear) on a soil surface can simulate much larger impacts where an explosion of 1 kg of TNT corresponds to 10^6 cal = 4.186×10^6 J.

Apart from the limitation relating to the energetic scale, the most important difference between laboratory studies and events in space is that Earth gravity can not be removed (although note that for small impacts its effect can be modified by means of experiments in centrifuges). So gravity is always acting on the flux of impact ejecta and on the crater formation. A further important complication is that, at laboratory energy scales, the energy density on the impact surface is sufficient to remove shattered material from brittle targets as ejecta. However for planetary scale impacts the energy density is such that much of the shattered material cannot escape the surface and thus crater growth even on brittle materials is via crater excavation and bulk movement of material rather than by its ejection into space.

6.2 Expected Output from Laboratory Experiments

- Crater morphology (crater dimensions and shape).
- Scaling laws for crater depth H, diameter D, and volume V versus impactor momentum mv or versus energy $mv^2/2$. This can then lead to other, e.g. non-dimensional, scaling parameters. Target material properties as well as impact

angle are the key parameters. In most of the cases the material of the impactor is not very critical. However, the response of very low density targets on impact strongly depends on projectile density (Kadono 1999).

- Criteria for catastrophic disruption of targets (Mizutani et al. 1990; Nolan et al. 2001; Arakawa et al. 2002).
- Studies of angular distributions of ejecta by number and by mass (Burchell et al. 1998b).

6.3 Particular Cases of Targets of Planetological Importance

Impact experiments on rocky targets in the form of solid blocks or loose granular material (e.g., sand) are of major importance for studies of the craters on the rocky planets. Accordingly there are well-known laboratory studies of hypervelocity impact experiments on rocks and sands (Gault 1973, Gault and Wedekind 1978 and Burchell and Whitehorn 2003 are typical examples) as well as lengthy treatments of large impacts on rocky planetary surfaces (e.g. Melosh, 1989, or Spudis, 1993). However, water ice (not rock) is the best recognised material on the surfaces of the icy satellites as well as on cometary nuclei. So, for this group of bodies separate experiments are required, performed with pure water ice to provide reference results. However, in the Solar System water ice is admixed by other frozen volatiles and by minerals. Indeed, complex organic compounds could be present as well, so further experimentation beyond simple water ice targets is also required.

Historically there was relatively little impact cratering data for hypervelocity impacts on ices. Here, for convenience, hypervelocity is taken as being impact speeds greater than 1 km s^{-1}. The earliest experiments were those of Croft (1981), who reported on impacts in ice (7 impacts) and ice saturated sand (7 impacts) at 1.3, 2.3 and 6.2 km s^{-1}. The crater depths and diameters were given as a function of energy (10 to 5000 J) for all 14 impacts. Polyethylene projectiles (density 950 kg m^{-3}) were use for the impacts on water ice, and pyrex projectiles (density 2400 kg m^{-3}) for those on water ice saturated sand. Unfortunately, the raw data itself was not given, nor were the results of any fits to the data. The paper did report that the fast ejected material from the impact was ejected not at 45°, (typical for impacts in rocks) but at 60-65° to the target surface. It can be noted that Arakawa et al. (2002) observed that the ejection angle for icy targets increased from (27° – 35°) to (48° – 58°) when the impact velocity increased from 160 m s^{-1} to 660 m s^{-1}. For fixed impact speeds the ejecta angle was larger when porosity was smaller. General crater ejecta scaling laws are discussed by Housen et al. (1983).

For many years this represented the sum total of hypervelocity impact data. The next published data were those of Frisch (1992). Glass beads (18 to 124 μm in diameter) were fired at water ice targets at speeds of between 1.80 and 9.62 km s^{-1}. The excavated crater mass was found, not by direct measurement, but by measuring crater depth and diameter and assuming a geometric shape for the crater. This was then fitted and given as a function of impact kinetic energy. See Table 7 for details. It was also reported that the fast ejected material was highly concentrated perpendicular to the target surface.

At approximately the same time Eichhorn and Grün (1993) reported impacts of dust particles onto water ice. They used micrometre-sized iron particles (with masses 10^{-17} to 10^{-13} kg) at impact speeds of 1 to 50 km s^{-1}. They did not measure any craters directly, but obtained the excavated mass by inference from the pressure change in the target vacuum chamber at the moment of impact. They assumed this was due to production of water vapour at the impact crater. They combined their data for excavated crater volume at low energies (10^{-7} J), with that of Frisch (1993) (10^{-2} J) and data from Lange and Ahrens (1987) at higher energy (10^{2} J). By doing this they obtained a relation between crater volume and impact energy that ranged over 9 orders of magnitude in energy (10^{-7} to 10^{2} J, see Table 7). It should be noted however that the data of Lange and Ahrens (1987) were not from hypervelocity impacts, but from lower speed impact events.

The next published data were that of Burchell et al. (1998a). They reported on 9 impacts of 1 mm diameter metal spheres (aluminium, titanium and stainless steel) onto water ice and 11 onto CO_2 ice. All impacts were at approximately 5 km s^{-1}. Thus for the first time hypervelocity impacts were reported onto a non-water ice. Also, inside a single experiment, data were available for the effect of impacts by a wide range of projectile densities. Using their data, Burchell *et al.* (1998a) found that crater size was smaller in CO_2 ice than in water ice (see Table 7 for details). They also found that there was at best only a weak dependence of crater shape (depth/diameter) on the density of the projectile. This contradicts previous beliefs (e.g., Croft 1981) which used low velocity data with lead bullets combined with higher speed data with lower density projectiles, and which found a stronger dependence on projectile density.

A more complete discussion of impact cratering on CO_2 ice, including studies of the ejecta, was given in Burchell et al. (1998b). There, data were given for 14 impacts at 5 km s^{-1} of stainless steel spheres (diameter ranging from 0.4 to 2.0 mm) onto CO_2 blocks. Their results for crater size and shape are given in Table 7. They reported that craters in water ice were 2.2 times larger in diameter and 8.9 times greater in volume than those in CO_2 ice. They also found that the ejecta was preferentially emitted at more than 60° from the target surface and gave size distributions of the ejecta as a function of angle of elevation.

More data is now available for hypervelocity impact craters in water ice. Burchell et al. (2001) give data for 23 impacts into water ice. The data cover a range of projectile densities (aluminium to stainless steel), projectile sizes (0.4 to 2.0 mm diameter spheres) and impact speeds 1 to 7 km s^{-1}. The data were used to explore the dependence of the crater size and shape on the projectile properties (density, size, speed). In addition, they also repeated the exercise of Eichhorn and Grün (1993) and fitted crater volume vs. impact energy, using Eichhorn and Grün's 1993 data, Frisch's 1992 data, their own data and that of Lange and Ahrens (1987) (which was at lower impact speed). This again covers some nine orders of

Table 7. Comparison of hypervelocity impact craters produced in icy targets in the laboratory. (Impact Energy *E* is in J).

Target	Impactor velocity *v*, energy *E*, type[1] and size *d* ranges	Diameter D (cm)	Depth H (cm)	H/D (v in km s^{-1})	Volume V (cm^3)
H_2O ice (Frisch 1992)	$v = 1.8 - 9.6$ km s^{-1} $E = 10^{-4} - 10^{-1}$ J. Glass, d = 18 – 124 µm	-	-	-	$V = 0.01\,E^{0.84}$
H_2O ice (Eichhorn and Grün 1993)	$v = 1 - 50$ km s^{-1} $E = 10^{-7} - 10^{2}$ J. Iron, d = 1 µm	-	-	-	$V = 0.061\,E^{0.98}$
H_2O ice (Burchell et al. 1998a)	$v = 5.2 \pm 0.2$ km s^{-1} E = 15 – 60 J. Al, Ti, St. st. d = 1 mm	$D = 1.88\,E^{0.39}$	$H = 0.56\,E^{0.25}$	0.16 ± 0.05	$V = 0.14\,E^{1.34}$
H_2O ice (Burchell et al. 1998b)	$v = 5.2 \pm 0.2$ km s^{-1} E = 60 J. St. st., d = 1 mm	-	-	0.15 ± 0.01	-
H_2O ice (Burchell et al. 2001)	$v = 1 - 7$ km s^{-1}. E = 1 - 100 J Various, d = 0.4 – 2 mm	-	-	-	$V = 0.087\,E^{1.06}$
H_2O ice (Shrine et al. 2002)	$v = 1.9 - 7.3$ km s^{-1} E = 1 – 40 J. Al., d = 1 mm	$D = (19 \pm 3) \times E^{0.36 \pm 0.04}$	$H = (4.4 \pm 0.5) \times E^{0.24 \pm 0.03}$	$(0.24 \pm 0.02) - (1.5 \pm 0.4) \times 10^{-2} v$	$V = (0.16 \pm 0.09) \times E^{1.2 \pm 0.2}$
H_2O-silicate ice mixture (Koschny and Grün 2001a)	$v = 1 - 12$ km s^{-1} $E = 10^{-4} - 70$ J. Nylon, d = 0.02 - 4 mm	-	-	0.28 ± 0.20	$V = 6.69 \times (0.0149)^{C} E^{1.23}$ C = fractional content of silicate by mass

Al = Aluminium, Ti = Titanium, St. St. = Stainless Steel

Table 7. (cont.)

Target	Impactor velocity *v*, energy *E*, type[1] and size *d* ranges	Diameter D (cm)	Depth H (cm)	H/D (v in km s^{-1})	Volume V (cm^3)
CO_2 ice (Burchell et al. 1998a)	v = 4.9 ± 0.3 km s^{-1} E = 15 - 60 J. Al, Ti and St. st., d = 1 mm	D = 1.79 $E^{0.20}$	H = 0.033 $E^{0.82}$	0.18 ± 0.06	V = 0.022 $E^{1.17}$
CO_2 ice (Burchell et al. 1998b)	v = 4.6 – 5.2 km s^{-1} E = 2.7 - 380 J. St. st., d = 0.4 - 2 mm	D = 0.95 $E^{0.38}$	H = 0.178 $E^{0.40}$	0.20 ± 0.05	V = 0.035 $E^{1.13}$
NH_3-H_2O ice (50% ammonia by weight) (Grey et al. 2001b)	v = 4.8- 4.9 km s^{-1} E = 96 J. St. st., d = 1 mm Target temperature 155 K	6.1 ± 1.0	0.96 ± 0.15	0.16 ± 0.04	V = 15.9 ± 2.5

Al = Aluminium, Ti = Titanium, St. St. = Stainless Steel

magnitude in energy, but now at the higher energies there is data in the hypervelocity regime as well as low velocity data. The result is given in Table 7. More detailed studies of the dependence of cratering in ice on impact speed are given in Shrine et al. (2000 and 2002).

Given that most impacts in space are at non-normal incidence (statistically 45° is the most probable angle for an impact, e.g. see Pierazzo and Melosh, 2000, for a derivation) it would seem important to study the effects of oblique impacts in the laboratory. Experiments have recently been reported concerning the evolution of crater shape in ice with the angle of impact. Grey et al. (2002) give details of the evolution of crater shape for impacts on water ice of 1 mm diameter aluminium spheres at 5.2 km s^{-1}. Only at extreme angles of incidence (greater than 70° from the normal) does the crater start to acquire a noticeably non-circular appearance when seen from above. However, crater volume decreases immediately upon non-normal incidence of the impact, decreasing as $(\cos\theta)^{1.4\pm0.3}$ (where θ is measured from the normal to the target surface).

Data are also now available for the first time concerning the dependence of the crater in an impact on the temperature of the ice target (Grey et al., 2001a, 2003). Impacts of 1 mm diameter stainless steel spheres at 5 km s^{-1} onto water ice were reported. Crater depth and volume decreased sharply as the ice temperature was lowered from 255 K (the usual temperature in all previous work) to 100 K.

Data for hypervelocity impacts on other ices, e.g., ammonia-water ice targets (Grey et al. 2001b), are also becoming available, but not yet in quantities to permit development of scaling laws. What is reported however is that crater volume in ice targets which are 50:50 mixtures of ammonia and water have decreased by a factor of 2 in volume compared to those in pure water ice targets.

For other targets, e.g. for rocky or regolith-type targets, the scaling equations are usually written in a similar mathematical form to that in Table 7. Direct approach of any of the equations in Table 7, to planetary scale must be taken with caution: only the result of an impact i.e. the crater is known. However, nothing is known about the particular interplanetary impactor (i.e. its mass and velocity and therefore its energy remains unknown). Furthermore, the problems of extrapolating over so many orders of magnitude in energy remain unsolved. Also requiring attention is the issue of how at laboratory scales crater growth is influenced by surface spallation, whilst at planetary scales it is bulk movement of material which forms the crater. Finally, the collapse of the transient crater in the modification stage of craters in planetary impacts has to be allowed for.

All of the preceding relate to impacts on solid pure ices. Impacts on rocky-icy mixtures (granular, porous, layered and slightly sintered due to irradiation by flux of light) are probably more meaningful for simulation of impacts on comet nuclei for example. Some experimental laboratory data exist. Experiments were performed relating to studies of impacts onto highly porous samples simulating the outer layers of cometary nuclei (Arakawa et al. 2000) or cratering records in regoliths such as Mars (e.g., Croft et al. 1979, and Lange and Ahrens 1982). However, an extrapolation from laboratory scale to, for example, cometary scale is even more difficult than that for the impacting of the icy or rocky uniform targets, since the relation in sizes of the laboratory layered structures to the structures on the nuclei is an additional parameter. Also, all these experiments were carried out at low velocity. The only data in the hypervelocity regime is that of Koschny et al. (2001). They report on impacts at up to 4 km s^{-1} of mm sized nylon projectiles into porous ice (with no silicate content).

Recently, more data has appeared for impacts in ice-silicate mixtures. Koschny and Grün (2001a) give crater size for normal incidence impacts of projectiles of mass 10^{-11} to 10^{-9} kg at speeds of 1 to 12 km s^{-1} into water ice targets with 5-20% silicate content by mass. The ice targets were at 250 K. They found that crater yield (excavated mass) falls as the silicate content increases. They also looked at the ejecta (Koschny and Grün, 2001b), finding that the ejecta was emitted at a peak angle of elevation of 70° to the target surface. They found that the fastest ejecta was also emitted at this angle was travelling faster, and was ejected at up to ¼ of the impact speed.

A sufficiently energetic impact onto a solar system body leads to disruption of that body. Criteria for catastrophic break up can be established on the basis of laboratory experiments (providing that an extrapolation to planetary scale holds). Some representative experimental papers are these of Fujiwara et al. (1989), Mizutani et al. (1990) and Arakawa (1999). A particularly useful discussion of the role of mechanical strength and target gravity as regards impact generated disruption is presented by Holsapple (1994).

Hereafter an example of extrapolation from laboratory to planetary scale is presented for icy bodies (Leliwa-Kopystynski 2002). It is based on the experiments of Arakawa et al. (2002). They studied impact disruption of pressure and thermally sintered fine-grained icy-silicate targets. The porosity of the targets was ϕ and the mass ratio of silicate to total was C. The equations below are for accretion velocity $v_{i\text{-acc}}$ (for impact velocity $v < v_{i\text{-acc}}$ the impactor is absorbed by the target) and for disruption velocity $v_{i\text{-dis}}$ (for $v > v_{i\text{-dis}}$ the target is broken up; however, it is not dispersed):

$$v_{i\text{-acc}} = 3.09 \times 10^{-4} \times [1\text{-}0.6867C]^{-1/2} \times [(1\text{-}\varphi)^{1/2}\, 10^{1.65\varphi}] \times R, \text{ (m s}^{-1}\text{)} \tag{4}$$

$$v_{i\text{-dis}} = (41.4 + 349.2C)^{1/2} \times (1\text{-}\varphi)^{-0.8+3.9C} \times (\rho/\rho_{imp})^{1/2} \times (R/r)^{3/2}, \text{ (m s}^{-1}\text{)} \tag{5}$$

Here R is the target radius, r is the impactor radius, ρ is the target density, and ρ_{imp} is the impactor density. The target density ρ is related to C, ϕ, and to the densities of the various components as follows:

$$(1 - \varphi)/\rho = (1 - C)/\rho_{ice} + C/\rho_{rock} \tag{6}$$

The numerical coefficients in equations (4) and (5) result from the adopted values of the densities of the icy and silicate components ρ_{ice} = 940 kg m^{-3}, and ρ_{rock} = 3000 kg m^{-3}. Equations (4) and (5) are based on experiments performed for

$C = 0$, $C = 0.5$, and for $0 < \phi < 0.55$. Extrapolation holds up to $C = 1$ (formally; however, physically it is rather doubtful to use these equations for ice-less bodies) and up to $\phi = 0.8$. The important role of the target porosity and composition can be seen best by considering the term $(1 - \phi)^{-0.8 + 3.9C}$ in equation (5). For a low silicate content ($C < 0.205$) the exponent is negative. It means that disruption of the higher porous targets needs a larger impact velocity than disruption of targets with lower porosity. For a concentration of silicates $C > 0.205$ the more porous targets are easier to disrupt than the lower porosity ones. Impact velocities with the limits fixed by equations (4) and (5) can be presented in 2D form on the different planes: v_i versus C, or ϕ, or R, or r. Three, or four at most, areas can be distinguish on each of these planes:

1. The area below the curve $v_{i\text{-acc}}$ as well as below the curve $v_{i\text{-dis}}$. This is a cratering and regolith formation area, without escape of the fragments (an accretion area).
2. The area above the curve $v_{i\text{-acc}}$ and below the curve $v_{i\text{-dis}}$. This is the cratering and regolith formation area with escape of some ejecta.
3. The area above the curve $v_{i\text{-dis}}$ and below the curve $v_{i\text{-acc}}$. This is the rubble pile formation area.
4. The area above both the curves $v_{i\text{-acc}}$ and $v_{i\text{-dis}}$ that is the catastrophic mass loss area.

In addition to laboratory experiments on ices, hydrocode simulations of impacts on ices are also now appearing. For example, Turtle and Pierazzo (2001) investigated impact crater shapes on Europa. Using their simulations and the reported observation of peak ring structures in some large Europan craters, they showed that the ice surface on Europa must be at least 3 to 4 km thick.

7 Conclusions

The study of impact cratering on ices has its origins in the Voyager missions to the outer planets which produced the first close up images of the natural satellites of these planets, several of which were ice covered and showed signs of cratering. More recently, the subject has gained new impetus with the Galileo Extended Mission to Jupiter. Galileo's long tour of the Jovian system has produced a wealth of images of the surfaces of the icy satellites. The Cassini spacecraft's mission to Saturn (arrival 2004) will similarly revolutionise our knowledge of the icy satellites of that planet (along with so much else pertaining to Saturn and the Saturnian system). It therefore seems artificial to continue describing impacts as things that occur only on rocky surfaces. Accordingly, in this paper it has been shown how impacts on icy surfaces in the Solar System can be integrated into general discussions of impact phenomena. Examples of how impacts on icy bodies can serve as tools to explore the history of those bodies have been given alongside better known examples relating to rocky planets. The result is that it can be seen that it is quite natural to treat impacts as a Solar System wide phenomenon.

In future, as well as impacts on planetary surfaces and the surfaces of natural satellites, it will also be necessary to understand impacts on cometary nuclei. These bodies pose a novel problem for impact studies as they are small, porous bodies, whose composition may well be inhomogenous on size-scales comparable to the craters. Several missions have now been launched to visit comets. Stardust (launched 1999) will fly past a comet in 2004 and return captured cometary material to Earth in 2006. Contour (launched 2002) will make a tour past several comets. Future launches will include Rosetta (2004), which will orbit a comet and launch a lander onto the nucleus. Finally, Deep Impact (launch 2004) will not only fly past a comet, but will deliberately launch a 347 kg copper projectile at the comet at ≈ 10 km s^{-1}. It will thus deliberately create its own impact crater. The results of current laboratory experiments concerning impacts into such targets are summarised here, and the problem is shown to be a complex one.

The developing field of laboratory studies of hypervelocity impacts on ices has also been summarised. There is a rapidly growing body of data from such experiments, what is required now is to see how the insights developed by such experimentation can be applied to impacts on icy solar system bodies.

Acknowledgements

This work was partially supported by grant No. 5-PO3D-028-20 from the Polish Academy for Scientific Research.

References

Arakawa M (1999) Collisional disruption of ice by high-velocity impact. Icarus 142: 34 - 45

Arakawa M, Higa M, Leliwa-Kopystynski J, Maeno N (2000) Impact cratering of granular mixture targets made of H_2O ice – CO_2 ice – pyrophylite. Planetary and Space Science 48: 1437 - 1446

Arakawa M, Leliwa-Kopystynski J, Maeno M (2002) Impact experiments on porous icy-silicate cylindrical blocks and the implication for disruption and accumulation of small icy bodies. Icarus 158: 516 - 531

Bergstralh JT, Miner ED (1991) The Uranian System: an overview. In: Bergstralh JT, Miner ED, Matthews MS (eds) Uranus. The University of Arizona Press, Tucson, pp 3 - 25

Bergstralh JT, Miner ED, Matthews MS (eds) (1991) Uranus. The University of Arizona Press, Tucson, 1076 pp

Burchell MJ, Whitehorn L (2003) Oblique incidence hypervelocity impacts on rock. Monthly Notices of the Royal Astronomical Society 341: 192 - 198

Burchell MJ, Leliwa-Kopystynski J, Vaughan B, Zarnecki J (1998a) Comparison of H_2O and CO_2 ices under hypervelocity impact. In: Schmidt SC, Dandekar DP, Forbes JW (eds) CP429 Shock Compression of Condensed Matter – 1997. The American Institute of Physics, pp 949 - 952

Burchell MJ, Brooke-Thomas W, Leliwa-Kopystynski J, Zarnecki JC (1998b) Hypervelocity impact experiments on solid CO_2 targets. Icarus 131: 210 - 222

Burchell MJ, Grey IDS, Shrine NRG (2001) Laboratory investigations of hypervelocity impact cratering in ice. Advances in Space Research 28: 1521 - 1526

Burns JB, Matthews MS (eds) (1986) Satellites. The University of Arizona Press, Tucson, 1021 pp

Cameron AGW, Ward WR (1976) The origin of the Moon [abs.]. Lunar and Planetary Science 7: 120 - 122

Cameron AGW (1997) The origin of the Moon and a single impact hypothesis. Icarus 126: 126 - 137

Cameron AGW (2000) Higher-resolution simulations of the giant impact. In: Canup RM, Righter K (eds) Origin of the Earth and Moon. The University of Arizona Press, Tucson, pp 133 - 144

Chapman CR, McKinnon WB (1986) Cratering of planetary satellites. In: Burns JB, Matthews MS (eds) Satellites. The University of Arizona Press, Tucson, pp 492 - 580

Cole GHA, Woolfson MM (2002) Planetary science: The science of planets around stars. Institute of Physics, London, 508 pp

Croft SK, Kieffer SW, Ahrens TJ (1979) Low velocity impact craters in ice and ice saturate sand with implications for Martian crater count ages. Journal of Geophysical Research B14: 8023 - 8032

Croft SK (1981) Hypervelocity impact craters in icy media [abs.]. Lunar and Planetary Science 12: 190 - 191

Cruikshank DP (ed) (1996) Neptune and Triton. The University of Arizona Press, Tucson, 1249 pp

Czechowski L, Leliwa-Kopystynski J (2002a) Solid state convection in the icy satellites: discussion of its possibility. Advances in Space Research 29: 751 - 756

Czechowski L, Leliwa-Kopystynski J (2002b) Solid state convection in the icy satellites: numerical results. Advances in Space Research 29: 757 - 762

Eichhorn K, Grün E (1993) High-velocity impacts of dust particles in low-temperature water ice. Planetary and Space Science 41: 429 - 433

Frisch W (1992) Hypervelocity impact experiments with water ice targets. In: McDonnell JAM (ed) Hypervelocity Impacts in Space. University of Kent, Kent UK, pp 7 – 14

Fujiwara A, Cerroni P, Davis D, Ryan E, Di Martino M, Holsapple K, Housen K (1989) Experiments and scaling laws for catastrophic collisions. In: Binzel RP, Gehrels T, Matthews MS (eds) Asteroids II. The University of Arizona Press, Tucson, pp 240 - 265

Gault DE (1973) Displaced mass, depth, diameter, and effects of oblique trajectories for impact craters formed in dense crystalline rocks. The Moon 6: 32 - 44

Gault DE, Wedekind JA (1978) Experimental studies of oblique impact. Proceedings of the 9th Lunar and Planetary Science Conference, pp 3843 – 3875

Grey IDS, Burchell MJ (2003) Hypervelocity impact cratering on water ice targets at temperatures ranging from 100 K to 253 K. Journal of Geophysical Research 108: 5019, doi:10.1029/2002JE001899

Grey IDS, Burchell MJ, Shrine NRG (2001a) Laboratory investigations of the temperature dependence of hypervelocity impact cratering in ice. Advances in Space Research 28: 1527 – 1532

Grey IDS, Burchell MJ, Johnson E (2001b) Hypervelocity impact craters in ammonia-water ice [abs.]. Lunar and Planetary Science 32: 1200 (CD-ROM)

Grey IDS, Burchell MJ, Shrine NRG (2002) Scaling of hypervelocity impact craters in ice with impact angle. Journal of Geophysical Research 107: 5076, doi:10.1029/2001JE001525

Grieve RAF, Shoemaker EM (1994) The record of past impacts on the Earth. In: Gehrels T (ed) Hazard due to comets and asteroids. The University of Arizona Press, Tucson, pp 417 - 462

Hartmann WK, Davies DR (1975) Satellite-sized planetesimals and lunar origin. Icarus: 24: 504 - 514

Head JW, Murchie S, Mustard JF, Pieters CM, Neukum G, McEwen A, Greeley R, Nagel E, Belton MJS (1993) Lunar impact basins: new data for the western limb and far side (Orientale and south pole Aitken basins) from the first Galileo flyby. Journal of Geophysical Research 98: 17149 - 17182

Holsapple KA (1994) Catastrophic disruptions and cratering of solar system bodies: a review and new results. Planetary and Space Science 42: 1067 - 1078

Holsapple KA, Schmidt RM (1980) On the scaling of crater dimensions. 1. Explosive processes. Journal of Geophysical Research 85: 7247 - 7255

Holsapple KA, Schmidt RM (1982) On the scaling of crater dimensions. 2. Impact processes. Journal of Geophysical Research 87: 1849 - 1870

Holsapple KA, Schmidt RM (1987) Point source solutions and coupling parameters in cratering mechanics. Journal of Geophysical Research: 92: 6350 - 6376

Housen KR, Schmidt RM, Holsapple KA (1983) Crater ejecta scaling laws: fundamental forms based on dimensional analysis. Journal of Geophysical Research 88: 2485 - 2499

Kadono T (1999) Hypervelocity impact into low density material and cometary outburst. Planetary and Space Science 47: 305 - 318

Koschny D, Grün E (2001a) Impacts into ice-silicate mixtures: Crater morphologies, volumes, depth-diameter ratios and yield. Icarus 154: 391 - 401

Koschny D, Grün E (2001b) Impacts into ice-silicate mixtures: Ejecta mass and size distributions. Icarus 154: 402 - 411

Koschny D, Kargl G, Rott M (2001) Experimental studies of the cratering process in porous ice targets. Advances in Space Research 28: 1533 – 1537

Kring DA, Cohen BA (2002) Cataclysmic bombardment throughout the inner solar system 3.9 – 4.0 Ga. Journal of Geophysical Research 107: doi:10.1029/2001JE001529

Lange MA, Ahrens TJ (1982) Impact cratering in ice- and ice-silicate targets: An experimental assessment [abs.]. Lunar and Planetary Science 13: 415 - 416

Lange MA, Ahrens TJ (1987) Impact experiments in low-temperature ice. Icarus 69: 506 - 518

Leliwa-Kopystynski J (2002) Impact break-up of the cometary nuclei – conclusions from the impact experiments. Earth, Moon, and Planets 90: 283 - 291

Leliwa-Kopystynski J, Taniguchi T, Kondo K, Sawaoka A (1984) Sticking in moderate velocity oblique impact: application to planetology. Icarus 57: 280 - 293

McKinnon WB, Chapman CR, Housen KR (1991) Cratering of the Uranian satellites. In: Bergstralh JT, Miner ED, Matthews MS (eds) Uranus. The University of Arizona Press, Tucson, pp 929 - 692

Melosh HJ (1989) Impact Cratering. Oxford University Press. Oxford, 245 pp

Melosh HJ, Ivanov BA (1999) Impact Crater Collapse. Annual Review of Earth and Planetary Sciences 27: 385 - 415

Mizutani H, Takagi Y, Kawakami S (1990) New scaling law on impact fragmentation. Icarus 87: 307 - 329

Moore JM, Asphaug E, Sullivan RJ, Klemaszewski JE, Bender KC, Greeley R, Geissler PE, McEwan AS, Turtle EP, Phillips CB, Tufts BR, Head JW, Pappalardo RT, Jones KB, Chapman CR, Belton MJS, Kirk RL, Morrison D (1998) Large impact features on Europa. Icarus 135: 127 - 145

Moore, JM, Asphaug E, Belton MJS, Bierhaus B, Breneman HH, Brooks SM, Chapman CR, Chuang FC, Collins GC, Giese B, Greeley R, Head JW, Kadel S, Klassen KP, Klemaszewski JE, Magee KP, Moreau J, Morrison D, Neukum G, Pappalardo RT, Phillips CB, Schenk PM, Senske DA, Sullivan RJ, Turtle EP, Williams KK (2001) Impact features on Europa. Icarus 151: 93 - 111

Neukum G, Ivanov BA (1994) Crater size distributions and impact probabilities on Earth from lunar, terrestrial planet, and asteroid cratering data. In: Gehrels T (ed) Hazards Due to Comets and Asteroids. The University of Arizona Press, Tucson, pp 359 - 416

Nolan MC, Asphaug E, Greenberg R, Melosh HJ (2001) Impact on asteroids: Fragmentation, regolith transport, and disruption. Icarus 153: 1 - 15

Pater I, Lissauer JJ (2001) Planetary Surfaces. Cambridge University Press, Cambridge, 544 pp

Passey QR, Shoemaker E (1982) Craters and Basins on Ganymede and Callisto; Morphological indicators of crustal evolution. In: Morrison D (ed) Satellites of Jupiter, University of Arizona, Tuscon, pp 379 – 434

Pierazzo E, Melosh HJ (2000) Understanding oblique impacts from experiments, observations, and modeling. Annual Review of Earth and Planetary Sciences 28: 141 - 167

Rampino MR, Haggerty BM (1994) Extraterrestrial impacts and mass extinctions of life. In: Gehrels T (ed) Hazards Due to Comets and Asteroids. The University of Arizona Press, Tucson, pp 827 - 857

Safronov VS (1969) Evolution of the Protoplanetary Cloud and Formation of the Earth and Planets (in Russian), Nauka, Moscow. Also NASA TTF-677 (1972)

Schmidt RM, Holsapple KA (1982) Estimates of crater size for large-body impact: gravity-scaling results. Geological Society of America. Special Paper 190: 93 - 101

Schmidt RM, Housen KR (1987) Some recent advances in the scaling of impact and explosion cratering. International Journal of Impact Engineering 5: 543 - 560

Shrine NRG, Burchell MJ, Gray IDS (2000) Velocity scaling of impact craters in water ice with relevance to cratering on icy planetary surfaces [abs.]. Lunar and Planetary Science 31: 1696 (CD-ROM)

Shrine NRG, Burchell MJ, Grey IDS (2002) Velocity scaling of impact craters in water ice over the range 1 to 7.3 km s^{-1}. Icarus 155: 475 - 485

Spudis PD (1993) The Geology of Multi-Ring Impact Basins. Cambridge University Press, United Kingdom, 263 pp

Squyres SW, Howell C, Liu MC, Lissauer JJ (1997) Investigation of crater "saturation" using spatial statistics. Icarus 125: 67 - 85
Thomas PC (1999) Large craters on small objects: occurrence, morphology and effects. Icarus 142: 89 - 96
Turtle EP, Pierazzo E (2001) Thickness of a Europan ice shell from impact crater simulations. Science 294: 1326 - 1328
Uchupi E, Emery KO (1993) Morphology of the rocky members of the solar system. Springer-Verlag Berlin Heidelberg, 394 pp
Veverka J, Brown RH, Bell JF (1989) Uranus satellites: surface properties. In: Bergstralh JT, Miner ED, Matthews MS (eds) Uranus. The University of Arizona Press, Tucson, pp 528 - 560
Wetherill GW (1980) Formation of the terrestrial planets. Annual Review of Astronomy and Astrophysics 18: 77 - 113
Wetherill GW (1998) Contemplation of things past. Annual Review of Earth and Planetary Sciences 26: 1 - 21
Yoder CF, Synnot SP, Salo H (1989). Orbits and masses of Saturn's co-orbiting satellites, Janus and Epimethus. Astronomical Journal 98: 1875 - 1889

Paleomagnetism and $^{40}Ar/^{39}Ar$ Age Determinations of Impactites from the Ilyinets Structure, Ukraine

Lauri J. Pesonen[1*], Dieter Mader[2], Eugene P. Gurov[3], Christian Koeberl[2*], Kari A. Kinnunen[4], Fabio Donadini[1], and Robert Handler[5]

[1]Division of Geophysics, University of Helsinki, P.O. Box 64, FIN-00014 Helsinki, Finland
[2]Department of Geological Sciences, University of Vienna, Althanstrasse 14, A-1090 Vienna, Austria
[3]Institute of Geological Sciences, Ukrainian Academy of Sciences, Kiev, Ukraine
[4]Geological Survey of Finland, P.O. Box 96, FIN-02151 Espoo, Finland
[5]Institute of Geology and Paleontology, University of Salzburg, Hellbrunnerstrasse 34, A-5020, Salzburg, Austria

*corresponding authors: Lauri.Pesonen@helsinki.fi; christian.koeberl@univie.ac.at

Abstract. Oriented hand samples were collected at the Ilyinets impact structure, western Ukraine, for paleomagnetic and petrographic studies. The samples consist of suevites, melt-breccias, impact melt rocks, autochthonous granite breccias, fractured granites, and unfractured gneisses. Three melt-breccias and one impact glass sample were used for laser $^{40}Ar/^{39}Ar$ dating. The characteristic remanent magnetization (ChRM) of the impact rocks has been acquired during the post-shock cooling, presumably in the presence of hydrothermal fluids. The alternating field demagnetization data indicate that this "impact" ChRM is of dual polarity and distinct from the ca. 1830-Ma-old pre-impact "target" remanence in the unfractured basement rocks. Some fractured autochthonous granites give evidence for impact overprinting, suggesting a fully positive paleomagnetic impact test. Paleomagnetic results suggest that no large-scale post-impact structural tilting of the melt sheets have taken place. The data allow a minor ~5° tilting towards the north. The combined paleomagnetic and $^{40}Ar/^{39}Ar$ dating results suggest an age of ca. 445 ± 10 Ma for the impact. The petrographic observations show that some secondary hydrothermal alteration took place in the rocks after the impact, which may have been responsible for some of the slightly younger Ar-Ar ages. The results show that the Ukrainian Shield was in contact with Baltica 445 Ma ago.

1 Introduction

The terrestrial impact crater database (at June 2003) contains about 170 structures with ages from 2.3 Ga to present (Grieve and Pesonen 1996; Abels et al. 2002). Only about 20 % of the structures are accurately dated. This is mainly because impactites are difficult to date with high-resolution isotope techniques (e.g., U-Pb SHRIMP), or because there are no impact materials available for dating due to erosion or to sedimentary cover. Accurate ages for impact events are needed for several reasons. The age data are important in linking impact events to mass extinctions (e.g., Rampino 1999), in searching for periodicity in the impact record (Alvarez and Muller 1984; Grieve and Pesonen 1996), for estimating the amount of erosion of impact structures and to confirm the existence of twin craters (e.g. Ries-Steinheim, Clearwater West-Clearwater East, Suvasvesi North-Suvasvesi South; Bottke and Melosh 1996; Lehtinen et al. 2002). The Ukrainian Shield has (so far) seven confirmed impact structures (Fig. 1, inset) but, unfortunately, most of them are poorly dated (Gurov et al. 1998; Grieve et al. 1995). The age of the Ilyinets structure in the western part of Ukraine is assumed to be 395-500 Ma, based on previous biostratigraphic and K-Ar age determinations (Gurov et al. 1998, and references therein), but no reliable isotope or paleomagnetic age data have been available so far. The purpose of this paper is to provide new age data for the Ilyinets impact event, using integrated results from paleomagnetic and $^{40}Ar/^{39}Ar$ dating. We also apply the paleomagnetic "impact test" of Pesonen et al. (2001) to demonstrate that the characteristic remanent magnetization (ChRM) in the impactites is related to the impact event or to the impact generated processes, which follow the shock event. Using the paleomagnetic tilt test we estimate the amount of northerly tilt that the melt sheet experienced in post-impact tectonic processes (Gurov et al. 1998). We also present new petrographic and petrophysical descriptions of the impactite samples, which, together with paleomagnetic and ^{40}Ar-^{39}Ar data, show that the Ilyinets impact structure was subjected to hydrothermal fluid activity causing secondary alterations in the rocks. We argue that a few of the somewhat younger ^{40}Ar-^{39}Ar ages are results of these secondary alterations.

2 The Ilyinets Structure

2.1 Location and Impact Origin

The Ilyinets impact structure is located in the southwestern part of the Ukrainian Shield (centered at 49° 07′N; 29° 06′E; Fig. 1), ca. 45 km southeast of the town of Vinnitsa, within the basin bounded by the Sobik and Sob Rivers (Gurov et al.

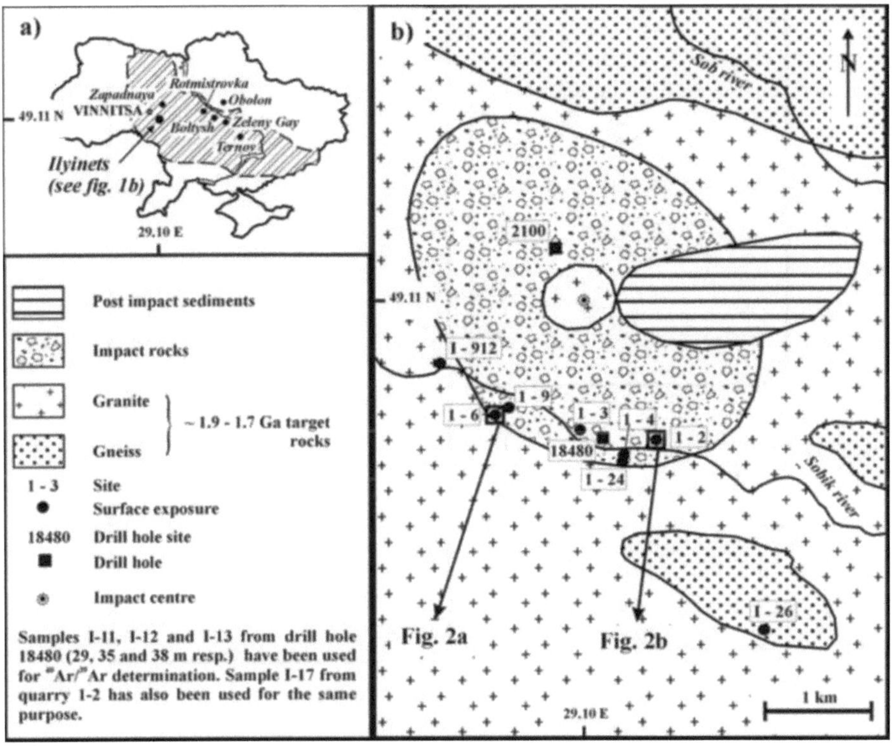

Fig. 1. Schematic outline of the geology of the Ilyinets structure, with the inset showing the locations of the seven recognized impact structures in Ukraine (modified after Gurov et al. 1998). The NW-SE trending ellipsoid, which contains the various impactite rocks, traces the present erosional level of the structure. The central uplift is marked within the main structure. The horizontally hatched ellipsoid shows the area where post-impact sediments have been mapped. The numbers (I-2, I-3, I-4, I-6, I-8, I-9, I-24) refer to the sampling locations of impact rocks (suevites, melt breccias, and impact melts) along the Sobok river. Site I-912 and I-913 are fractured granite sites and site I-26 is the unfractured target gneiss site. The Ukrainian Shield is marked with diagonal lines.

1998). Most of the structure is buried under Quaternary sediments. The impact origin of this structure has been described by Masaitis (1973) and Valter and Ryabenko (1973, and references therein), and is based on data from several drill holes, as well as on studies of natural outcrops.

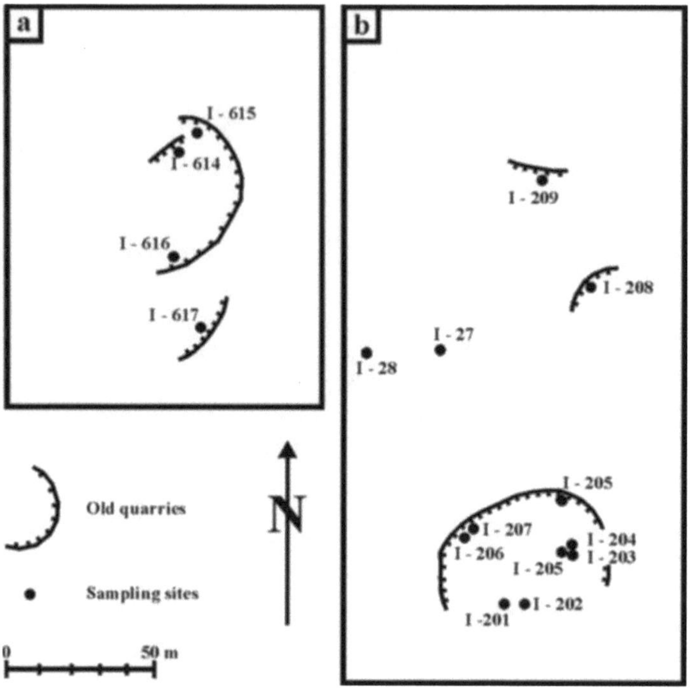

Fig. 2. Close-up of sampling sites for the present study in some quarries at the Ilyinets structure (locations are indicated in Fig. 1).

2.2 Geological Background

The structure is strongly eroded, with a present-day diameter of about 4 kilometers (Gurov et al. 1998). The crater basement is composed of Archean and ca. 1.9-1.7 Ga Proterozoic crystalline rocks of the Ukrainian Shield (Figs. 1 and 2). Biotite granites are the predominant target rocks, whereas gneisses, amphibolites, and schists are less voluminous. Fragments of argillites and sandstones occur in the upper suevites (a mixture of target rocks which suffered impact effects, were lifted into the air and then redeposited in the crater bowl), but have not been preserved *in situ* on the surface of the basement (Gurov and Gurova 1991). The structure is generally buried under some tens of meters of Cenozoic sediments. The subsurface structure of the crater is known on the basis of drill core data. The crater has a central uplift, which is surrounded by an annular trough filled with allochthonous breccias, suevites, and impact melt rocks (Gurov et al. 1998). Melt rocks occur within the suevite bodies in the form of several sheet-like bodies or melt dikes. Melt rocks occur in the southern part of the crater within the suevites in the form of a sheet-like body that dips gently northwards. A variety of impactites and weakly shocked autochthonous basement rocks are also exposed in quarries dating from the third and fourth century A.D. to present, and in natural outcrops in the Sobik river valley in the southern part of the crater (Figs. 1 and 2). Unfortunately, no geophysical data are available to study the extent of the structure below the overburden.

Based on data from several boreholes it is clear that the Ilyinets crater is a complex impact structure with a present oval shape with dimensions of 3.4 x 4.1 km and the major axis oriented northwest to southeast. The crater floor dips gently from its edge to the base of the central uplift. The depth of the annular depression is ca. 300 m at its deepest part to the north of the central uplift. Allochthonous impactites fill the crater annulus. Preservation of post-impact sediments at a distance of up to 3 km from the crater center indicates that the minimal diameter of the structure was about 7.5 km (Gurov and Gurova 1991, Valter and Ryabenko 1977).

Detailed geological, petrographic, and geochemical data of these samples, as well as of various drill core samples, were previously presented by Gurov et al. (1998) and Pesonen et al. (2000).

2.3 Stratigraphy of the Structure

The cross section (see Gurov et al. 1998) outlines the stratigraphy of the Ilyinets impact structure. The target rocks consist of unfractured (unshocked) granites, gneisses, and other rocks (e.g., site I-26; Fig. 1). On top of the unshocked basement lies fractured granite, followed by autochthonous monomict granitic breccia, up to 70 m thick, which forms the crater floor. Two exposures of this fractured granitic breccia occur on both slopes of the Sobik river (sites I-912 and I-913; Fig. 1) valley. The structure is filled with allochthonous impactites including, from bottom to top, lithic breccia, melt breccia and suevite. The transition from allochthonous impactites to autochthonous granitic breccia of the crater floor is gradual within a zone up to 30 m thick. The suevites form an annular layer around the central uplift. The lower horizon of the suevite layer is a lithic breccia and glass-poor suevite with a glass content of less than 10 vol%, whereas glass-rich rocks with glass contents up to 40 vol% form the upper part of the suevite lens. The transition from glass-poor suevite to glass-rich suevite is also gradual. The maximum thickness of the suevite layer is up to 300 m.

2.4 Previous Age Determinations

In the eastern part of the crater post-impact lacustrine sediments of about 30-40 m thickness (mainly clay, siltstone, argillite) are preserved. A probable Early Devonian age for these sediments is suggested by palynological studies (Andreeva and Sergeeva, cited in Bistrevskaya et al. 1974; Masaitis et al. 1980; Valter and Ryabenko 1977), which represents a minimum age of the Ilyinets structure.

Previously K-Ar studies of Ilyinets impact melt rocks yielded ages of about 395-400 Ma (Nikolsky 1975), later, unpublished K-Ar analysis of bulk samples of clast-free impact melt (A.K. Boiko 1997, pers. comm. to E.P. Gurov) yielded ages of about 390 Ma (sample location I-4) and 410 Ma (drill hole 2100, 255 m depth). Thus, no precise and coherent age information has so far been available for this impact structure.

3 Sampling

Twenty-four oriented hand samples were collected from old quarries and outcrops along the Sobik River in the southern part of the structure for paleomagnetic and petrophysical study. The sampling sites are shown in Figs. 1 and 2. Several of the impactite types occur at several of the quarry sites (Fig. 2). The investigated samples consist of suevites, melt-breccias, impact melt rocks, granite breccias, and fractured target granites (autochthonous), as well as unfractured target gneisses (see Table 1). Three impact breccias and one impact glass sample were used for the $^{40}Ar/^{39}Ar$ determinations. Impact melt breccias (I-11, I-12, I-13) from a coherent, about 30-m-thick melt body in the southern part of the impact structure, were obtained from drill core 18480 (at depths of 29, 35, and 38 m, Gurov et al. 1998; Fig. 1) The impact glass sample I-17 was obtained from an outcrop (I-2 in Gurov et al. 1998), also located in the southern part of the structure (Fig. 1).

The orientations were measured with a magnetic compass, and the local magnetic declination (4°E) was corrected for using the IGRF 1980.0 global magnetic maps by Pesonen et al. (1994). The hand samples were further drilled in the Geological Survey of Finland (GSF in Espoo) to obtain ca. 1x1-inch cylinders (height to diameter ratio of 0.9), which were used in paleomagnetic, petrographic, and petrophysical measurements.

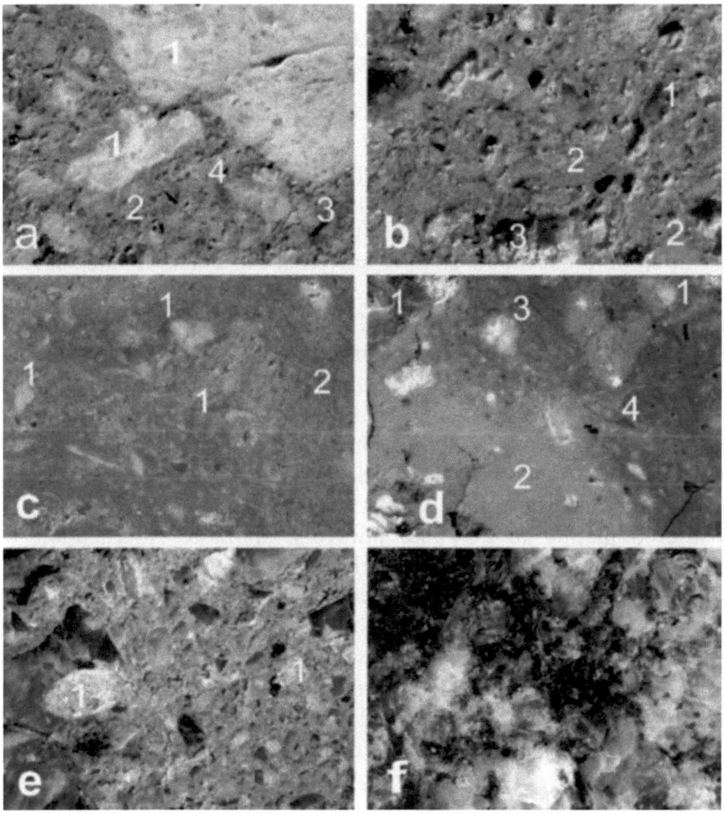

Fig. 3. Examples of Ilyinets rock types, following (from a to f) the stratigraphy from top to bottom. All examples are in the same scale (width of image 10 mm). (a) Suevite with altered glass fragments (sample I206-1a) and voids filled by secondary siliceous materials, including agates. Clasts are composed of granitoid fragments and fine-grained sedimentary rocks. Glass fragments and matrix are hydrothermally altered, showing indication of silicification in the voids. Clasts are low to moderately shocked (no diaplectic glasses). (b) Glass-poor suevite (sample I208-1a). Clasts are composed of granitoid fragments and fine-grained sedimentary rocks. The rock is more hydrothermally altered than the sample shown in Fig. 3a. (c) Melt breccia (sample I615-1a). Clasts are composed mainly of granitoid-derived mineral fragments. Matrix glass is devitrified and hydrothermally altered. Clasts are highly shocked. This sample has a much lower porosity than the samples shown in Figs. 3a and b. (d) Impact melt rock (sample I410-1b). Clasts are composed mainly of granitoid-derived mineral fragments. Matrix glass is devitrified and hydrothermally altered. Clasts are highly shocked. Silicification (e.g., agate in cavities) is observed among hydrothermal alteration products. This sample has a relatively low porosity (2.2 vol%). (e) Monomict brecciated target granite (sample I912-2a). Breccia matrix is hydrothermally altered and shows light brownish coloration. (f) Unfractured target gneiss (sample I26-2a). No shock features and no indication of hydrothermal alterations are visible. The sample has basically no porosity.

4 Petrography and Shock Features

Information on the petrography of Ilyinets rocks has been given before (see Gurov et al. 1998, and references therein); we just add a few additional observations. Fig. 3 shows examples of the petrographic appearance of impactites (a-d) and target rocks (f-g), respectively. These figures delineate visually the various types of clast fragments seen in the impactites. The overall light brownish color reflects the secondary alteration that the rocks have suffered. For example, the suevite sample in Fig. 3b (I208-1a), which contains only few glass fragments, is distinctly more altered than the suevite sample I206-1a of Fig. 3a.

Petrographic evidence of shock metamorphism in the clasts includes shatter cones, kink-banding in biotite, PDFs in quartz, and impact-induced high-pressure phases, such as coesite and diamond (Gurov et al. 1998; Koeberl et al. 1996). It should be noted that there is no indication of lechatelierite, although the presence of impact diamond shows that the shock pressure may have exceeded 35 GPa. Accordingly, glass in most Ilyinets suevites, melt breccias, and melt rocks is strongly altered mainly to clay-like minerals; sporadic silicification was also observed in some impact rocks. The silicification is expressed in two main types: silcrete-like silicification of earlier, mainly matrix minerals, and void-filling agates with diameters up to 32 cm (Gurov et al. 1998). The silicification is interpreted to have occurred during low-temperature post-impact hydrothermal activity corresponding to the formation of agates in the Sääksjärvi impact crater in southern Finland (Kinnunen and Lindquist 1998). At present, Ilyinets and Sääksjärvi are the only craters known to contain agate-filled voids in their impact melt rocks.

5 Experimental Methods

5.1 Petrophysics

Petrophysical properties, such as weak field magnetic susceptibility and its anisotropy, density, and porosity were measured at the Laboratory for Paleomagnetism of the GSF using standard techniques as explained in Pesonen et al. (1992; 1999). The porosities were measured using the Archimedean technique, where the samples are first soaked in tap water for 3 days and then heated in an oven (105° C) for two days (e.g., Kivekäs 1993). Mass and volume of each sample were measured after both steps. Some of the specimens appeared to take up a large quantity of water during soaking, resulting in very high porosity values (Table 1).

5.2 Paleomagnetism

Paleomagnetic measurements were done with the superconducting 2G-SQUID magnetometer at the GSF with an associated 3-axis alternating field (a.f.) demagnetizer. We used a.f. demagnetization steps of 5-10 mT up to 160 mT. Thermal demagnetization was conducted only for a few samples with the Schonstedt TSD-1 furnace. These experiments were terminated due to complex behavior of the samples, probably due to laboratory-induced magneto-mineralogical changes in the samples during heating.

Many of the samples showed multicomponent remanence. Five techniques were applied to separate the components: (i) the 3-D "tube-find" technique for orthogonal vector plots by Leino (1991), (ii) the vector difference method, (iii) the stable end point technique, (iv) the principal component analysis (Kirschvink 1980), and (v) the technique involving intersection of great circles (Halls 1979). In the majority of cases, two to three remanent magnetization components were isolated in the specimens, but there are a few cases where only a single component forms the ChRM.

5.3 $^{40}Ar/^{39}Ar$ Dating

The impact melt breccias contain highly shock-metamorphosed clasts, which are partially to completely shock melted. The aphanitic to fine-grained crystalline impact melt consists of quartz, feldspar, and mafic minerals in a devitrified glassy matrix. Samples I-11 and I-12 are black colored, fine-grained, clast-poor and devitrified impact melt breccias; sample I-13 is more clast-rich than the other two samples (granite derived, biotite-rich) with a schlieric, partly aphanitic and vesicular gray matrix; sample I-17 is a light brown, clast-poor (mainly quartz) devitrified impact glass.

The four samples were prepared by first wrapping the rocks in plastic foil, and splitting off small chips with a hammer. Subsequently, these splits were hand-crushed in a mortar to millimeter size in order to avoid altered parts. After decanting with distilled water the samples were repeatedly washed in an ultrasonic bath with distilled water and acetone for 10 minutes each. After drying at 60° C, some of the most unaltered and clast-free melt matrix and glass pieces (one to ten pieces per sample, with sizes of ~0.5 - 1 mm) were handpicked under a binocular microscope for radiometric dating.

Table 1. Physical properties of rocks for the Ilyinets crater.

Spec.	d	n/N	ρ	φ	χ	NRM	J20/J	Q	comments
	km		kgm^{-3}	%	$10^{-6}SI$	mAm^{-1}			
				Suevite					
I-311	1.52	2	1967	15.5.	627	69	0.8	2.8	altered, with PDFs in quartz
I-614	1.63	1	2298	10.6	214	44	0.94	5.2	glass poor
I-617	1.65	1	2272	9.2	556	38	0.70	1.7	fluidal features
I-209	1.68	2	2028	15.7	528	81	0.93	3.8	altered, clayish
I-208	1.72	2	1947	14.6	584	45	0.76	2.0	poor of glass fragments, more altered than I-206
I-28b	1.72	1	1897	18.3		16	1.03		glass rich, vesicular
I-27P	1.72	1	1897		1017	90	1.07	2.2	altered, glass rich
I-205	1.80	2	2010	20.4	934	119	0.66	3.2	clast rich
I-206	1.80	2	2097	18.5	899	77	0.89	2.4	hydrothermally altered glass fragments,PDFs
I-207	1.80	2	2159	11.9	798	94	1.00	3.1	altered, PDFs
Mean	1.70	10/16	2113	15.4	743	82	0.84	2.8	
				Melt breccia					
I-615	1.63	1	2463		515	50	0.81	2.4	clast (granitoid) rich, strongly shocked
I-616	1.64	1	2274	7.0	408	44	0.78	2.7	PDFs
I-202	1.82	2	2302	8.0	316	19	0.72	1.5	glass poor, PDFs
I-203	1.82	2	2222	10.6	853	57	0.85	1.7	altered, clayish, glass rich, PDFs
I-204	1.82	2	2115	13.5	468	97	0.97	1.8	clasts of shocked granite
I-25P	1.82	1	2221		443	30	0.73	1.7	granite clasts, empty vesicles
Mean	1.76	6/9	2289	11.1	553	54	0.86	1.7	
					Impact melt				
I24-P	1.71	1	2404		71	0.5	0.93	0.18	brecciated granite, kink bands, altered (brownish)
I-410	1.71	1	2592	0.8	72	0.1	1.24	0.05	brecciated granite, brownish, no shock features
I-913	1.85	1	2390	10.0	161	1.55	0.66	0.24	fractured target granite, kink bands in biotite
Mean	1.76	3/3	2462	5.2	101	0.72	0.92	0.16	
			brecciated/fractured target granite (autochtonous breccia)						
I-912	1.71	1	2404		71	0.5	0.93	0.18	brecciated granite, kink bands, altered (brownish
I-913	1.71	1	2592	0.8	72	0.1	1.24	0.05	brecciated granite, brownish, no shock features
I-23P	1.85	1	2390	10.0	161	1.55	0.66	0.24	fractured target granite, kink bands inbiotite
Mean	1.76	3/3	2462	5.2	101	0.72	0.92	0.16	
			unfractured target gneiss (basement						
I-26P	4.59	1	2684	0.0	12249	741	0.50	1.93	unshocked target gneiss
I-26P1	4.59	1	2615	0.8	6390	239	0.24	0.87	unshocked target gneiss
Mean	4.59	2/2	2650	1.1	9319	490	0.37	1.40	

spec — specimen, see Figs. 1-2 ,d — distance from the impact center (km)
n/N — number of samples/specimens , ρ, φ — laboratory bulk density, porosity
χ, NRM, J_{20}/J — weak field susceptibility, intensity of NRM after 20 mT alternating field demagnetization/original NRM
Q — Koenigsberger ratio

Table 2. Paleomagnetic results for the Ilyinets structure (49.07°N, 25.06°E)

				Impact components		Target components
Rock type	**d**	**n/N**	**PEF**	**ChRMR**	**ChRMN**	**SvF**
	km		D,I	D,I	D,I	D,I
Suevites	1.70	10/14	309,51	6,48	145,54	
Melt breccias	1.76	6/9	87,7	10,54		
Impact melts	1.81	3/5	33,52	22,44	33,46	
Fractured target granites	1.76	2/3		34,36		342,33
Unfractured target granites	4.59	2/3				347,57
Mean			15,62	19,46		345,49
				k=8		k=25
				β=4		β=3
				α95=15°		α95=25°
Paleopoles				Plat. = 11.7°		Plat. = 67.7°N
				Plon. = 191.3°		Plon. = 245.3°E
				dp = 11°		dp = 22°
				dm = 17°		dm = 33°
				A95 = 13°		A95 =27°

rock type	see Table 1, Fig. 1-2 and text.
d	approximate distance (meters) of the rocks from the impact center
B/N/n	number of sites/samples/specimens
PEF	remanence component probably related to present Earth's magnetic field contamination by weathering
$ChRM_R$ ($ChRM_N$)	characteristic remanence components of NRM isolated by

The individually handpicked sample concentrates were packed in aluminum-foil, and encapsulated in sealed quartz vials. The sealed quartz vials were irradiated in the MTA KFKI reactor (Budapest, Hungary) for 32 hours. Correction factors for interfering isotopes have been calculated from 10 analyses of two Ca-glass samples and 22 analyses of two pure K-glass samples, and are: $^{36}Ar/^{37}Ar_{(Ca)}$= 0.00026, $^{39}Ar/^{37}Ar_{(Ca)}$= 0.00065, and $^{40}Ar/^{39}Ar_{(K)}$= 0.01566. Variation in the flux of neutrons within the irradiation package was monitored with B4M white mica standard (Flisch 1982), for which an $^{40}Ar/^{39}Ar$ plateau age of 18.6 ± 0.4 Ma has been reported (Burghele 1987). After irradiation the minerals were unpacked from the quartz vials and the aluminum-foil packets, and handpicked into 1-mm-diameter holes within one-way Al-sample holders.

$^{40}Ar/^{39}Ar$ analyses were carried out at the Institute for Geology and Paleontology at the University Salzburg, using an UHV Ar-extraction line equipped with a combined MERCHANTEKTM UV/IR laser ablation facility, and a VG-ISOTECH^{TM} NG3600 Mass Spectrometer. Stepwise heating analyses of samples were performed using a defocused (~1.5-mm-diameter) 25 W CO_2-IR laser operating in Tem_{00} mode at wavelengths between 10.57 and 10.63 µm. Clean-up of the extracted gas was performed using one hot and one cold Zr-Al SAES getter. Gas admittance and pumping of the mass spectrometer and the Ar-extraction line were computer controlled using pneumatic valves. The NG3600 is an 18 cm-radius 60° extended geometry instrument, equipped with a bright Nier-type source operated at 4.5 kV. Measurements were performed on an axial electron multiplier in static mode, with peak jumping of the magnet controlled by a Hall probe. For each increment the intensities of ^{36}Ar, ^{37}Ar, ^{38}Ar, ^{39}Ar, and ^{40}Ar were measured, and the baseline readings on mass 35.5 are automatically subtracted. Linear or polynomial fits were used to back-extrapolate 16 measured peak intensities (per measurement) to the time of gas admittance. Intensities were corrected for system blanks, background, post-irradiation decay of ^{37}Ar, and interfering isotopes. Ages and errors were calculated following suggestions by McDougall and Harrison (1999) and decay factors reported by Steiger and Jäger (1977).

6 Results

6.1 Petrophysics

The petrophysical data are summarized in Table 1 and plotted in bivariate diagrams in Fig. 4. The results show that the physical properties including the susceptibility, the Koenigsberger Q-ratio, and the porosity, show a decreasing trend when moving from suevites through melt-breccias to melt rocks, fractured granites, and finally to unfractured target gneisses. These trends in physical property data of the Ilyinets structure are similar to those observed in many other impact structures (e.g., Lappajärvi, Mien, Popigai, Bosumtwi), which contain an impact breccia lens with impact melt bodies or sheets (e.g., Pesonen et al. 1999; Pilkington et al. 2002; Plado et al. 2000). Figure 4 shows that susceptibility (with the exception of unfractured target gneiss, which has a very high susceptibility), Q-value, and porosity decrease with increasing density (decreasing shock), as is

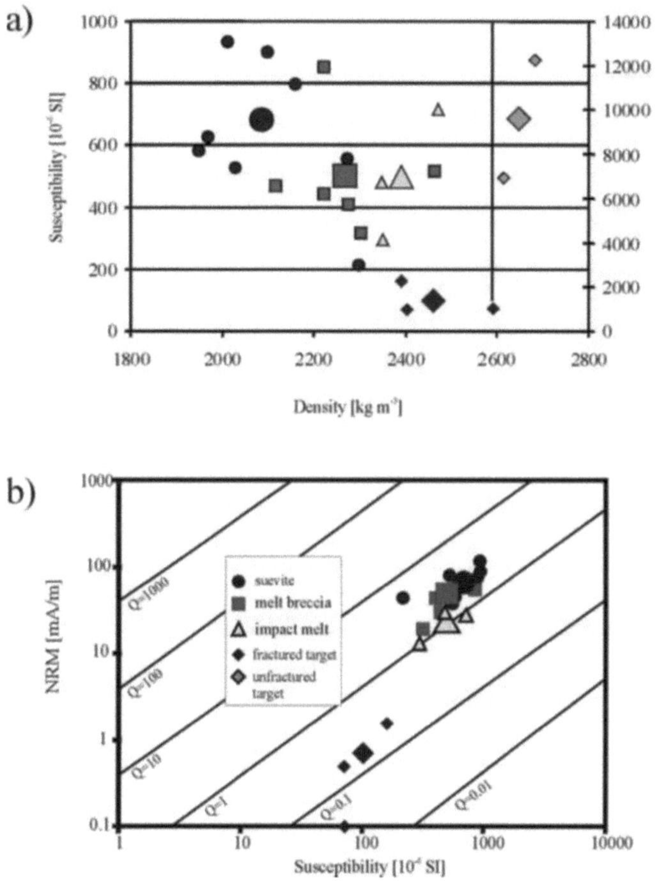

Fig. 4. Petrophysical data of the rocks plotted in bivariate diagrams. (a) susceptibility vs. density, (b) NRM vs. density, (c) Q-ratio vs. density, and (d) porosity vs. density. The density (see Table 1) is the bulk density in laboratory conditions (a value roughly between the wet bulk and dry bulk density. Note that the unfractured target gneisses have very high susceptibilities (to be read in the right vertical axis). Larger symbols denote mean values for the particular rock type.

the case for example in the Lappajärvi complex impact structure (Pesonen et al. 1992). High porosities in suevites and melt breccias (from 20% to 7%) have also been observed in some other impact structures (e.g., Lappajärvi and Sääksjärvi; Pesonen et al. 1992; Kivekäs 1993).

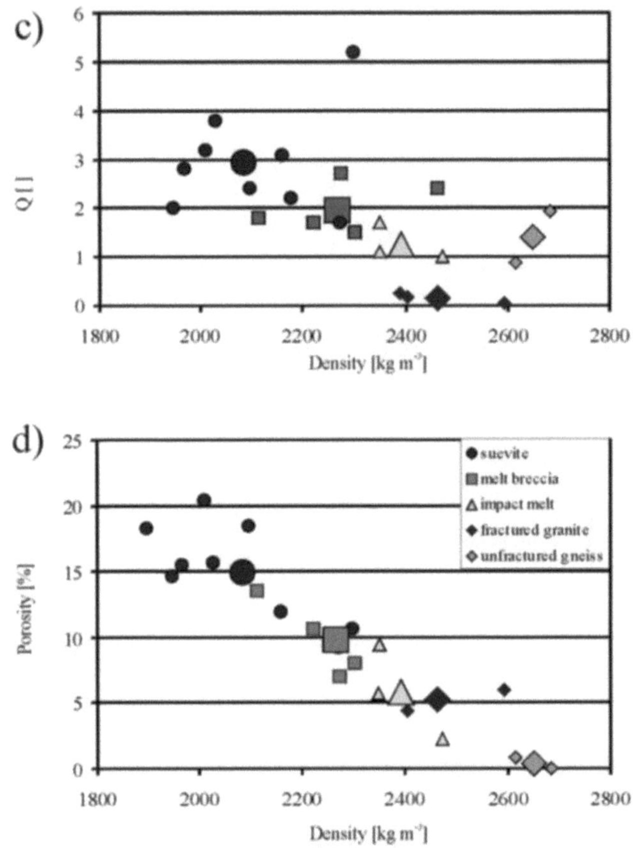

Fig. 4 cont. Figure text see opposing page.

As noted above, the Ilyinets impactites underwent secondary alteration, which resulted in void fillings by secondary products, such as silica (agate), calcite, and zeolite minerals (Gurov et al. 1998). However, these void fillings have not led to a marked decrease of the porosity of the impactites. This is either a real feature in the Ilyinets suevites and melt-breccias or a consequence of the laboratory procedure used to determine the porosity, where the soaking and heating steps drive out the secondary clay minerals from the pores and, thus, result in porosity values that are too high.

Unfortunately, no unshocked target granites were available for petrophysical measurements. The unfractured target rocks are represented here by samples from the gneissic body south of the impact structure (Fig. 1). These gneisses (e.g., Fig. 3f) have distinctly higher susceptibility and lower porosity than the fractured granites and other impact rocks. The fractured or brecciated granites have relatively high porosities (from 1 to 10%) consistent with their petrographic characteristics (Fig. 3e), which shows that they have suffered mild shock metamorphism and fracturing.

6.2 Paleomagnetism

6.2.1 General

Paleomagnetism, when coupled with rock magnetic data, can be used to date impact events (e.g., Pesonen et al. 1999; 2000). The paleomagnetic dating method is based either on the Apparent Polar Wander Path (APWP) technique (e.g., Manicouagan, Dellen, Suvasvesi North; Werner et al. 2001) or on magnetostratigraphy, which is used when oriented samples are not available (e.g., Chicxulub, Ries, Manson, e.g., Pesonen et al. 1996 and references therein). Paleomagnetism has been most successful when applied to impact melts or suevites, where the NRM is a thermoremanent magnetization (TRM) acquired at the time of post-impact cooling (e.g., Dellen, Manicouagan), or a thermochemical remanence (TCRM), acquired at a time when hydrothermal fluids percolated in the structure (e.g., Siljan; Elming and Bylund 1991). In about 80 % of the cases where paleomagnetism has been used for dating impact events with the APW technique, the paleomagnetic age agrees (within the errors limits) with the radiometric age. Failures in paleomagnetic dating involve post-impact structural tilting (e.g., Lappajärvi; Pesonen et al. 1992), non-averaged secular variation (e.g., Lake Mistastin), and scattered NRM directions caused by tectonic movements during the cratering processes (e.g., Scott et al. 1997).

6.2.2 Impactites

Figure 5 shows typical examples of the behavior of the rocks under the a.f. demagnetization treatments. In each case the intensity decay curves, the directions on stereo plots and the orthogonal vector diagrams are shown. The examples reveal that the ChRM of the impactites is generally hard and stable and consists mainly of two or three components. This is the case in particular for suevites (Fig.

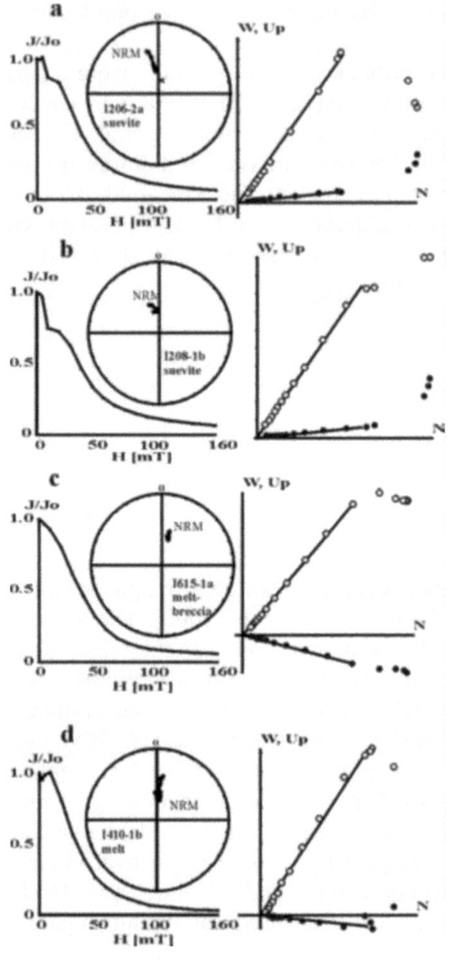

Fig. 5. Eight examples of a.f. demagnetization characteristic for the samples. Left: intensity decay curve; center: equal area stereo plot; right: orthogonal (Zijderveld) vector plot. Open (closed) symbol denotes upward (downward) direction, respectively. (a) Specimen I206-1a (suevite), (b) I208-1b (suevite), (c) I 615-1a (melt breccia), (d) I410-1b (impact melt rock), (e) I912-2a (fractured target granite). Note that the weak cleaning fields isolate a possible "impact" component in sample (e), superimposed with the characteristic "target" remanence. (f) I26-4a (unfractured target gneiss), showing the "target" remanence direction. (g) I617-1a unfractured target granite showing an odd direction. (h) I203-1a (melt breccia) showing hints of a possible reversed polarity direction (see Fig. 6).

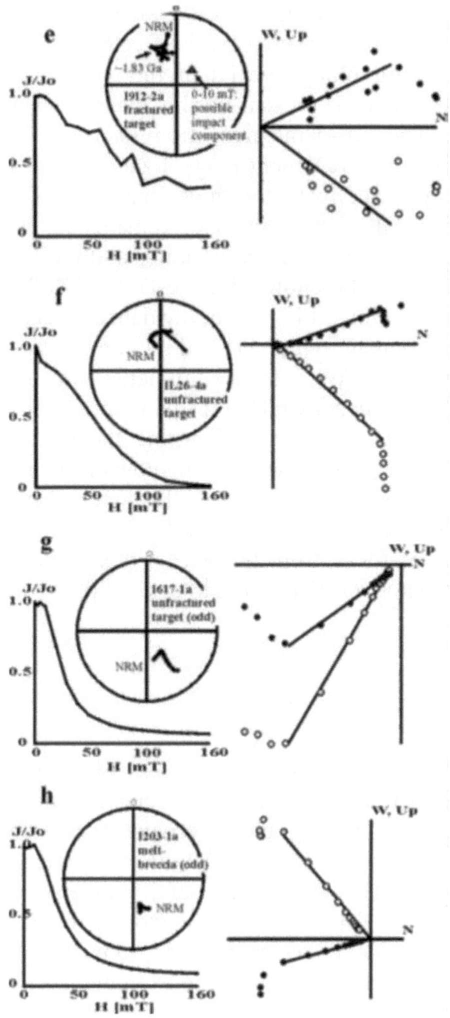

Fig. 5 cont. Figure text see opposing page.

5a), impact melt rocks (Fig. 5d), and melt breccias (Fig. 5c). The fractured target granite (Fig. 5e) shows less stable behavior and somewhat unstable intensity decay. A noticeable feature in all the data is a small but visible soft component, which is removed by 5-20 mT cleaning fields. Vector difference analysis and the study of orthogonal vector plots reveal that this component is often (but not always, see sample 5e) a viscous present-Earth's magnetic field component (called "PEF" in Table 2). The a.f.-demagnetization data show that the remaining component, the characteristic remanent magnetization (ChRM) in the impactites, has a moderate steep upward direction with declination to N or NNE (Figs. 5 a-d). This ChRM, hereafter called the "impact" remanence, is different from the component in the partially shocked target granites and unshocked gneisses, which reveal the target remanence direction (Fig. 5 e-f). Study of demagnetization data reveal two further observations. First, some evidence exists in melt breccias and suevites of another component, which has a direction nearly opposite direction to the impact ChRM (Fig. 5 h). We interpret this latter direction to be the reversed polarity component of the impact ChRM. If this interpretation is correct, we have evidence of a geomagnetic field reversal during the time when the impactites acquired their characteristic remanence. Fig. 6 shows further examples of a.f. data in the form of great circles, which delineate the superimposed N and R polarity components in some specimens. The situation here resembles that observed in the paleomagnetic studies of rocks from the Ries crater (14.8 Ma; Pohl 1977) and from the Bosumtwi structure (1.07 Ma; Plado et al. 2000).

6.2.3 *Target Rocks*

The fractured autochthonous granite samples (Fig. 5e) and the unfractured target gneiss (Fig. 5f, 5g) yield very different remanent magnetization directions compared to the impactites. The a.f. characteristics of these samples are also less stable, but collectively, they point to a characteristic remanent magnetization, which has a northwesterly declination and moderate shallow downward inclination (D 345°, I 59°; Table 2). It is noteworthy, however, that at least two autochthonous granitic breccia specimens yield some evidence of the "impact" remanence in their demagnetization data. One example is shown in Fig. 5e, where the fractured granite has two superimposed remanence components, the ChRM component typical of unfractured granites (the "target" component) and the component isolated in cleaning fields of 0-10 mT (the "impact" component).
Although only two specimens yield this "impact" remanence, the paleomagnetic data collectively provide a fully positive impact test (Pesonen 2001), which confirms that the "impact" remanence is acquired during the time when impact-related processes (e.g., post-impact cooling) took place and was not the result of remagnetization.

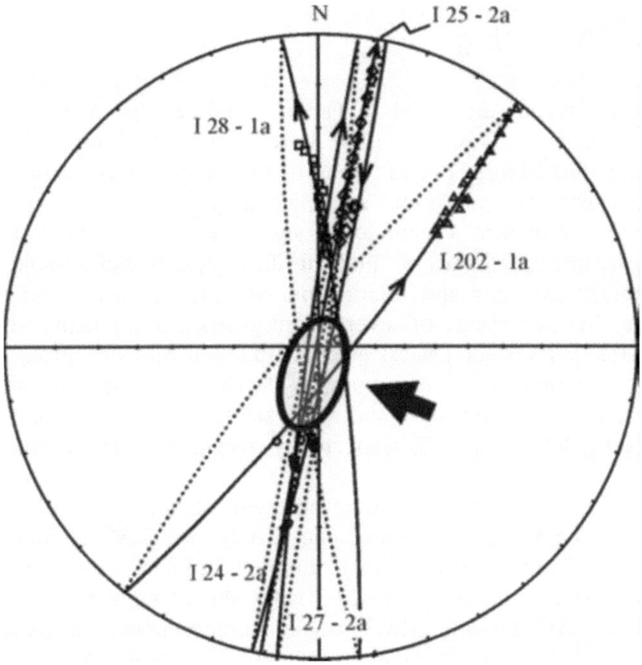

Fig. 6. Examples of great circles in five specimens. The intersections of the great circles define the characteristic N polarity direction in the upper (northerly) hemisphere, which is also obtained by vector plots and seen in stereo plots. Note, however, that the great circles extend away from this N polarity direction southwards and give a hint of another, probably R polarity component in the southern hemisphere (see also specimen I203-1a in Fig. 5h).

6.2.4 Type of NRM

The generally very smooth and stable (after removal of the PEF) a.f.-demagnetization behavior of many impactite specimens and the moderate high Q-values in impactites suggest that the remanence is of thermal (TRM) or thermochemical (TCRM) origin, as is the case with many other impact structures containing melt and suevite bodies (e.g., Pesonen 1996). We suggest that the ChRM in the Ilyinets impactites is either TRM or TCRM acquired during cooling of the hot hydrothermal fluids, for which there is distinct petrographic evidence (Fig. 3; see above). The acquisition process of the characteristic remanent magnetization may last thousands of years, allowing a field reversal to be recorded in the impactites.

7
Results of $^{40}Ar/^{39}Ar$ Dating

The $^{40}Ar/^{39}Ar$ analytical results are listed in Table 3, and are displayed as age spectra in Fig. 7. Increments with less than 1% ^{39}Ar released were eliminated. The samples show rather disturbed age spectra. None of the samples yield a flat age spectrum, which could strictly be interpreted as a plateau age.

None of these spectra represent a statistically significant plateau; thus, aside from the total gas ages, integrated ages of "plateau-like" steps of each sample are also given. These integrated ages and their errors are calculated as weighted means on cumulative ^{39}Ar percentage of several contiguous heating steps, which comprise the most stable parts of the gas release pattern. No inverse isochrons can be calculated for the examination of excess argon effects, as the samples show high radiogenic ^{40}Ar contents, which show too much scatter (low $R^2 = 0.68$, 0.89, 0.02, 0.03 for I-11, I-12, I-13, and I-17, respectively, where R is the correlation coefficient).

Samples I-11 and I-12 show hump-shaped spectra, with sample I-11 yielding a total gas age of 398.6 ± 3.6 Ma. An integrated age of 410.1 ± 3.7 Ma is obtained by only using steps 8-20 (70.55 % of total ^{39}Ar). Sample I-12 yields a total gas age of 410.2 ± 3.7 Ma, and an integrated age of 424.7 ± 3.8 Ma is obtained from steps 6-14 (57.18 % of total ^{39}Ar). Sample I-13, showing a more concordant pattern, yields a total gas age of 431.1 ± 3.8 Ma and an integrated age of 439.8 ± 3.9 Ma by the steps 2-20 (71.30 % of total ^{39}Ar). A fairly well concordant spectrum is shown by sample I-17, which yields a total gas age of 438.3 ± 3.9 Ma and an integrated age of 444.6 ± 3.9 Ma for the steps 2-23 (90.55 % of total ^{39}Ar).

According to Deutsch and Schärer (1994) more than 90 % of impact-shocked rocks preserve their pre-shock age. The studied samples, however, do not show higher apparent ages at the last increments, which would indicate the presence of pre-shock phases. All samples display young apparent ages at the first increments, which are interpreted to be caused by argon loss during post-impact low temperature (hydrothermal) processes. In general, the patterns show a rather homogeneous composition of the analyzed grains and do not imply incomplete degassing or the presence of non-equilibrated relic phases. Two of the samples (I-11, I-12) have several lower temperature increments with younger apparent ages, which comprise, however, only 4.49 to 5.11 % of released total ^{39}Ar. This is in the same range as at the first increment of I-13 and of I-17. These first release steps (in all samples) have higher total $^{37}Ar/^{39}Ar$ ratios, which roughly indicate a higher Ca/K ratio for the first incremental steps. The $^{37}Ar/^{39}Ar$ ratio in all higher incremental steps are fairly constant. Thus, the younger age increments may be interpreted as rejuvenated ages of Ca-rich phases by thermal resetting of the K-Ar

Table 3. $^{40}Ar/^{39}Ar$-analytical data from multi-grain incremental heating analysis on impact melt from the Ilyinets impact structure. Ukraine. [a] measured; [b] corrected for post-irradiation decay of ^{37}Ar (35.1 days half-life); [c] ($^{40}Ar_{tot}$ - $^{36}Ar_{atmos}$ x 295.5) / $^{40}Ar_{tot}$; J-value for all samples: 0.036564; integrated age and error: weighted on cumulative % ^{39}Ar

Sample:	I-11						
increment	$^{36}Ar/^{39}Ar^{a}$	$^{37}Ar/^{39}Ar^{b}$	$^{40}Ar/^{39}Ar$	$\%^{39}Ar$	$\%^{40}Ar^{*c}$	age [Ma]	± 1σ [Ma]
1	0.026331	0.247584	11.083225	1.48	29.80	205.9	2.5
2	0.015559	0.187219	8.993319	1.67	48.88	268.9	2.7
3	0.005343	0.098946	7.127619	1.34	77.85	332.8	3.1
4	0.004279	0.077992	7.727589	3.39	83.64	382.2	3.5
5	0.002725	0.086103	7.420443	1.92	89.15	390.4	3.5
6	0.002775	0.091941	7.348831	3.94	88.84	385.8	3.5
7	0.001865	0.059485	7.308530	5.72	92.46	397.8	3.6
8	0.001240	0.046542	7.375807	6.17	95.03	411.0	3.7
9	0.000932	0.043068	7.314431	4.19	96.23	412.6	3.7
10	0.000798	0.045577	7.476598	2.98	96.85	423.1	3.8
11	0.000757	0.047913	7.289924	5.20	96.93	414.0	3.7
12	0.000725	0.043633	7.350132	4.22	97.09	417.6	3.7
13	0.000710	0.046052	7.269573	7.10	97.11	413.7	3.7
14	0.000757	0.043395	7.262694	4.90	96.92	412.6	3.7
15	0.000812	0.044888	7.167677	7.82	96.65	406.7	3.6
16	0.000882	0.045237	7.248665	5.55	96.40	409.9	3.7
17	0.000901	0.045138	7.164176	4.19	96.28	405.1	3.6
18	0.001182	0.056181	7.268092	7.35	95.19	406.3	3.6
19	0.001118	0.055219	7.246403	5.52	95.44	406.1	3.6
20	0.001732	0.051411	7.330904	5.35	93.02	401.0	3.6
21	0.001480	0.057678	7.028098	5.29	93.78	388.9	3.5
22	0.001753	0.060698	7.085359	3.03	92.69	387.7	3.5
23	0.002484	0.069333	7.127616	1.67	89.70	378.5	3.4
total	0.001995	0.060025	7.362852	100	91.99	398.6	3.6
integrated	(steps 8-20)			70.6		410.1	3.7

Sample:	I-12						
increment	$^{36}Ar/^{39}Ar^{a}$	$^{37}Ar/^{39}Ar^{b}$	$^{40}Ar/^{39}Ar$	$\%^{39}Ar$	$\%^{40}Ar^{*c}$	age [Ma]	± 1σ [Ma]
1	0.034269	0.265677	13.622063	1.18	25.66	217.3	2.9
2	0.015028	0.138404	10.231908	2.02	56.60	346.3	3.4
3	0.003778	0.076246	7.604436	1.91	85.32	383.5	3.5
4	0.002634	0.073078	7.600292	3.23	89.76	401.2	3.6
5	0.002107	0.070616	7.711983	5.72	91.93	415.3	3.7
6	0.001369	0.053630	7.713544	6.21	94.76	426.7	3.8
7	0.000993	0.051746	7.660581	7.78	96.17	429.7	3.8
8	0.000993	0.049993	7.590535	7.27	96.14	426.1	3.8

9	0.001200	0.051013	7.528225	9.45	95.29	419.6	3.7
10	0.001159	0.055170	7.620530	7.62	95.51	425.1	3.8
11	0.001346	0.055644	7.600830	5.05	94.77	421.2	3.8
12	0.001347	0.061958	7.690735	5.48	94.82	425.9	3.8
13	0.001353	0.051978	7.621024	3.44	94.76	422.1	3.8
14	0.001360	0.057632	7.677079	4.87	94.76	425.0	3.8
15	0.001561	0.055366	7.398291	4.88	93.76	407.2	3.7
16	0.001808	0.058330	7.260704	4.13	92.64	396.1	3.6
17	0.002159	0.063980	7.216629	1.31	91.16	388.3	3.6
18	0.001743	0.059177	7.096148	1.84	92.74	388.4	3.5
19	0.001892	0.057537	7.129410	2.46	92.16	387.8	3.5
20	0.001855	0.062800	7.274036	4.07	92.46	396.1	3.6
21	0.001975	0.062244	7.115258	2.23	91.80	385.8	3.5
22	0.002089	0.066507	7.294749	4.88	91.54	393.6	3.6
23	0.002628	0.068200	7.464169	1.88	89.59	394.1	3.6
24	0.002034	0.057936	7.259945	1.08	91.72	392.6	3.6
total	0.002227	0.064260	7.651111	100	91.40	410.2	3.7
integrated	(steps 6-14)			57.2		424.7	3.8

Sample:	I-13						
increment	$^{36}Ar/^{39}Ar$ [a]	$^{37}Ar/^{39}Ar$ [b]	$^{40}Ar/^{39}Ar$	% ^{39}Ar	% $^{40}Ar^*$ [c]	age [Ma]	± 1σ [Ma]
1	0.003183	0.042726	5.101236	4.77	81.56	254.8	2.4
2	0.001142	0.015061	7.537871	4.61	95.52	420.9	3.8
3	0.000812	0.014904	7.695608	4.07	96.88	434.2	3.9
4	0.000586	0.013367	7.916736	3.79	97.81	449.0	4.0
5	0.000391	0.010693	7.835644	1.59	98.53	447.8	4.0
6	0.000499	0.012981	7.995384	3.54	98.15	454.4	4.0
7	0.000609	0.011556	7.932812	10.96	97.73	449.5	4.0
8	0.000632	0.010501	7.722959	4.05	97.58	438.3	3.9
9	0.000657	0.010498	7.774708	5.51	97.50	440.6	3.9
10	0.000739	0.010578	7.731921	5.23	97.18	437.2	3.9
11	0.000907	0.008969	7.844424	6.37	96.59	440.4	3.9
12	0.000878	0.009286	7.682306	3.06	96.62	432.4	3.8
13	0.000924	0.008294	7.643859	2.17	96.43	429.7	3.8
14	0.000994	0.009089	7.859042	6.38	96.26	439.8	3.9
15	0.001389	0.007444	7.898333	7.66	94.80	435.8	3.9
16	0.001790	0.009281	8.026781	2.30	93.41	436.3	3.9
17	0.001649	0.008619	7.794522	3.48	93.75	426.4	3.8
18	0.001401	0.009828	7.834313	3.00	94.72	432.3	3.9
19	0.001493	0.008657	7.898147	3.50	94.42	434.2	3.9
20	0.001719	0.007487	8.148552	13.95	93.76	443.7	3.9
total	0.001158	0.011658	7.738501	100	95.58	431.1	3.8
integrated	(steps 2-20)			71.3		439.8	3.9

Sample:	I-17						
increment	^{36}Ar/^{39}Ar^{a}	^{37}Ar/^{39}Ar^{b}	^{40}Ar/^{39}Ar	% ^{39}Ar	% ^{40}Ar*^{c}	age [Ma]	± 1σ [Ma]
1	0.001456	0.030748	6.793469	9.45	93.67	376.7	3.4
2	0.000489	0.005766	7.879595	6.66	98.16	448.6	4.0
3	0.000495	0.005108	7.807605	4.02	98.13	444.8	3.9
4	0.000492	0.005762	7.811087	3.21	98.14	445.0	3.9
5	0.000513	0.006096	7.871208	9.26	98.07	447.8	4.0
6	0.000712	0.006593	7.819860	3.43	97.31	442.1	3.9
7	0.000566	0.007168	7.779465	2.96	97.85	442.2	3.9
8	0.000537	0.007277	7.838912	2.39	97.98	445.7	3.9
9	0.000513	0.006638	7.902290	5.96	98.08	449.4	4.0
10	0.000503	0.007541	7.846073	3.88	98.11	446.6	4.0
11	0.000609	0.007082	7.920803	2.74	97.73	448.9	4.0
12	0.000590	0.008381	7.788663	2.17	97.76	442.4	3.9
13	0.000618	0.007391	7.716887	2.07	97.63	438.2	3.9
14	0.000613	0.007899	7.777276	1.74	97.67	441.4	3.9
15	0.000669	0.006369	7.839981	4.24	97.48	443.8	3.9
16	0.000633	0.005842	7.835498	3.68	97.61	444.1	3.9
17	0.000713	0.006857	7.834224	2.98	97.31	442.8	3.9
18	0.000785	0.006292	7.926480	4.59	97.07	446.5	4.0
19	0.000862	0.006670	7.889037	3.34	96.77	443.4	3.9
20	0.000959	0.006117	7.793848	5.03	96.36	437.0	3.9
21	0.001078	0.005201	7.929655	3.73	95.98	442.2	3.9
22	0.001250	0.005725	7.942830	5.04	95.35	440.2	3.9
23	0.002187	0.005415	8.321566	7.44	92.24	445.5	3.9
total	0.000862	0.007961	7.790592	100	96.73	438.3	3.9
integrated	(steps 2-23)			90.6		444.6	3.9

isotope system. Samples I-11 and I-12, with their younger apparent ages, generally show higher ^{37}Ar/^{39}Ar ratios, implying degassing of a Ca-rich phase, probably due to hydrothermal alteration. Chemical analyses of Gurov et al. (1998) showed a higher CaO content of these samples compared to the other impactites. Gurov et al. (1998) also reported anomalously low Na and high K abundances in the upper part of the impact melt rocks, most likely due to post-impact hydrothermal mobilization. In addition to argon loss a higher potassium content due to secondary alteration could be a cause for the younger ages. Samples I-11 and I-12 (6.22 and 6.19 wt% K_2O) with the lower ages, however, show lower potassium abundances than samples I-13 and I-17 (10.74 and 10.93 wt% K_2O). Thus, argon loss seems to be the main reason for the younger ages. Nevertheless, it should be noted that the K/Ca ratios, calculated from the Gurov et al. (1998) bulk compositions, yield higher values for samples I-13 and I-17 (27.1 and 24.9) than for I-11 and I-12 (6.75 and 6.42). The slight decrease of apparent ages in the last increments of samples I-11 and I-12, together with a slight increase of the $^{37}Ar/^{39}Ar$ ratio, could be explained by recoil of ^{39}Ar into potassium-poor groundmass phases and/or microclasts.

The glass sample (I-17) shows the best, almost concordant, spectrum, and it gives the oldest apparent age. As glass is commonly completely degassed and rapidly quenched, this may indicate the cooling age of the impact melt. The fairly concordant spectrum precludes mobilization of potassium and argon (and, thus, a possible anomalous old age). This spectrum can be interpreted either by homogeneous gain of potassium during hydrothermal post-impact event(s), or by shock-induced trapping of argon gas from potassium-rich phases. The first may not be the case as the apparent age should be lowered due to the gain of potassium relative to the radiogenic produced argon since the impact induced resetting of the K-Ar system. Moreover, argon loss during (hydro)thermal alteration process(es) should have increased the fractionation between the radiogenic argon and potassium contents. A fairly concordant $^{37}Ar/^{39}Ar$ ratio probably precludes recoil redistribution of ^{39}Ar as a cause for the higher apparent age of this sample compared to the impact melt breccia samples. Furthermore, a general ^{39}Ar loss due to irradiation-caused recoil of the glass sample may be excluded due to the fact that the first increment yields a younger apparent age. Argon solubility is thought to be very low and may preclude strong incorporation of excess argon from hydrous fluids (Kelley 2002), which would increase the apparent age. And excess argon, in, e.g., vesicular inclusions, would also yield high apparent ages at the initial incremental steps.

Thus, we assume that argon loss due to secondary (hydro)thermal alteration may have played a major role in the K-Ar isotopic system of the studied samples. It is probable that alterations, such as silification (agate), calcite formation, and growth of zeolites and clay minerals have taken place at relatively low temperatures (less than 150 °C), which, however, depending on the length of the temperature increase, could be enough to partially reset the Ar ages.

As excess argon in the glass sample cannot be completely excluded, the age of I-17 should be regarded as a maximum age of the impact event. The two most concordant integrated ages of the samples I-13 and I-17 are interpreted to yield the apparent age of the impact (339.8 to 444.6 ± 3.9 Ma).

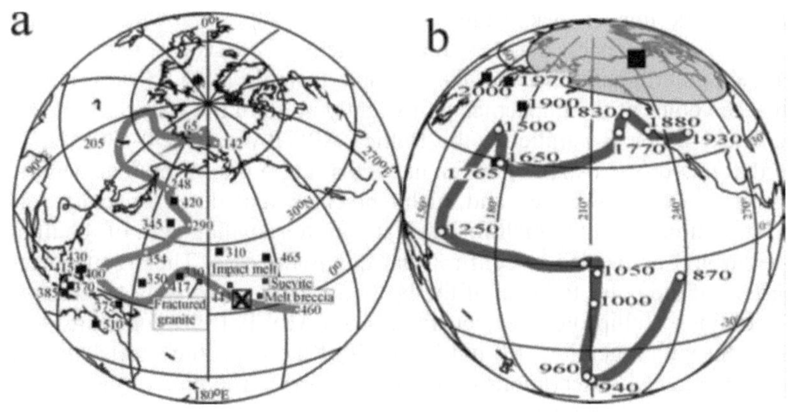

Fig. 7. (a). The mean paleomagnetic poles of Table 2 on the APWP of Baltica. The gray squares denote the "impact" poles of suevites, melt breccias, melt rocks, and fractured granites. The crossed square indicates the mean pole with the 95%-confidence oval. The black squares are Ukrainian poles from the same time period taken from the GPDB, but their scattered nature does not allow them to be used for paleomagnetic APWP dating. The square with a cross indicates the mean pole if a 5° structural tilt to the north is applied (see text). (b) The mean paleomagnetic pole (closed square) as defined for the unshocked target rocks in Table 2 plotted on the APWP of Baltica for 1930-870 Ma (modified after Elming et al. 2001 and Pesonen et al. 2003). The smaller squares denote the available Ukrainian poles of 2000-1765 Ma in age, but due to their large scatter only the Baltica APWP was used to define the paleomagnetic age for Ilyinets target (host-rock) magnetization, which is interpolated to be ca. 1830 Ma.

8 The Paleomagnetic Age

Table 2 shows the mean directions and paleomagnetic poles for impactites and target rocks. The mean "impact" direction is D = 19°, I = -46° (k 48, $\alpha95$ = 13°). These values are defined by four rock types (suevites, melt breccias, melt rocks, and fractured target rocks) from 10 sites and are considered a reliable estimate of the cooling following the impact event. The mean paleomagnetic pole is plotted on the APW path of Baltica for the period of 460-65 Ma in Fig. 7. The Ukrainian poles (shown in Fig. 7) are too scarce or too scattered to allow the APWP of Ukraine to be used for this period. However, the use of Baltica APWP is reasonable in so far that Elming et al. (2001) have shown that Baltica and Ukraine amalgamated together some time during the period of 1750 to 1800 Ma (see also Pesonen et al. 2003) and have most likely remained connected ever since. The mean pole for the impact rocks suggest an age of ~ 445 Ma for the time of the acquisition of the impact magnetization, which is consistent with the best $^{40}Ar/^{39}Ar$ age determination of the glass sample (sample I-17). However, Fig. 7 shows that the poles of the impactites plot slightly north of the Baltica APWP and define a track, which is parallel to the Baltica APWP during 450-430 Ma. It is possible that this reflects slightly different acquisition times of the remanences in these rocks, with the suevites and melts breccias recording somewhat older magnetization times than the melt rocks. Many more sampling sites of the impact lithologies would be needed to confirm that this is the case. The significance of the departure of the poles to north of the Baltica APWP is probably fortuitous, as the 95% confidence ellipsoid intersects the Baltica APWP. If the departure is real, it is consistent with a gentle (~5°) northerly tilt of the impactite sheets suggested from some geological observations (Gurov et al. 1998). The pole of the fractured target granites is based only on two specimens and cannot be regarded as reliable.

The formation of hydrothermal clay minerals is a process, which takes a considerable time (millions of years) when temperatures decrease from high to ambient values. The scatter in the Ar-Ar age spectra is consistent with this and shows a trend from a well-defined cooling age (sample I-17, age 445 Ma) to somewhat more complex ages from 430 to 400 Ma, respectively. However, the paleomagnetic poles do not spread considerably and do not support a prolonged acquisition mechanism, but rather indicate a magnetization time of 450-440 Ma (Fig. 7). This probably indicates that the ChRM of the impactites is a TRM (or TCRM) reflecting thermal cooling, rather than a prolonged CRM.

Data for target rocks are sparse. Based on the well defined APWP of Baltica during 1930-870 Ma, and on the less well defined Ukrainian pole data (Fig. 7), the very tentative "target" pole (only few samples available) suggests an age of ~1830 Ma for the acquisition time of the target remanence.

9
Summary and Conclusions

1. The integrated paleomagnetic and $^{40}Ar/^{39}Ar$ results show that the Ilyinets impact probably occurred at 445 ± 10 Ma. The "impact" magnetization occurred during the time of post-shock cooling, while hydrothermal activity was ongoing in the structure, and is probably a TRM or a TCRM. Some of the Ar-Ar age spectra reveal some evidence of later disturbance. As there is no paleomagnetic evidence of another separate significant thermal event in the area at about 400 Ma, we propose that the somewhat younger Ar-Ar ages in some of the samples were caused by hydrothermal fluids; these also resulted in the secondary alterations observed in the petrographic data. The lock-in of the impact-related characteristic magnetization probably took place at the time of cooling of the melts and suevites. However, the lock-in process may have been long enough so that a field reversal is recorded in the impactite rocks. The "impact test" is positive, suggesting that

the impactites have acquired their magnetization during impact-related processes and have not been remagnetized during later geological events.

2. Paleomagnetic data do not indicate that large scale post-impact structural tilting has taken place. A minor (less than 5°) northerly tilt of the melt sheet is possible. Paleomagnetic data of the impact rocks suggest that the Ukrainian Shield was in contact with the Baltica at the time of the impact event 445 Ma ago.

3. The secondary alteration features seen in petrographic thin sections, such as silicification, calcification, and the formation of clay minerals, are related to post-impact hydrothermal activity.

4. Petrophysical properties of the impactites reveal a trend of decreasing susceptibility, Q-value, and porosity from suevites to melt breccias, to melt rocks, and to brecciated target rocks. The unshocked target gneisses have very different petrophysical characteristics compared to the impactites and to fractured or target rocks.

Acknowledgments

We are grateful for M. Kuulusa and A. Kataikko for carrying out the paleomagnetic and petrophysical measurements at the GSF Paleomagnetic Laboratory. Work in Vienna was supported by the Austrian Science Foundation, project Y58-GEO (to C. K.) and by a research scholarship from the University of Vienna (to D. M.).

References

Abels A, Plado J, Pesonen LJ, Lehtinen M (2002) The Impact Cratering record of Fennoscandia - A new look at the database. In: Plado J, Pesonen LJ (eds) Impacts in Precambrian Shields, Impact Studies vol. 2, Springer Verlag, Berlin, Heidelberg, pp 1-58

Alvarez W, Muller R (1984) Evidence from crater ages for periodic impact on Earth. Nature 308: 718-720

Bistrevskaya SS, Zemskova GA, Vinogradov GG (1974) New data of the Ilyinets paleovolcano structure on the Ukrainian Shield [in Russian]. Geological Journal (Kiev) 34: 123-126

Bottke WF, Melosh HJ (1996) Binary asteroids and the formation of doublet craters. Icarus 124: 372-391

Burghele A (1987) Propagation of error and choice of standard in the ^{40}Ar-^{39}Ar technique. Chemical Geology 66: 17-19

Deutsch A, Schärer U (1998) Dating terrestrial impact events. Meteoritics 29, 301-322

Elming S-Å, Bylund G (1991) Palaeomagnetism and the Siljan impact structure, central Sweden. Geophysical Journal International 105: 757-770

Elming S-Å, Mikhailova NP, Kravchenko S (2001) Palaeomagnetism of Proterozoic rocks from the Ukrainian Shield: new tectonic reconstructions of the Ukrainian and Fennoscandian shields. Tectonophysics 339: 19-38

Fisher RAF (1953) Dispersion on a sphere. Proceedings of the Royal Society (London) A 217: 293-305
Flisch M (1982) Potassium-argon analysis. In: Odin GS (ed) Numerical dating in stratigraphy. Wiley and Sons, Chichester, New York, Brisbane, pp 151-158
Grieve RAF, Pesonen LJ (1996) Terrestrial impact craters: their spatial and temporal distribution and impacting bodies. Earth, Moon and Planets 72: 357-376
Grieve RAF, Rupert J, Smith J, Therriault A (1995) The record of terrestrial impact cratering. GSA Today 5: 189, 194-196
Gurov EP, Gurova EP (1991) Geological structure and rock composition of impact structures [in Russian]. Naukovka Dumka, Kiev, Ukraine, 160 pp
Gurov EP, Koeberl C, Reimold WU (1998) Petrography and geochemistry of target rocks and impactites from the Ilyinets Crater, Ukraine. Meteoritics and Planetary Science 33: 1317-1333
Halls HC (1979) The Slate Islands meteorite impact site: a study of shock remanent magnetization. Geophysical Journal of the Royal Astronomical Society 59: 553-591
Järvelä J, Pesonen LJ, Pietarinen H (1995) On palaeomagnetism and petrophysics of the Iso-Naakkima impact structure, southeastern Finland: Open File Report Q19/29.1/3232/95/1, Laboratory for Paleomagnetism, Geological Survey of Finland, Espoo, 43 pp
Kelley S (2002) Excess argon in K-Ar and Ar-Ar geochronology. Chemical Geology 188, 1-22
Kinnunen KA, Lindqvist K (1998) Agate as an indicator of impact structures: An example from Sääksjärvi, Finland. Meteoritics and Planetary Science 33: 7-12
Kirschvink J (1980) Least squares line and plane and the analysis of paleomagnetic data. Geophysical Journal of the Royal Astronomical Society 62: 699-718
Kivekäs L (1993) Density and porosity measurements at the petrophysical laboratory of the Geological Survey of Finland. In: Autio S (ed) Current Research 1991-1992, Geological Survey of Finland, Special Paper 18: 119-127
Koeberl C, Reimold WU, Gurov EP (1996) Petrology and geochemistry of target rocks, breccias, and impact melt rocks from the Ilyinets crater; Ukraine [abs.]. Lunar and Planetary Science 27: 681-682
Lehtinen M, Pesonen LJ, Moilanen J (2002) Impactites from Lake Suvasvesi impact structures, a possible double impact crater in Finland [abs.]. In: Jakes P (ed) Impacts: a geological and astronomical perspective, Programme and Abstracts, 9th Workshop of the European Science Foundation Impact Programme, October 12-16, 2002, Prague, Czech Republic, pp 43-45
Leino M (1991) Paleomagneettisten tulosten monikomponenttianalyysi painotetulla pienimmen neljösumman menetelmällä. GSF internal report Q29.1/91/2 (in Finnish).
Masaitis VL (1973) The geological consequences of falls of crater-forming meteorites [in Russian]. Nedra Press, Leningrad, Russia, 17 pp
Masaitis VL, Danilin AN, Maschak MS, Raykhlin AI, Selivanovskaya TV, Shadenkov EM (1980) Geology of Astroblemes [in Russian]. Leningrad, Nedra Press, 231 pp
McDougall I, Harrison MT (1999) Geochronology and thermochronology by the $^{40}Ar/^{39}Ar$ method, 2nd edition. Oxford University Press, Oxford, New York, 269 pp
Nikolsky AP (1975) The meteorite explosion craters of the Ukrainian Shield near Vinnitsa [in Russian]. Geological Journal (Kiev) 35: 76-86
Pesonen LJ (1996) The geophysical signatures of terrestrial impact craters [abs.]. In: Abstracts, International Workshop "The Role of Impact Processes in the Geological

and Biological Evolution of Planet Earth", September 27 - October 2, 1996, Ivan Rakovec Institute of Palaeontology, Scientific Research Centre SAZU, Postojna, Slovenia, pp 61-62

Pesonen LJ (1999) Pegmatites and impactites - Useful rocks for paleomagnetism. In: Abrahamsen N (ed.) Aarhus Geoscience 8, Geology Department, Aarhus University, pp 91-94

Pesonen LJ (2001) Primary magnetization associated with meteorite impact: the field test and case histories [abs.]. In: Geo Abstracts, XXVI EGS General Assembly, Nice, France, April 25-30, 2001 (CD-ROM).

Pesonen L (2003) Paleomagnetic configuration of supercontinents during the Proterozoic. Tectonophysics (in press)

Pesonen LJ, Marcos N, Pipping F (1992) Paleomagnetism of the Lake Lappajärvi impact structure, western Finland. Tectonophysics 216: 123-142

Pesonen LJ, Järvelä J, Sarapää O, Pietarinen H (1996) The Iso-Naakkima meteorite impact structure: physical properties and palaeomagnetism [abs.]. Meteoritics and Planetary Science 31: A105-A106

Pesonen LJ, Elo S, Lehtinen M, Jokinen T, Puranen R, Kivekäs L (1999) The Lake Karikkoselkä impact structure, central Finland - new geophysical and petrographic results. In: Dressler BO, Sharpton VL (eds) Large meteorite impacts and planetary evolution. Geological Society of America Special Paper 339: 131-139

Pesonen LJ, Mader D, Gurov E, Koeberl C, Kinnunen KA (2000) Paleomagnetism and ^{39}Ar-^{40}Ar age determinations of impactites from the Ilyinets Impact Structure, Ukraine [abs.]. In: Plado J, Pesonen LJ (eds) Meteorite impacts in Precambrian Shields, Programme and Abstracts, the 4th Workshop of the European Science Foundation Impact Programme, Lappajärvi-Karikkoselkä-Sääksjärvi, Finland, May 24-28, 2000. Geological Survey of Finland and University of Helsinki, pp 39-40

Pesonen LJ, Kuulusa M, Donadini F (2001) Impact structures and a new field test for palaeomagnetism [abs.]. In: Martinez-Ruiz F (ed) Abstracts, ESF Workshop on "Stratigraphic Record of Impact Events", May 19-26, 2001, Granada, Spain, pp 85-86

Pilkington M, Grieve RAF (1992) The geophysical signature of terrestrial impact craters. Reviews of Geophysics 30: 161-181

Pilkington M, Pesonen LJ, Grieve RAF, Masaitis V (2002) Geophysics, petrophysics and paleomagnetism of the Popigai impact structure, Siberia. In: Plado J, Pesonen LJ (eds) Impacts in Precambrian Shields, Impact Studies vol. 2, Springer Verlag, Berlin, Heidelberg, pp 87-108

Plado J, Pesonen LJ, Koeberl C, Elo S (2000) The Bosumtwi meteorite impact structure, Ghana: A magnetic model. Meteoritics and Planetary Science 35: 723-732

Pohl J (1977) Paläomagnetische und gesteinsmagnetische Untersuchungen an den Kernen der Forschungsbohrung Nördlingen. Geologica Bavarica 75: 329-348

Rampino MR (1999) Impact crises, mass extinctions, and galactic dynamics: the case for a unified theory. In: Dressler BO, Sharpton VL (eds) Large meteorite impacts and planetary evolution. Geological Society of America Special Paper 339: 241-248.

Scott RG, Pilkington M, Tanczyk EL (1997) Magnetic investigation of the West Hawk, Deep Bay and Clearwater impact structures, Canada. Meteoritics and Planetary Science 32: 293-308

Steiger RH, Jäger E (1977) Subcommission on geochronology: Convention on the use of decay constants in geo- and cosmochronology. Earth and Planetary Science Letters 36: 359-362

Torsvik TH, Smethurst MA, Meert JG, Van der Voo R, McKerrow WS, Brasier MD, Sturt BA, Walderhaug HJ (1996) Continental break-up and collision in the Neoproterozoic and Palaeozoic- a tale of Baltica and Laurentia. Earth-Science Reviews 40: 229-258

Valter AA (1975) Deciphering of the Ilyinets structure as astrobleme [in Russian]. Doklady Akademi Nauk SSSR 224: 1377-1379

Valter AA, Ryabenko VA (1973) Petrological evidence of shock metamorphic origin of Ilyinets structure [in Russian]. Geological Journal (Kiev) 33: 142-144

Valter AA, Ryabenko VA (1977) The impact craters of the Ukrainian Shield [in Russian] (Ed. VG Bondarchuk). Kiev, Naukova Dumka, 156 pp

Werner SC, Plado J, Pesonen LJ, Kuulusa M (2001) The two Suvasvesi lakes in Finland - a possible doublet impact structure [abs.]. Meteoritics and Planetary Science 36: A223-A224

Cathodoluminescence, Electron Microscopy, and Raman Spectroscopy of Experimentally Shock Metamorphosed Zircon Crystals and Naturally Shocked Zircon from the Ries Impact Crater

Arnold Gucsik[1], Christian Koeberl[1*], Franz Brandstätter[2], Eugen Libowitzky[3] and Wolf Uwe Reimold[4]

[1]Department of Geological Sciences, University of Vienna, Althanstrasse 14, A-1090 Vienna, Austria.
[2]Department of Mineralogy, Natural History Museum, P.O. Box 417, A-1014 Vienna, Austria.
[3]Institute of Mineralogy and Crystallography, University of Vienna, Althanstrasse 14, A-1090 Vienna, Austria
[4]Impact Cratering Research Group, School of Geosciences, University of Witwatersrand, Private Bag 3, P.O. 2050, Johannesburg, South Africa
[*]Corresponding author: christian.koeberl@univie.ac.at

Abstract. Thorough understanding of the shock metamorphic signatures of zircon will provide a basis for the application of this mineral as a powerful tool for the study or recognition of old, deeply eroded, and metamorphically overprinted impact structures and formations. This study of the cathodoluminescence (CL) and Raman spectroscopic signatures of naturally shocked (Ries Crater, Germany) zircon crystals and experimentally (at 38, 40, 60, and 80 GPa) shock-metamorphosed single crystals of zircon contributes to the understanding of the formation of microdeformation in zircon under very high, dynamic pressures.

Unshocked samples show crosscutting, irregular fractures in the backscattered electron (BSE) images. The 38 GPa sample exhibits a dense pattern of narrow-spaced lamellar features, in CL mode, which could represent the twinning effect noted in a 40 GPa sample by earlier workers and which was ascribed to partial conversion from the zircon-structure phase to the more voluminous scheelite-structure phase. The CL images of experimentally, at 40, 60, and 80 GPa, shocked samples show subplanar and nearly parallel microdeformations. All experimentally shocked zircon, as well as all investigated naturally shocked samples from the Ries crater, show an inverse relationship between the brightness of the BSE signal and the corresponding cathodoluminescence intensity of the zonation patterns. The CL spectra of unshocked and experimentally shock-deformed specimens and naturally shock-metamorphosed zircon samples are characterised by narrow emission lines and broad bands in the region of visible light and in the near-ultraviolet range. The emission lines likely result from rare earth element activators and the broad bands might be associated with lattice defects. Raman spectra reveal that the unshocked samples, as well as naturally shock-deformed zircon crystals from the Ries, represent zircon-structure material, whereas the 38 and 40

GPa samples yield additional peaks with relatively high peak intensities, which are indicative of the presence of the scheelite-type structure of zircon with zircon-structure relics. The 60 and 80 GPa samples display a Raman signature that is characteristic of only the scheelite-type phase. According to the Raman measurements, the naturally shock-deformed zircons might be related to the low-shock regime (<30 GPa), and do not represent the same shock stages indicated by whole-rock petrography.

The results show a relationship between the CL and Raman properties of zircon and shock pressure, which confirm the possible use of these methods as shock indicators.

1 Introduction

Zircon is a highly refractory and weathering-resistant mineral that has proven useful as a shock level indicator of shock metamorphism in the study of impact structures and formations that are old, deeply eroded, and metamorphically overprinted (e.g., Bohor at al. 1993; Kamo et al. 1996; Krogh et al. 1996; Reimold et al. 2002). Zircon has advantages compared to quartz or other shock-metamorphosed rock-forming minerals that have been widely used as impact indicators, but are far less refractory than zircon. Furthermore, U-Pb dating of zircon can provide constraints on the ages of impact events or deposition of impact formations (e.g., Deutsch and Schärer 1994; Kamo et al. 1996, and references therein).

The activation of cathodoluminescence (CL) in zircon, which may be due to, for example, the presence of trace elements or may result from defects in the crystal structure, is generally rather strong, in comparison with the CL activity of many other minerals (e.g., Hanchar and Miller 1993; Vavra 1993; Hartmann et al. 1997; Blanc et al. 2000; Kempe at al. 2000). The nature and CL effects of activators such as Tb^{3+} or Dy^{3+} can be relatively easily investigated through synthesis and analysis of element-doped zircon samples (e.g., Cesbron et al. 1995; Blanc et al. 2000). A distinct disadvantage of using zircon in shock studies of natural materials is that this mineral is generally not very abundant, in comparison with quartz or feldspar, and commonly occurs only as small grains. Furthermore, only relatively felsic crustal rocks contain zircon in accessory amounts.

Shock metamorphic effects in zircon have been described from a number of impact environments, including confirmed impact structures (e.g., Kamo and Krogh 1995; Krogh et al. 1996; Kamo et al. 1996; Gibson et al. 1997; Glass and Liu 2001), the Cretaceous-Tertiary boundary (Bohor et al. 1993; Kamo and Krogh 1995), and the Upper Eocene impact ejecta layer (e.g., Glass and Liu 2001), as well as from tektites (Deloule 2001; Glass and Liu 2001). Shock-induced microdeformation in experimentally shock-deformed zircon crystals has also been reported (Deutsch and Schärer 1990; Leroux et al. 1999; Gucsik et al. 2002).

Two different types of shock-deformation structures have been observed: (i) planar microdeformation features and (ii) granular texture (also called polycrystalline, microcrystalline, or strawberry texture). Some effort, especially by transmission electron microscopy (TEM), has been made to determine whether the planar microdeformation features discernable at the optical scale in shock-metamorphosed zircon represent bona fide planar deformation features (PDFs), well-known from many other shock-metamorphosed rock-forming minerals (e.g. French 1998; Stöffler and Langenhorst 1994; Grieve et al. 1996), or whether they represent planar fractures or some other type of microdeformation (Reimold et al. 2002). To date, this problem has not been solved. Leroux et al. (1999) and Reimold et al. (2002) also established that, on a nanometer scale, amorphous phases in the form of planar lamellae were formed in zircon experimentally shocked at 40 and 60 GPa. However, these authors were unable to confirm that these microlamellae, though resembling PDFs, indeed corresponded to the optically resolved, several μm wide, planar/subplanar microdeformations. Leroux et al. (1999) also observed numerous planar fractures and dislocation deformation bands. It can not be excluded that these features corresponded to the optically resolved planar microdeformation features.

Granular texture was first observed by Bohor et al. (1993) in zircon from the Cretaceous/Tertiary distal impact ejecta layer. Since then, it has been observed in zircon from a number of impact structures (e.g., Kamo et al. 1996), from a Late Eocene microkrystite layer, and from tektites (e.g., Glass and Liu 2001; Deloule et al. 2001). Baddeleyite, formed as a result of thermal decomposition of zircon (T >1676 °C), has been identified in impact-generated rocks (e.g., El Goresy 1966; El Goresy et al. 1968) and in Libyan Desert Glass (Kleinmann 1969), and its presence has been quoted as evidence for an impact origin of these enigmatic glasses. According to Reimold et al. (2002), granular shock texture has been interpreted as the result of recrystallization of zircon to aggregates of smaller crystals in response to high temperatures induced by the shock process.

The phase transformation from the zircon crystal structure ($ZrSiO_4$) to a scheelite ($CaWO_4$)-structure phase in shock-metamorphosed zircon was described by Kusaba et al. (1985) to begin at about 30 GPa and to be complete at around 53 GPa. These observations were confirmed by Leroux et al. (1999) through their TEM investigations of experimentally shocked zircon. More recently, according to Scott et al. (2002), high-pressure X-ray data show that a small amount of residual zircon-structured material remained at 39.1 GPa on $ZrSiO_4$ scheelite-structure material. Glass et al. (2002) found the scheelite-type phase in zircon samples from marine sediments from an upper Eocene impact ejecta layer sampled near New Jersey and Barbados. They named this mineral phase 'reidite' after Alan F. Reid, who first produced this high shock-pressure polymorph of zircon (Reid and Ringwood 1969).

This study of the cathodoluminescence and Raman spectroscopic properties of experimentally shock-metamorphosed zircon crystals and naturally shock-deformed zircon specimens from the Ries (Germany) impact structure contributes to the understanding of high-pressure microdeformation in zircon, and continues our previous work (Gucsik et al. 2002). We present the results of cathodoluminescence and backscattered electron (BSE) imaging, as well as of Raman spectroscopy, performed on experimentally shocked (shock pressures from 38 to 80 GPa) zircon specimens, which were cut perpendicular and parallel to, and at 45° to the c-axis, and on naturally shock-deformed zircon samples from the Ries impact crater (southern Germany). The purpose of this investigation is to further investigate the capability of the SEM-CL technique and Raman spectroscopy to document shock deformation and to determine whether specific CL or Raman effects in zircon/scheelite-structure can be utilised to determine particular shock pressure stages.

2
Samples and Experimental Procedures

Two natural zircon crystals of about 1 cm length and 0.5 and 0.7 cm width from Australia (sample A) and from Sri Lanka (sample B) were experimentally shock deformed at shock pressures between 38 and 80 GPa. The zircon samples had been cut into thin plates of about 1 mm thickness, parallel to and at 45° to their crystallographic c-axes.

Shock recovery experiments were performed on such plates using the shock reverberation technique at the Ernst-Mach-Institute, Germany (e.g., Deutsch and Schärer 1990; Stöffler and Langenhorst 1994). For the present study, polished thin sections, produced from the experimentally shocked zircon plates were utilized. However, only the specimens from the A sample were selected for further CL- and Raman spectroscopic investigations, because the sections are of better quality than those from sample B. The samples were first examined under a petrographic microscope. Digital optical images were stored without further computer enhancement. The samples were then examined with an Oxford Mono-CL system attached to a JEOL JSM 6400 scanning electron microscope (SEM). Operating conditions for all SEM-CL investigations were 15 kV accelerating voltage and 1.2 nA beam current; backscattered-electron (BSE) and cathodoluminescence (CL) images were obtained. The BSE and CL images were captured digitally. CL spectra were recorded in the wavelength range of 200-800 nm, with 1 nm resolution. The grating of the monochromator was 1200 lines/mm.

Raman spectra were obtained with a Renishaw RM1000 confocal micro-Raman spectrometer with a 20 mW, 632 nm He-Ne laser excitation system and a thermoelectrically cooled CCD detector. The power of the laser beam on the sample was approximately 3 mW. Spectra were obtained in the range 100-1200 cm^{-1}, with approximately thirty seconds total exposure time. The spectral resolution (apparatus function) was 4 cm^{-1}. Areas of approximately 450 x 450 μm were selected for CL and BSE imaging. Raman spectra were taken from 3 μm^3 sample volume and CL spectra were obtained from approximately 35 x 45 μm areas. In general, three CL spectra and ten Raman spectra were acquired per sample. Figures 1a and b show backscattered-electron images as overview images of experimentally (parallel and 45°) and naturally shock-deformed zircon samples (parallel and perpendicular). In

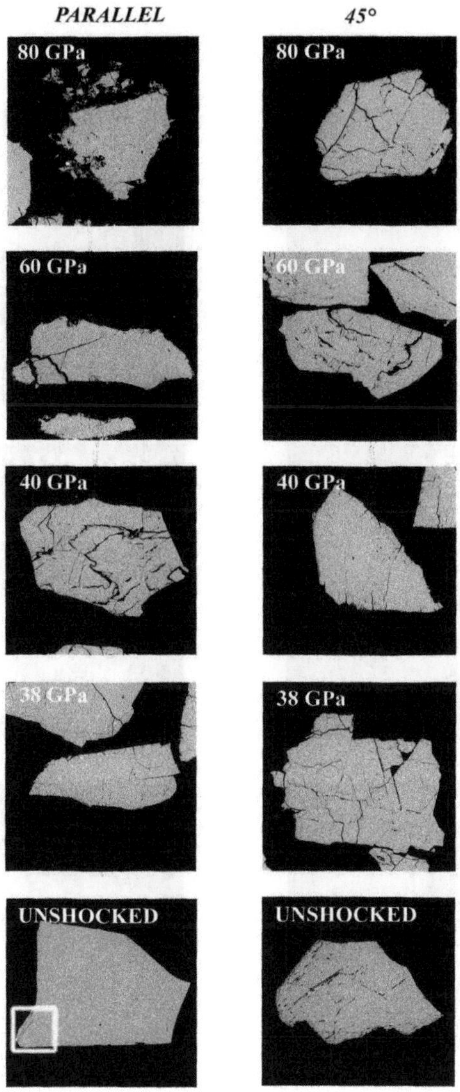

Fig. 1a. Backscattered-electron (BSE) images of experimentally ("parallel"; "45°") and naturally shock-deformed zircon at various shock stages. Grain sizes of experimentally shock-metamorphosed zircon samples vary between 0.8 and 1.4 mm. The BSE images exhibit grain-pervasive irregular fractures at all shock stages and show zonation patterns mainly in the naturally shock-deformed zircon samples. In general, 400x400 μm areas were scanned for CL images with higher magnification from approximately the center of all grains (except the parallel unshocked grain, which is marked by a white rim).

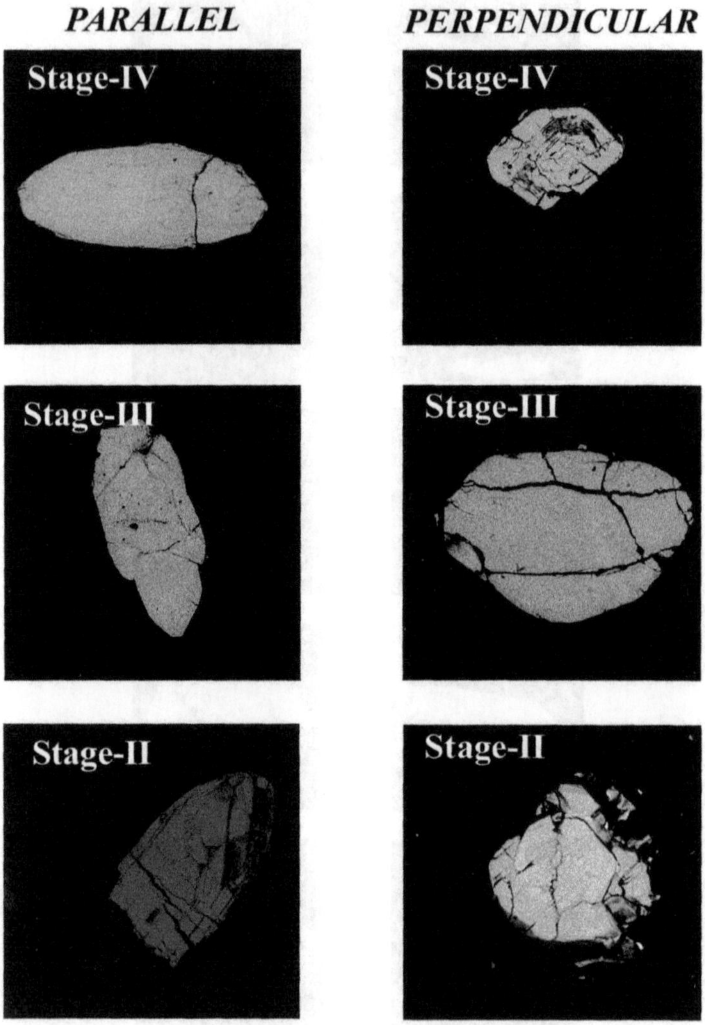

Fig. 1b. The grain sizes of naturally shock-metamorphosed zircon grains from the Ries impact structure range from 0.5 to 0.9 mm. The BSE images exhibit grain-pervasive irregular fractures at all shock stages. (Shock stages refer to the overall shock degree of the host rock sample).

this paper, the terms ‘perpendicular’, ‘parallel’ and ‘45°’ refer to direction with respect to the crystallographic c-axis.

For this study, zircon separates were obtained from Dr. R. Schmidt and Dr. A. Greshake, Humboldt-Universität Berlin. The zircons had been separated from three rock samples from the Ries crater: (1) a glass bomb from suevite from the Aumühle site, which had been classified on the basis of shock metamorphic effects as a shock stage IV specimen; (2) a biotite gneiss sample from Appertshofen, classified as stage II, and (3) crystalline rock fragments from a suevitic sample obtained near Seelbronn, the shock stage of which was given as stage III (regarding definitions of shock stages, compare Table 1). These locations are shown in Figure 2. Crystals from these three zircon separates were cut both parallel and perpendicular to the c-axis, for the purpose of CL and Raman spectrometric analysis. Zircon separates from shock deformed rock samples from the Ries crater were obtained with standard methods, starting with milling of the rock samples in an agate mill, sieving of the product to different grain size fractions, and finally separation with heavy liquid. The grains were washed in ultra pure acetone, and then cut parallel and perpendicular to their crystallographic c-axis.

3 Background on the Geology of the Ries Impact Crater

The Ries (also called “Nördlinger Ries”) is a complex impact structure, located in Southern Germany (centered at: N 48°53’, E 10°37’), with a rim-to-rim diameter of about 26 km (e.g., Engelhardt 1990; Deutsch 1998). The well-preserved ejecta blanket (Geologische Karte des Rieses, Bayerisches Geologisches Landesamt, 1: 50 000, 1999) and intra-crater breccia lens offer excellent conditions for studies of terrestrial impact structures (e.g., Engelhardt 1990) (Fig. 2). According to the level of shock metamorphism, the ejecta from the Ries crater can be divided into two formations: (1) Low shock level (<10 GPa), represented by the Bunte Breccia (sediments from the upper 600 m of the target area), megablocks of sedimentary rocks (Upper Jurassic limestone blocks occurring in the Megablock zone), and megablocks and monomict breccias derived from the crystalline basement (from the surrounding area of the Inner Crater); and (2) high shock level (>10 GPa), represented by polymict crystalline breccias (are common in the Megablock Zone), fall-out suevites, crater suevite, and tektites (moldavites) (Engelhardt 1999). A 15 Ma age was determined for the Ries impact by the ^{40}Ar-^{39}Ar method on impact glasses from suevitic impact breccias (Staudacher et al. 1982). The present structure formed as the result of the collapse of the transient impact crater by subsurface readjustment, which caused upward movement of the initial crater bottom and slumping of parts of the primary rim into the crater, producing a compensating ring depression (the so-called Megablock Zone) (Engelhardt 1990). The pre-impact stratigraphy of the target region includes a crystalline basement of pre-Variscan gneisses and amphibolites, and Variscan granite. These crystalline rocks were overlain by sedimentary rocks of Upper Jurassic (limestone), Middle Jurassic

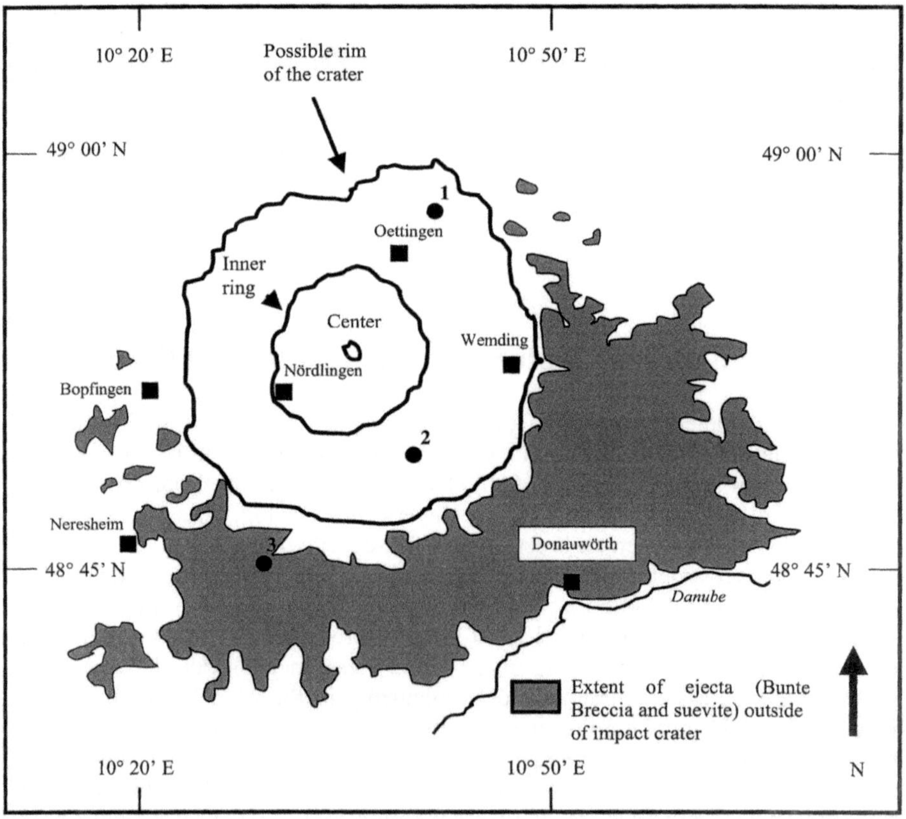

Fig. 2. Locality of the Ries impact structure in Germany and approximate extent of the occurrence of Bunte Breccia and suevite. Sample localities are indicated with a full black circle: 1=Aumühle, 2=Appertshofen, 3=Seelbronn. Outline of crater and extent of Bunte Breccia and suevite distribution redrawn from the 1:50 000 map of the Bayerisches Geologisches Landesamt (Geologische Karte des Rieses 1999). Scale: the diameter of the apparent rim of the crater is approximately 26 km.

(sandstone, marlstone, limestone), Lower Jurassic (sandstone, marlstone, limestone), Upper Triassic (sandstone, siltstone, marlstone, claystone) and Lower Triassic (sandstone) age. The southern part of the target area was covered by ~25 m of unconsolidated Upper Miocene sands, marls and clays (Engelhardt 1990).

The Bunte Breccia is an approximately 200-m-thick ejecta deposit occurring in an asymmetric distribution (extending farther to the south and east than to the north and west) pattern (Fig. 2). The breccia is composed of unshocked and moderately shocked rock and mineral fragments of sedimentary and crystalline origin, as well as megablocks and monomict breccias derived from the crystalline basement (Deutsch 1998). Suevite or suevite breccia is a polymict impact breccia with clastic matrix and mineral clasts of various stages of shock metamorphism including cogenetic impact melt particles which are in a glassy or crystallized state. Suevite types from the Ries crater can be classified into two groups: (1) The fallout suevite contains, on average, 4 vol% crystalline basement clasts, 0.4 vol% sedimentary rocks, 16 vol% glass bodies, and 79 vol% groundmass (matrix). It occurs in small isolated patches in the Megablock Zone, inside the morphological rim and up to 10 km beyond the rim. (2) The crater suevite contains the same crystalline rock types (62 vol% metamorphic and 38 vol% igneous rocks) that occur in the fallout suevites. The crater suevite differs from fallout suevite by a higher clast/glass ratio, by preponderance (65-95 %) of clasts, and by absence of aerodynamically shaped glass bodies (Engelhardt 1997).

Effects of high degrees of shock deformation (>10 GPa) in quartz and other rock-forming minerals (e.g., feldspars), such as planar deformation features (PDFs), were first described from shocked granite inclusions in suevite (Engelhardt and Stöffler 1965). Additionally, shock metamorphic indicators, such as high-pressure mineral phases (e.g., coesite, stishovite) (Chao 1968; Stöffler 1972), diaplectic quartz and feldspar glass (Stöffler 1966), and fused quartz glass (lechatelierite) (Stöffler 1974; French 1998) from suevite were also found in impact breccias from the Ries Crater. The Ries crater is the source of moldavite tektites of the Central European Strewn Field (e.g., Engelhardt 1990).

Recently, Boggs et al. (2001) identified planar deformation features (PDFs) in shocked quartz from the Ries Crater using a scanning cathodoluminescence imaging facility. These authors discussed some differences between these shocked quartz grains and planar microstructures in quartz associated with tectonic fracturing.

According to Stöffler (1974) and Engelhardt (1990), the stages of shock metamorphism in minerals from the impact formations of the Ries impact crater can be classified into six stages (known as 0, I, II, III, IV, and V) that are characterized by various elastic and plastic deformation phenomena as well as isotropization of minerals, the formation of high-pressure phases and the occurrence of mineral or bulk rock melting. Based on shock recovery experiments, shock pressure and post-shock temperature values can be assigned to these different stages, as listed also in Table 1.

Table 1. Stages of shock metamorphism in the Ries impact structure (from Engelhardt 1990)

Stage	Shock effects	Pressure (GPa)	Post-shock Temperature (°C)
0	Fragmentation; mosaicism;undulatory extinction; deformation bands in quartz; kinkbands in biotite; shatter cones	0-10	0-100
I	Planar elements in quartz; planar deformation lamellae in feldspar, amphibole and pyroxene; stishovite, coesite; kink bands in biotite	10-35	100-300
II	Diaplectic glasses of quartz and feldspar; deformation lamellae in amphibole and pyroxene; kinkbands in biotites	35-45	300-900
III	Selectively fused feldspar; diaplectic quartz glass; thermal decomposition of biotite and amphibole	45-50	900-1300
IV	Complete fusion of rocks (granitic composition); impact melts	60-80	1500-3000
V	Vaporization	>80	>3000

4 Background on Cathodoluminescence and Raman Effect

4.1 Origin of Cathodoluminescence

When an energetic electron beam interacts with a solid surface, its energy is dissipated as a series of physical effects, including emission of secondary electrons (SE), ejection of back-scattered electrons (BSE), electron absorption ("sample current"), characteristic X-rays, and cathodoluminescence (CL) emission (Marshall 1988). Eventually, a large amount of the total incident beam energy is transformed into heat, resulting in non-radiative emissions, such as phonons, which are tiny packets of vibrational energy associated with heat (Wolfe 1998). The penetration depth of electrons depends on their energy (generally in CL investigations, 10 - 20

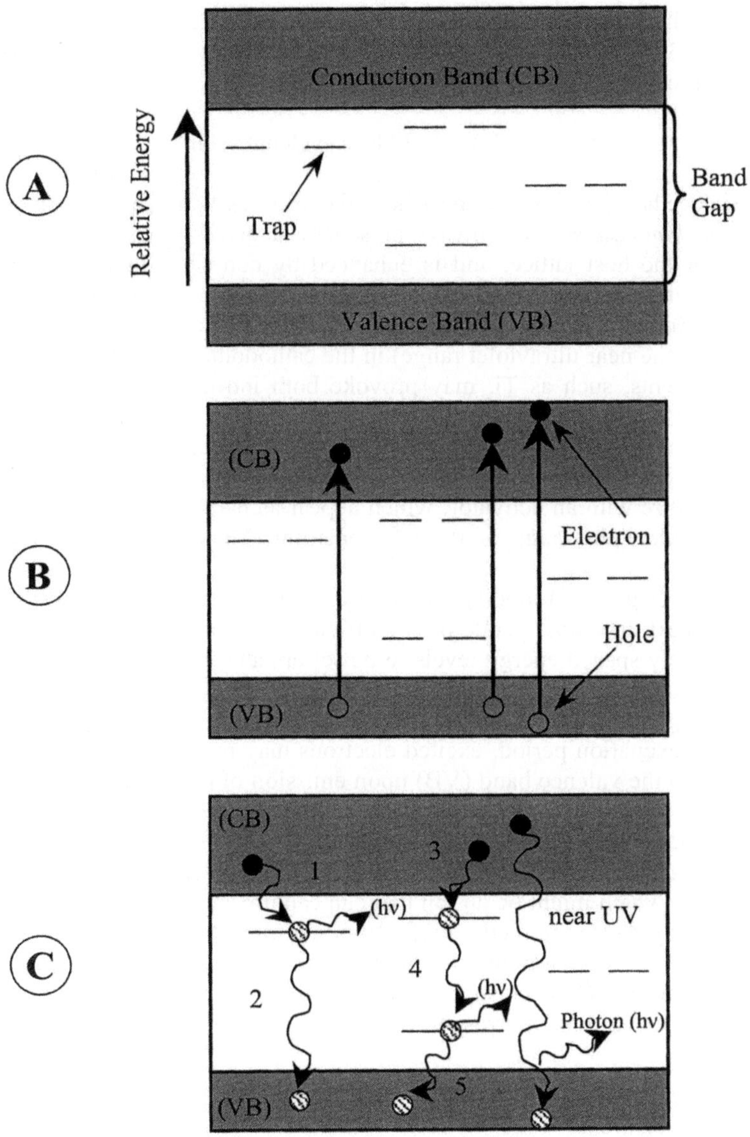

Fig. 3. Schematic representation of the luminescence effects produced by an electron beam interacting with a sample in the SEM. The origin of CL could be based on electrons, which may return back from the conduction band (CB) to the valence band (VB) (Figs. 3A and B), emitting photons in three different paths. They can move back directly (1), randomly (2), and encountering (3) a trap or a recombination center. Excited electrons may return back indirectly by first finding traps. They may do so through a single step (processes 1 and 2, Fig. 3C) or a series of steps (processes 3, 4, and 5 - Fig. 3C) (after Boggs et al. 2001).

keV) and is on the order of 2 to 8 μm (Marshall 1988, Hayward 1998). Cathodoluminescence represents visible radiation (also manifested in the ultraviolet range) in the wavelength range between 400 and 700 nm, corresponding to energies between 1.77 and 3.10 eV. The relationship between energy (eV) and wavelength (nm) can be expressed as follows (Rèmond et al. 2000):

$$\text{energy (eV)} = 1239.8/\ \text{wavelength (nm)} \qquad (4.1.)$$

The luminescence has three causes: intrinsic and extrinsic CL, and quenchers.

(1) *Intrinsic luminescence* (is always present in a given mineral species), is characteristic of the host lattice, and is enhanced by non stoichiometry (vacancies), structural imperfections (poor ordering, radiation damage, shock damage), and impurities (nonactivators), which distort the crystal lattice causing broad bands (mostly in the near ultraviolet range) in the cathodoluminescence spectrum. Some trace elements, such as Ti, may provoke both intrinsic and extrinsic CL (Marshall 1988, Hayward 1998).

(2) Impurities responsible for *extrinsic luminescence* are called activators. Elements are referred to as sensitizers when their presence is necessary to create a luminescence centre with an activator, which appear as narrow or sharp emission lines in the visible light range of the CL spectrum (Marshall 1988, Hayward 1998).

(3) *Quenchers* (e.g., Fe^{2+}) modify the energy level arrangement so that the luminescence process does not operate or is inefficient. The presence of a quencher causes new, closely spaced energy levels to be set up, and the electron can easily return to the ground state with the emission of a succession of low-energy photons or by losing energy to the lattice as heat (Marshall 1988, Hayward 1998) .

After a short excitation period, excited electrons may return from the conduction band (CB) to the valence band (VB) upon emission of photons, via three different pathways (Boggs et al. 1991) (Figs. 3A and B): (1) The promoted electrons fall directly back to their lower energy level; (2) they can move back randomly through the crystal structure, until they encounter a trap or a recombination center (i.e., an activator, which might be an ion that can capture an electron); and (3) excited electrons may return indirectly, by first finding traps, then vacating the traps with concomitant emission of photons. These paths are schematically illustrated in Figure 3C. A band gap (BG) exists between valence band and conduction band, which – in the case of zircon - corresponds to approximately 5.3 eV (Fig. 3A).

4.2 Background on Raman Spectroscopy

According to Roberts and Beattie (1995), when a monochromatic beam of light illuminates a medium (e.g., gas, liquid or solid), most of the light will traverse the sample without undergoing any changes. Approximately 10^{-3} of the incident intensity is scattered with the same frequency (ν_0) as that of the incident light source (elastic or Rayleigh scattering). Raman scattering ("Raman effect") occurs when, in addition to the Rayleigh scattering, about 10^{-6} of the incident intensity is scattered at new frequencies above ($\nu 0+\Delta\nu$) and below ($\nu 0-\Delta\nu$) the incident frequency. $\Delta\nu$ corresponds to vibrational energies in the medium, with "+" indicating a transition from an excited state to the ground state, and "-" the opposite transition (to the excited state). This effect was first documented by the Indian physicist Chandrasekhara Venkata Raman in 1928, who observed colour changes in the scattered light of focused sunlight. The shifts in frequency $\Delta\ \nu$ from that of the incident radiation are independent of the exciting radiation ν_0 and are characteristic of the species that gives rise to the scattering.

Raman transitions to lower frequency (i.e., red shifted bands) are referred to as Stokes lines, whereas transitions to higher frequency (i.e., blue shifted bands) are referred to as anti-Stokes lines. Stokes lines are normally much more intense than anti-Stokes lines, as the population of the ground state is usually very much greater than that of the excited state of the molecule. Not all molecular vibrations are Raman active. A Raman active vibration can be expected if the polarizability in a molecule is changed during the normal vibration (Roberts and Beattie 1995). The frequencies of vibrations depend upon the vibrating masses and the forces between them, including the anharmonic nature of interatomic and inter-and intramolecular interactions. If a phase transition occurs, the Raman selection rules, which ultimately depend on crystal and molecular symmetries, will also change (as well as forces) and new spectral features, characteristics for the new lattice, will appear (e.g., Williams and Knittle 1993; Nasdala 1995). Thus, this method is not only of great help in elucidating crystal structures (e.g., McMillan and Hofmeister 1988), but can also be used as a method of qualitative analysis (finger print), e.g., in determining the presence of specific phases of small thin section areas without destroying them.

5 Results

5.1 Backscattered-Electron (BSE) and Cathodoluminescence (CL) Image Observations

5.1.1 Experimentally Shock-Deformed Zircon Samples

The specimen of unshocked (parallel) zircon shows lack of any types of cracks or fractures in BSE mode. In contrast, the unshocked sample cut at 45° to the c-axis exhibits a number of open, irregularly shaped fractures in BSE mode, with a great variety of widths (Fig.1a). The CL images of both unshocked samples contain patchy areas of different CL brightness (Figs. 4a and b). These relatively bright and dark zones are interpreted as the result of variations in concentration and distribution of U and Th contents. Brighter fluorescing domains are relatively poorer

(a)

(b)

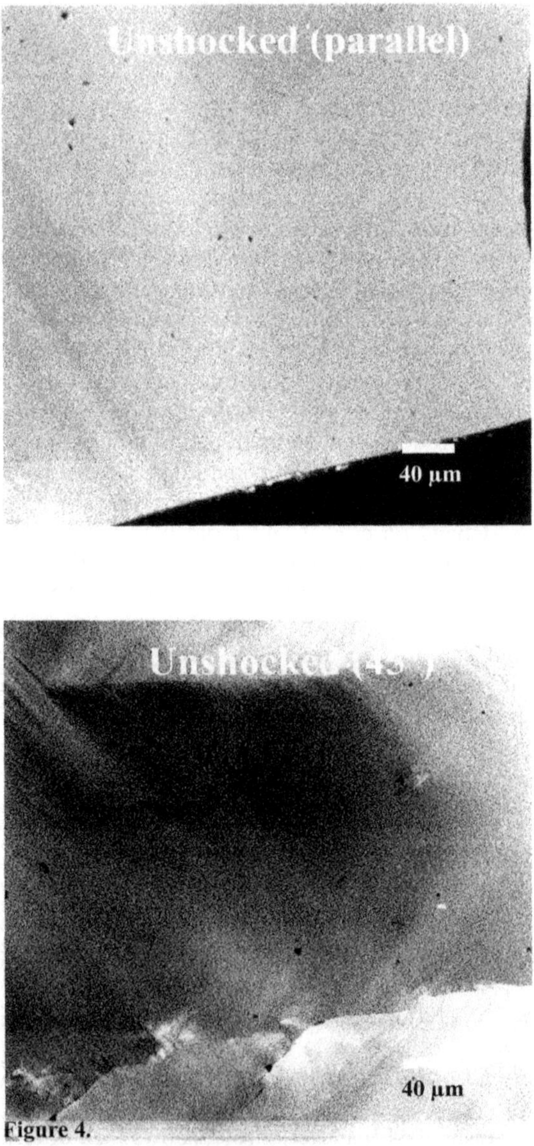

Fig. 4. CL images of the unshocked samples. The parallel-sample (the crystal was cut parallel to the crystallographic c-axis) does not show any fractures (Fig. 4a). The (45°) unshocked sample (the crystal was cut 45° to the crystallographic c-axis) exhibits a low density of individual, irregular fractures (Fig. 4b).

in U content than the weaker fluorescing (dark) domains (Rubatto and Gebauer 2000).

Gucsik et al. (2002) described open, irregularly shaped, fractures (widths typically ranging from 5 to 25 µm), which are easily visible in the BSE image of an unshocked sample. It was also noted that the unshocked grain displays a well-developed growth zonation parallel to the crystal edges and patchy areas exhibiting a distinct inverse relation between BSE and CL brightness. They concluded that the BSE contrast is sensitive to changes in abundance of elements, such as Hf, Y, Yb, U and Th, in natural zircon, causing a higher BSE yield. In contrast, in the CL images, these elements cause quenching of the CL signal.

Compared to the unshocked samples, the fracture density visible in BSE images (Fig. 1a) of the 38 GPa specimens is significantly higher, irrespective of crystallographic orientation of the specimen. The CL image of the parallel-sample shows 3-4 µm wide, curved and more or less parallel microdeformations spaced at 5-15 µm (Fig. 5a). A high density of much closer spaced, clearly crystallographically controlled features is visible in CL mode in the 45°-sample. These features are very narrow (widths of approximately 1-2 µm) and closely spaced (at approximately 1-3 µm) (Fig. 5b). Two sets of different orientation can be discerned (Fig. 5b).

Gucsik et al. (2002) noted in their 40 GPa specimen (cut perpendicular to the crystallographic c-axis) that this sample displays a high density of well-developed lamellae that are arranged in narrow-spaced sets (which are best visible in the CL images). These features (dark in the CL images) show a narrow range of widths (1-2 µm) and spacings (bright in the CL image) of 1-2 µm. In the CL image of the 40 GPa (45°) sample of this study, we observe narrow, closely-spaced planar microdeformations (widths: approximately 1-2 µm; spacing: approximately 1-3 µm) (Fig. 6b). The density of irregular fractures is similar to that in the 38 GPa samples. The subplanar, widely spaced fractures are easily visible in the CL images (Fig. 6b). In the parallel-cut 40 GPa specimen, the CL brightness is significantly higher than in the 45°-sample (Figs. 6a and b), which could be the result of relatively higher concentration of, for example, U or Th in this sample. The parallel-cut sample exhibits patchy areas with relatively higher CL intensity (Fig. 6a). Images for both samples were acquired under the same operating conditions. No computer enhancement was carried out on these images.

The long, irregular fractures and the narrower microfractures are very prominent in the 60 GPa samples (both, cut parallel and 45°) (Figs. 7a and b). In the CL images, subplanar, 3-8 µm wide and 15-20 µm spaced, apparently crystallographically controlled, lamellar microdeformations are observed between the irregularly shaped, cross-cutting fractures. They are best visible in the CL image of the 60 GPa, parallel-cut sample (Fig. 7a). Gucsik et al. (2002) observed grain-pervasive, relatively widely spaced (10-20 µm), clearly subparallel features, which are set into a network of <3 µm wide features in the CL images of the sample shocked at 60 GPa.

Compared to the parallel 80 GPa sample, the density of irregular fractures (width: approximately 2-8 µm) is much higher in the 45° specimen (Figs. 8a and b). The subplanar, widely spaced (width: 1-2 µm, spacing: 12-15 µm), crystal

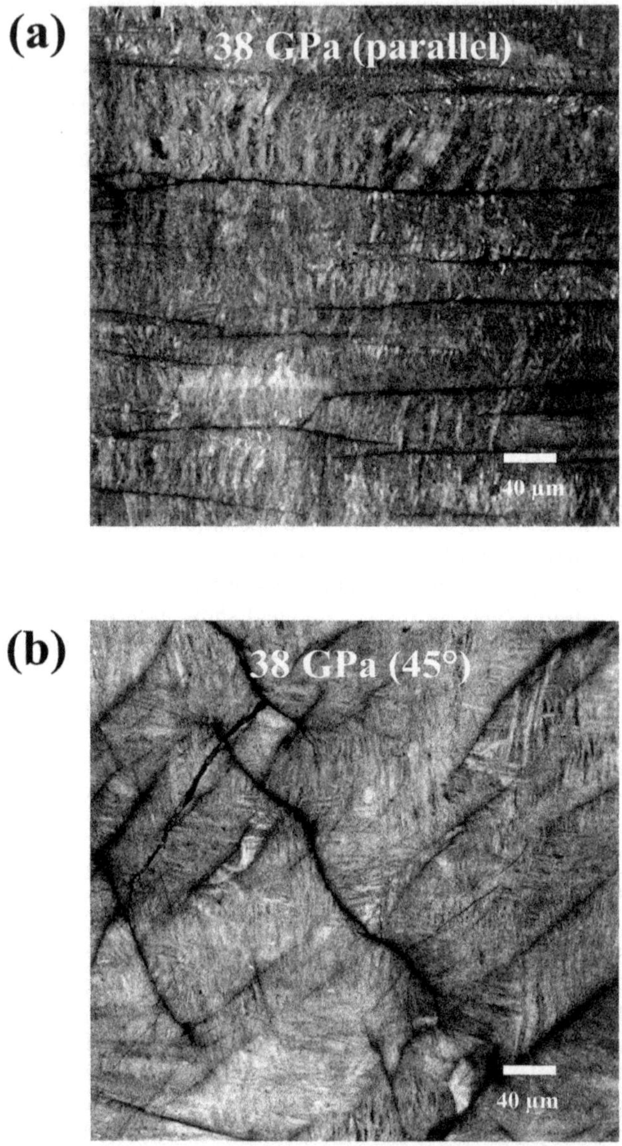

Fig. 5. The CL image of the 38 GPa parallel-sample shows irregular fractures and subplanar, nearly parallel microdeformations (Fig. 5a). The CL image of the 45°-sample is characterized by widely spaced irregular fractures, which are set into subplanar, mostly parallel microdeformations (Fig. 5b). The CL image of the 38 GPa 45°-sample exhibits planar microfractures as narrowly spaced, planar and parallel, dark lines (spacing: 1-3 µm) (Fig.5b).

(a)

(b)

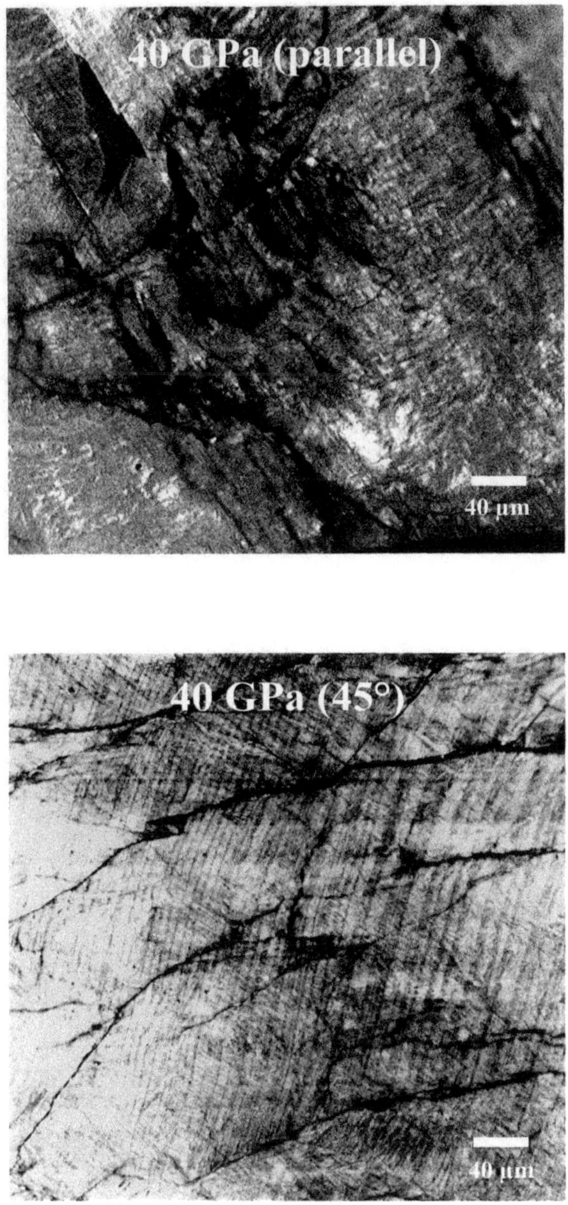

Fig. 6. Both CL images of this shock stage (parallel and 45°) show subplanar, mostly parallel microdeformations (Figs. 6a and b). In the CL image of the 40 GPa (45°) samples, dominantly narrow, dense, closely-spaced planar features (widths: approximately 1-2 μm; spacing: approximately 1-3 μm) can be seen (Fig. 6b).

(a)

(b)

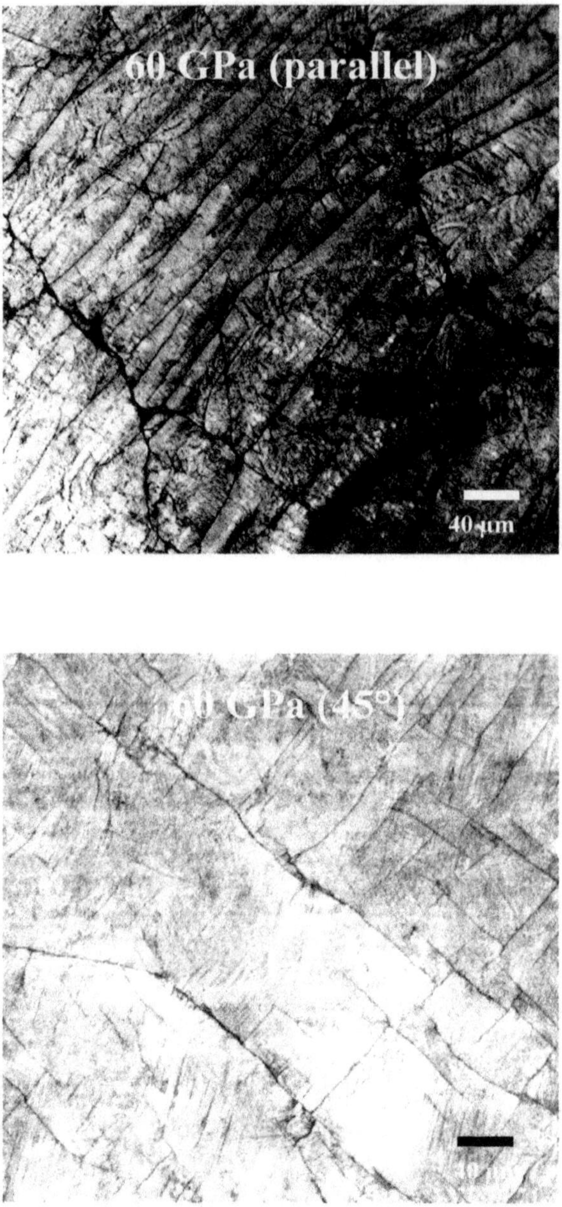

Fig. 7. CL images of the 60 GPa sample (Figs. 7a and b). The CL image of the parallel-sample is dominated by planar fractures (Fig. 7a). The 45° sample shows irregular, cross-cutting fractures (Fig. 7b).

(a)

(b)

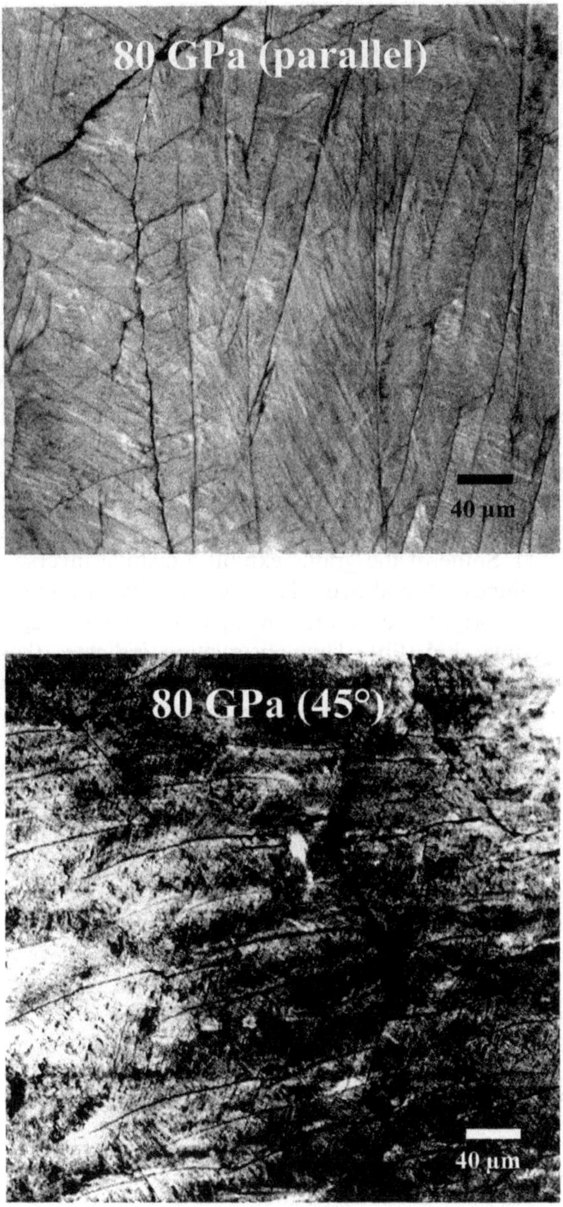

Fig. 8. The CL image of the parallel-sample shocked at 80 GPa is characterized by widely spaced irregular fractures, which are connected by subplanar, mostly parallel microdeformations (Fig. 8a). Irregular and crosscutting fractures are visible in the 45°-sample (Fig. 8b).

lographically controlled microdeformations are connected into the irregularly shaped, crosscutting fractures, which are best visible in the CL image of the parallel-sample in the 80 GPa specimen (Fig. 8a). The 45°-sample shows some areas (with relatively high CL intensity) without high density of irregular fractures or any types of planar microdeformations (Fig. 8b).

5.1.2
Naturally Shock-Deformed Zircon Crystals from the Ries Crater, Germany

The grain size of these specimens is relatively small (approximately 0.8 and 1.4 mm), in comparison with the experimentally shock-deformed zircon crystals. In general, all naturally shock-deformed zircon grains from the Ries Crater that were studied by us (in total, 12 grains) are characterized by the presence of low densities of open, wide, and irregularly-shaped fractures. Individual fractures typically have widths between 5 and 13 µm. Many of these grains display a well-developed growth zonation parallel to the crystal edges (e.g., Figs. 9-11). The widths of individual growth zones vary from 5 to 15 µm. These zones rarely exhibit weak variation of brightness in BSE imaging (Fig. 1b), but are generally well visible in CL mode (Figs. 9-11). Some of the grains exhibit a distinct inverse relation between BSE and CL brightness (see above). The high density of micro-lamellar features (two orientations of bright lines in the lower part of the CL images of a grain from the Stage-III sample: Fig.10b). This observation indicates that this grain was shocked at 40 GPa, as these grains were described by Gucsik et al. (2002). These authors found a high density of micro-lamellae in their 40 GPa sample (best visible in the CL mode), which is similar to the microdeformations of Stage-III. It was also suggested by these authors, according to Raman spectroscopy, that the zircon-structure was partly converted to the scheelite-type phase.

Whilst these three zircon fractions were obtained from three crystalline rock samples from the Ries crater, classified as shock stages II to IV on the basis of the respective variation of shock metamorphic effects in quartz and feldspar, the variation of deformation effects noted in this BSE and CL investigation of zircon grains is not compatible with these shock classifications. Clearly this is the result of shock distribution heterogeneity, resulting in a wide range of shock stages observed as microdeformations in different grains, as well as partially or completely melted phases.

5.2
Raman Spectroscopy

In the present study, we show only one representative spectrum per sample, because the other spectra were identical. Intensity variations due to polarization effects were not observed, as all samples were measured at approximately 45° to the

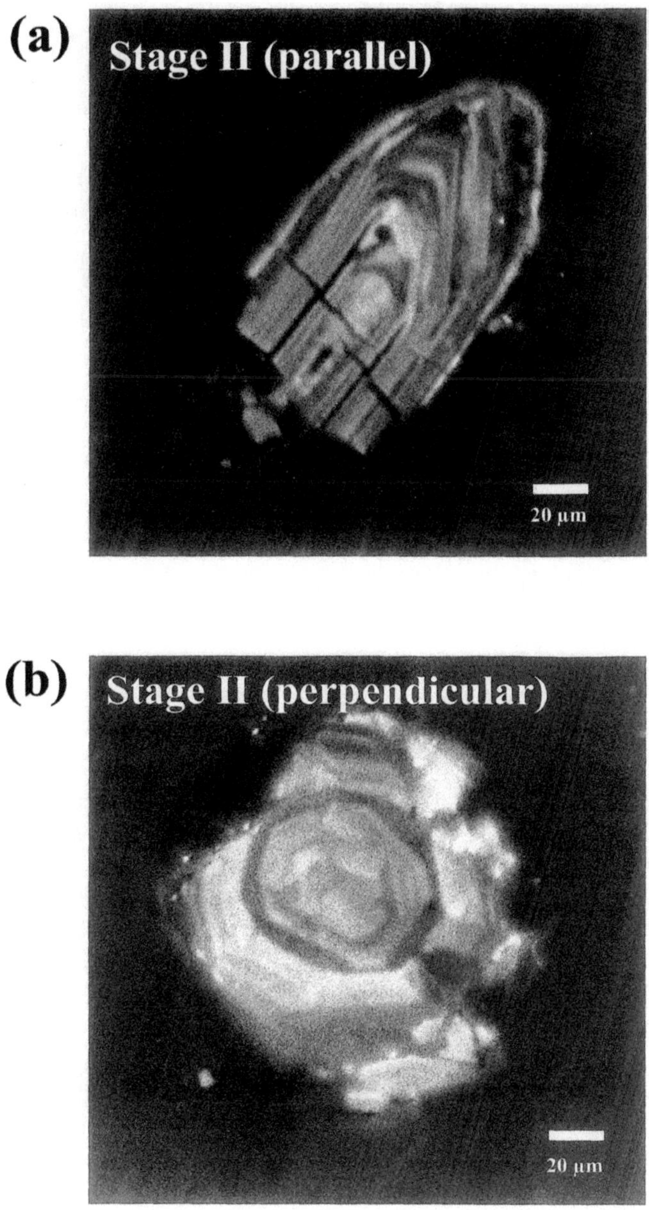

Fig. 9. CL images of both samples (specimens cut parallel and perpendicular to the c-axis) from a Stage-II rock sample from the Ries crater show grain-pervasive, irregular fractures (Figs. 9a and b). Both CL images show zonation patterns.

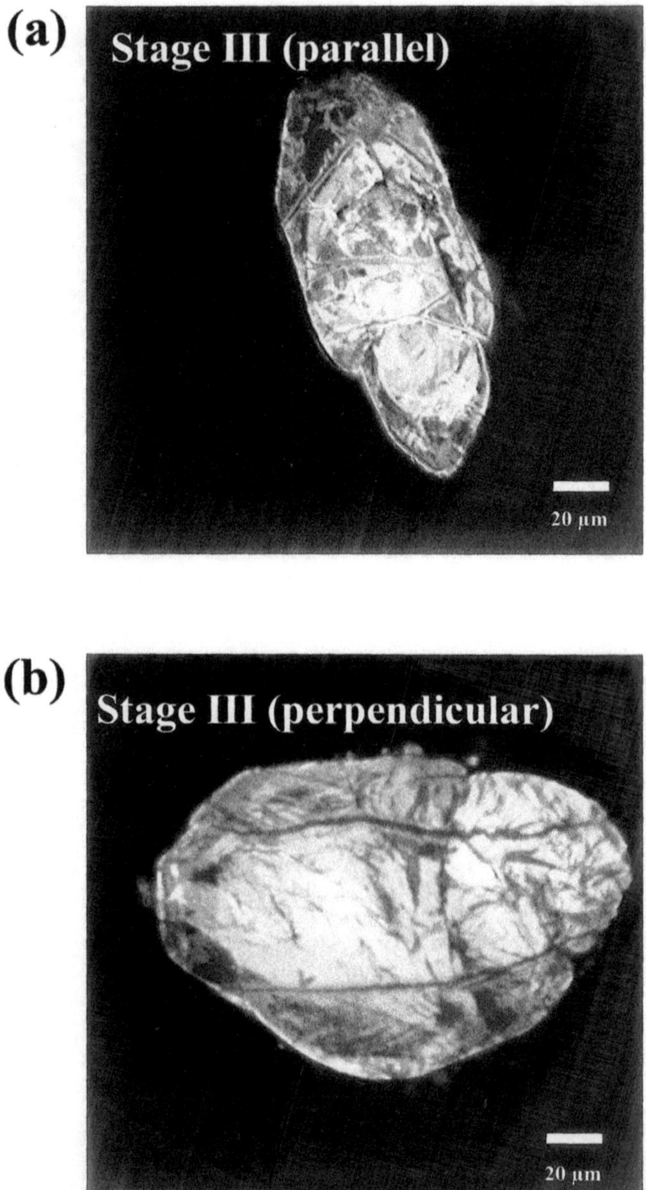

Fig. 10. CL images of zircon from a Stage-III sample from the Ries crater. Both CL images are characterized by zonation patterns (Figs. 10a and b). Shock-related lamellar structure is, for example, visible in the lower part of the image from the perpendicular sample (two sets of bright CL lines) (Fig. 10b).

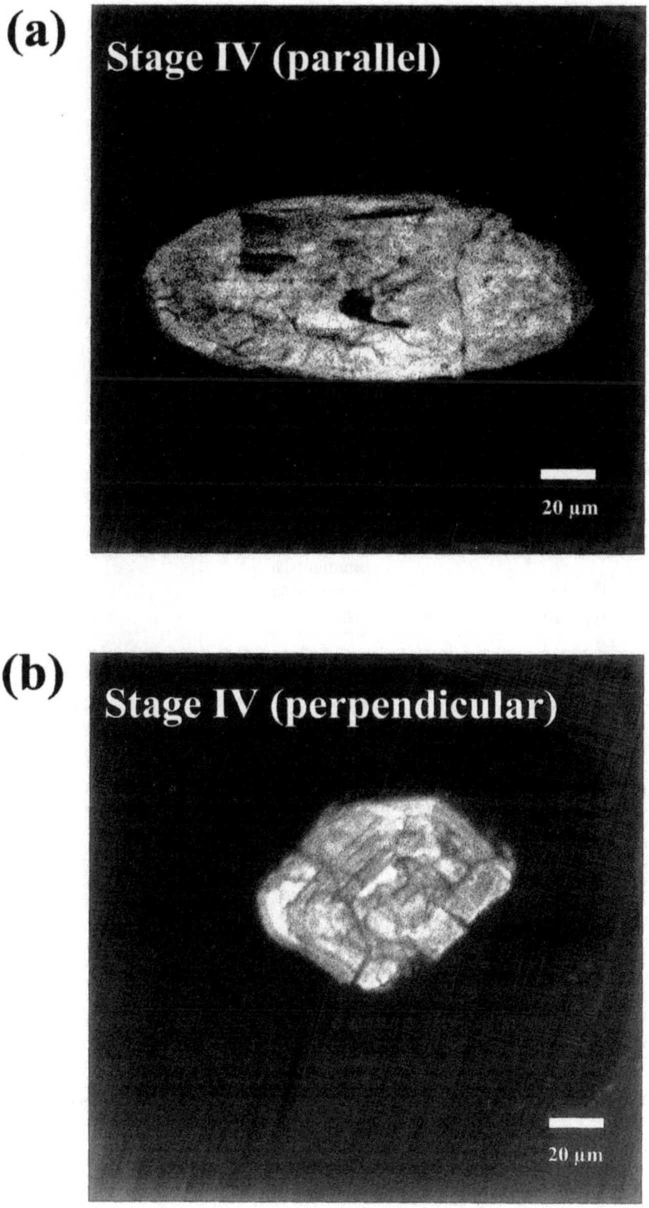

Fig. 11. CL images of zircon of the Stage-IV Ries crater sample are characterized by zonation patterns (Figs. 11a and b). The CL brightness is relatively weak in the CL image of the parallel-sample (Fig. 11a).

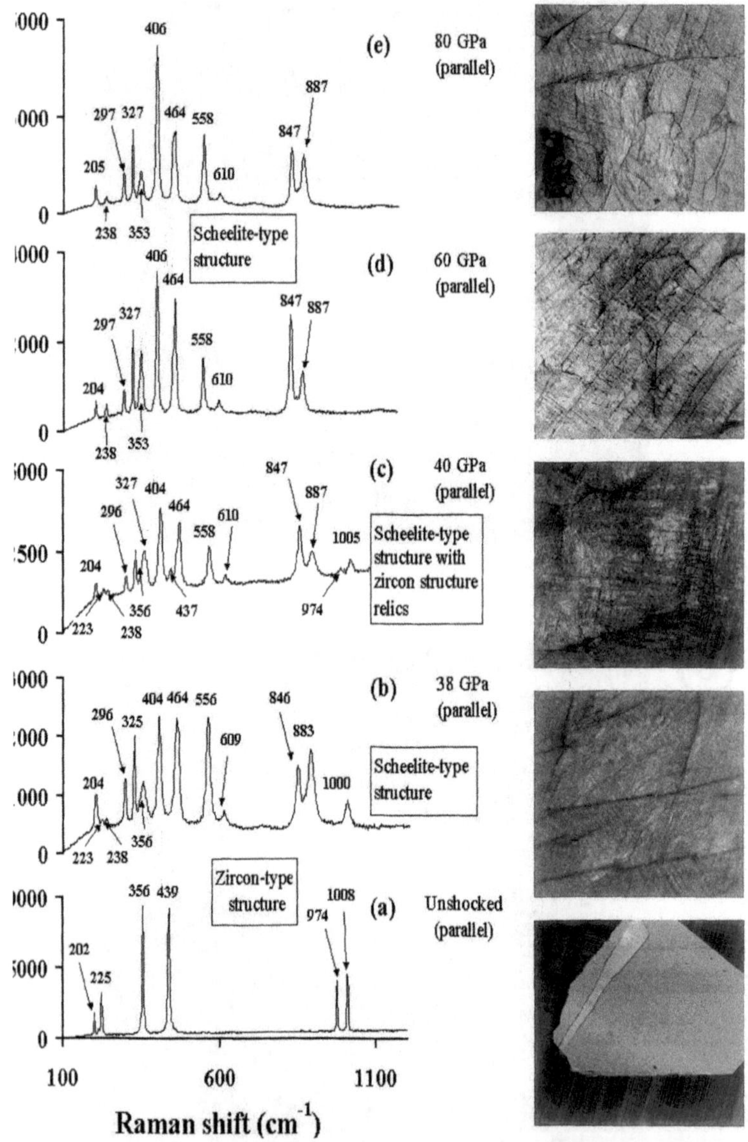

Fig. 12. Raman spectra of unshocked and experimentally shocked (38, 40, 60, 80 GPa) zircon samples, which were cut parallel to their c-axes. The peaks in the spectrum of the unshocked sample indicate zircon-structure and bands in the spectra of the shocked samples are characteristic of the high-pressure scheelite-type phase. The 38 and 40 GPa samples, however, exhibit some additional weak bands that indicate relics of zircon-phase (c). Numbers denote peak positions in [cm^{-1}].

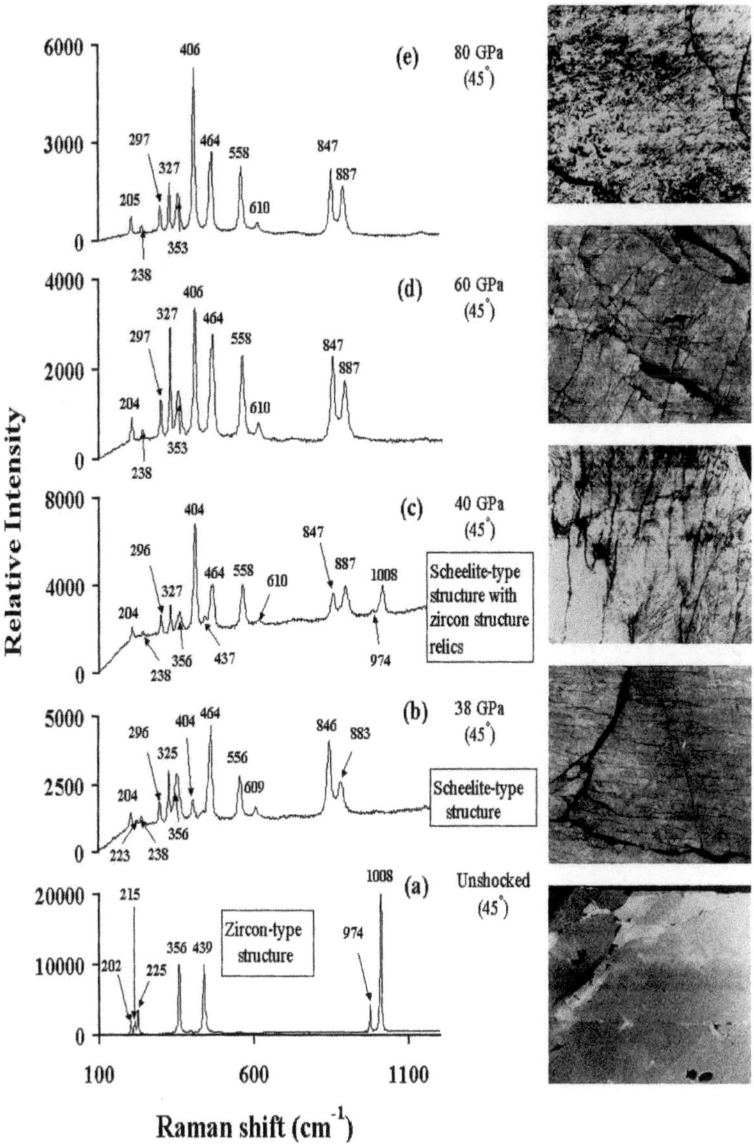

Fig. 13. Raman spectra of unshocked and experimentally shocked (38, 40, 60, 80 GPa) zircon samples, which were cut at 45° to the c-axis. The peaks in the spectrum of the unshocked sample indicate zircon-structure and bands in the spectra of the shocked samples are characteristic of the high-pressure scheelite-type phase. The 38 and 40 GPa samples, however, exhibit some weak bands that indicate relics of zircon-phase (c). Numbers denote peak positions in [cm^{-1}].

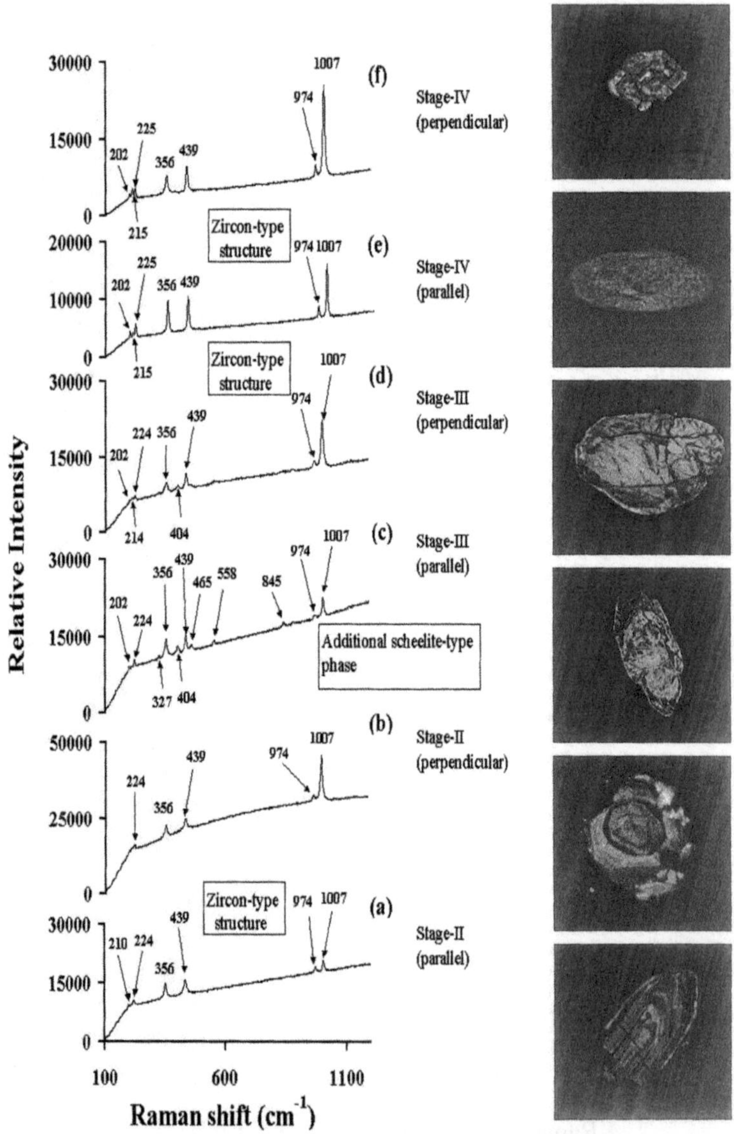

Fig. 14. Raman spectra and CL images of naturally shock deformed zircon specimens from Stage-II, Stage-III and Stage-IV rock samples from the Ries impact crater (Germany), which were cut parallel and perpendicular to the c-axis.The peaks in the spectra from all grains indicate the zircon structure. The bands in the Stage-III parallel-cut zircon sample indicate also the presence of scheelite-type phase (c). Numbers denote peak positions in [cm^{-1}].

crystallographic c-axis. The CL images (right columns of Figures 12-14) represent the areas where the Raman and CL spectra were taken.

5.2.1 *Raman spectroscopy of experimentally shock-deformed zircon samples*

The Raman spectra of the unshocked and experimentally shock-deformed zircon samples (38, 40, 60 and 80 GPa) show significant differences. The Raman spectra of the unshocked samples (parallel and 45°) contain seven peaks at 202, 215, 225, 356, 439, 974, 1008 cm^{-1} (Figs. 12a, 13a) indicating the zircon-type structure (Williams and Knittle 1993; Kolesov 2001). Whereas all these peaks are characteristic for the zircon-type structure, the bands at 356 and 439 cm^{-1} appear most useful to distinguish this phase from the scheelite-type structure (Williams and Knittle 1993); compare Figs. 13a, 14a. The narrow band widths of both unshocked samples indicate a highly crystalline structure without major zoning or defects.

The spectra of the 38 GPa samples (parallel and 45°) are dominated by a number of peaks at 204, 223, 238, 296, 325, 356, 404, 464, 556, 609, 846, and 883 cm^{-1}. These peaks are characteristic for the scheelite-type phase (Williams and Knittle 1993). The maximum peak intensities of the 45°- sample are higher than those of the parallel-sample (Figs. 12b,13b). The parallel-sample contains an additional peak at around 1000 cm^{-1} (Fig. 12b), which will be discussed below. Additional Raman spectra (available from the corresponding author) from the 38 GPa parallel-sample were obtained, for the acquisition of which this sample was rotated 45° with approximately thirty minutes total exposure time, which yielded a scheelite-type phase spectrum.

In both 40 GPa samples, typical scheelite-type peaks occur at 204, 238, 296, 327, 356, 404, 464, 558, 610, 847, and 887, (Figs. 12c, 13c). In addition, relatively weak peaks appear at 223, 437 and 1005 cm^{-1} in the spectrum of the parallel-sample (Fig. 12c) and at 437, 974, and 1005 cm^{-1} in the spectrum of the 45° sample (Fig. 13c). These features are discussed below. In general, maximum peak intensities of the 45^0-sample are higher than those of the parallel-sample (Figs. 12c, 13c).

The Raman spectra of the 60 GPa samples (parallel and 45°) show peaks at 204, 238, 297, 327, 353, 406, 464, 558, 610, 847 and 887 cm^{-1} (Figs. 12d, 13d). The peak intensities of these peaks in both samples are approximately equal, and all peaks are characteristic for the scheelite-type phase. These peaks are totally different from the corresponding findings of Gucsik et al. (2002), who observed an amorphous state without any vibration modes of Raman spectrum at 60 GPa. The Raman spectra of the 80 GPa samples exhibit peaks that are typical for the scheelite- type phase at 205, 238, 297, 327, 353, 406, 464, 558, 610, 847 and 887 cm^{-1} (Figs. 12e, 13e). The peak intensities of these bands in both samples are approximately equal. In both cases, the peak at 406 cm^{-1} is relatively strong.

5.2.2
Raman Spectroscopy of Naturally Shock-Deformed Zircon Samples from the Ries Crater

The Raman spectra of the naturally shock-deformed zircon samples from the Ries crater (Stage-II: 35-45 GPa, Stage-III: 45-50 GPa, Stage-IV: >50 GPa) cut parallel and perpendicular to the crystallographic c-axis do not exhibit significant differences. The fluorescence background and widths of the Raman bands in all samples are considerably larger than for the experimentally shock-deformed samples, which indicate lower crystallinity with major zoning and defects.

Both Stage-II (35-45 GPa) samples are characterized by five peaks at 224, 356, 439, 974 and 1007 cm^{-1}, indicating zircon-type structure (Williams and Knittle 1993; Kolesov et al. 2001) (Figs. 14a,b). Additionally, a weak peak at 210 cm^{-1} appears in the Raman spectra of the Stage-II parallel sample (Fig. 14a). The peak intensities of the perpendicular sample are higher than those of the parallel-samples. The peak at 1007 cm^{-1} is relatively strong in the perpendicular sample (Fig. 14b).

The Raman spectrum of the Stage-III (45-50 GPa) parallel-sample shows eleven peaks at 202, 224, 327, 356, 404, 439, 465, 558, 845, 974 and 1007 cm^{-1}, which indicate the presence of the scheelite-type phase among predominant zircon-type material (Fig. 14c). In contrast, the Stage-III perpendicular-sample contains only eight peaks at 202, 214, 224, 356, 404, 439, 974 and 1007 cm^{-1} showing pure zircon-type structure (Fig. 14d). A peak at 1007 cm^{-1} is relatively strong in the perpendicular-sample (Fig. 14d). In general, the fluorescence background in the parallel-sample is considerably higher than in the perpendicular-sample. In both cases, the peak intensities are similar.

The spectra of the Stage-IV samples (parallel and perpendicular samples) are characterized by seven peaks at 202, 215, 225, 356, 439, 974 and 1007 cm^{-1}, indicating zircon-type phase (Williams and Knittle 1993; Kolesov et al. 2001) (Figs. 14e,f). In both cases, a peak at 1007 cm^{-1} is relatively strong. The peak intensities of the perpendicular-sample are higher than those of the parallel-sample (Fig. 14f).

5.3. Cathodoluminescence Spectrometry

5.3.1
Cathodoluminescence Spectroscopy of Experimentally Shock-Deformed Zircon Samples

Figures 15 and 16 show a comparison of typical CL spectra of experimentally shock-deformed specimens (cut parallel and at 45° to the c-axis), representing the various shock stages (from unshocked to 80 GPa).

The CL spectra of all parallel-samples are dominated by four sharp emission lines at 406, 484, 491, and 580 nm, and a weak band at 548 nm in the visible light range (Fig. 15). A broad band in the unshocked sample is centered at 353 nm,

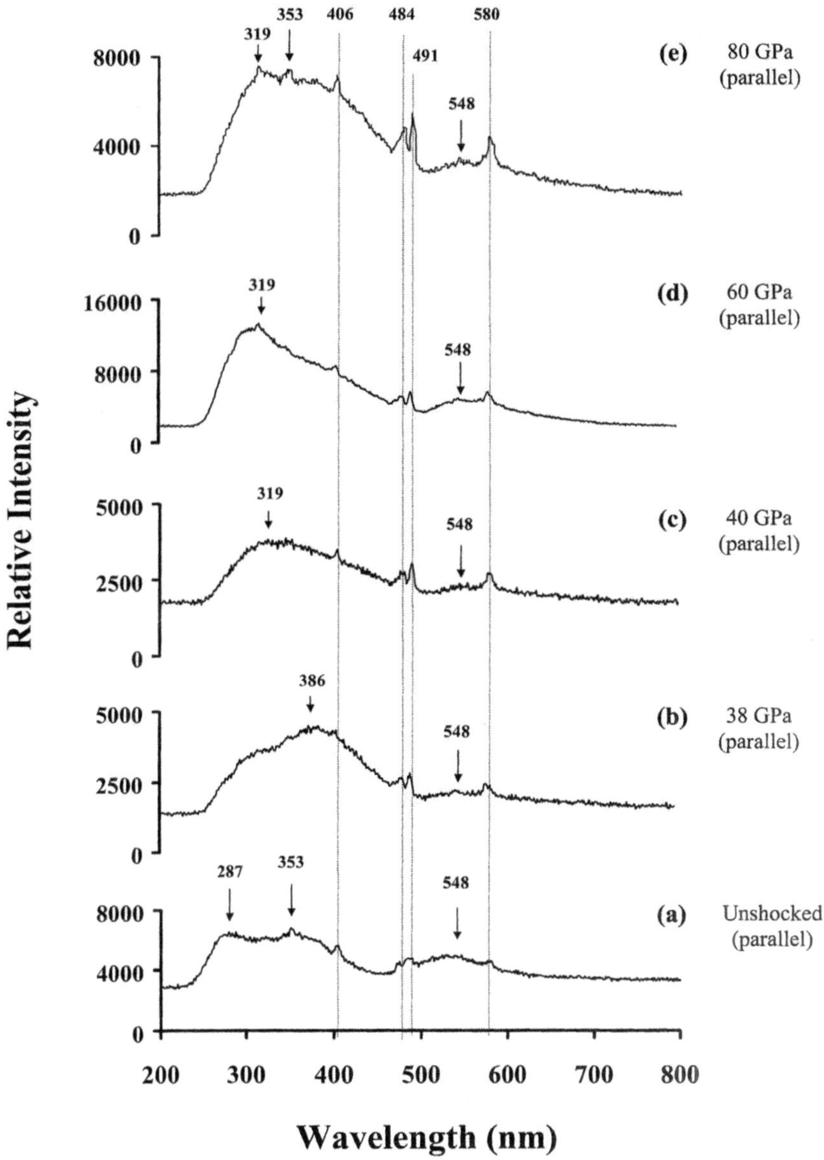

Fig. 15. Cathodoluminescence spectra of unshocked and experimentally shocked (38, 40, 60, 80 GPa) zircon samples, which were cut parallel to the c-axis, showing sharp emission lines in the visible light range and relatively high peak intensities of broad bands in the near-ultraviolet range. Reference lines of REE activators at: 406, 484, 491, and 580 nm. Numbers denote peak positions in [nm].

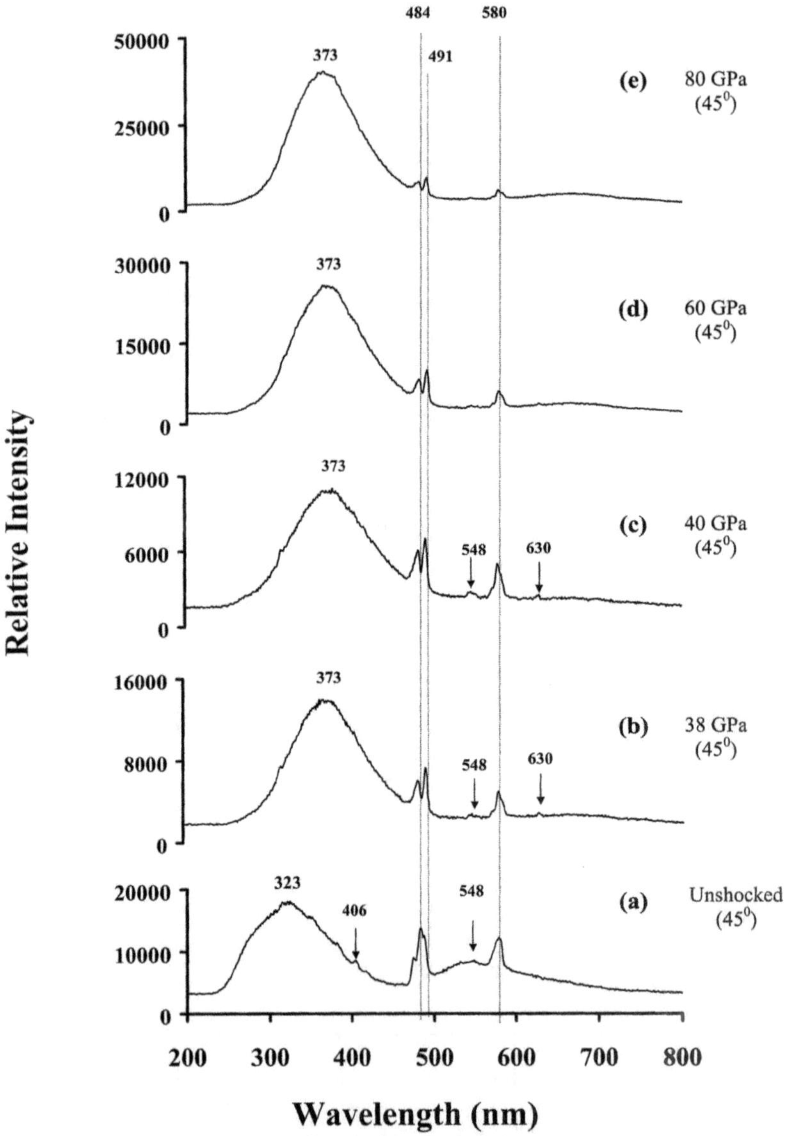

Fig. 16. Cathodoluminescence spectra of unshocked and experimentally shocked (38, 40, 60, 80 GPa) zircon samples, which were cut at 45° to the c-axis, showing sharp emission lines in the visible light range and relatively high peak intensities of broad bands in the near-ultraviolet range. Reference lines of REE activators at: 484, 491, and 580 nm. Numbers denote peak positions in [nm].

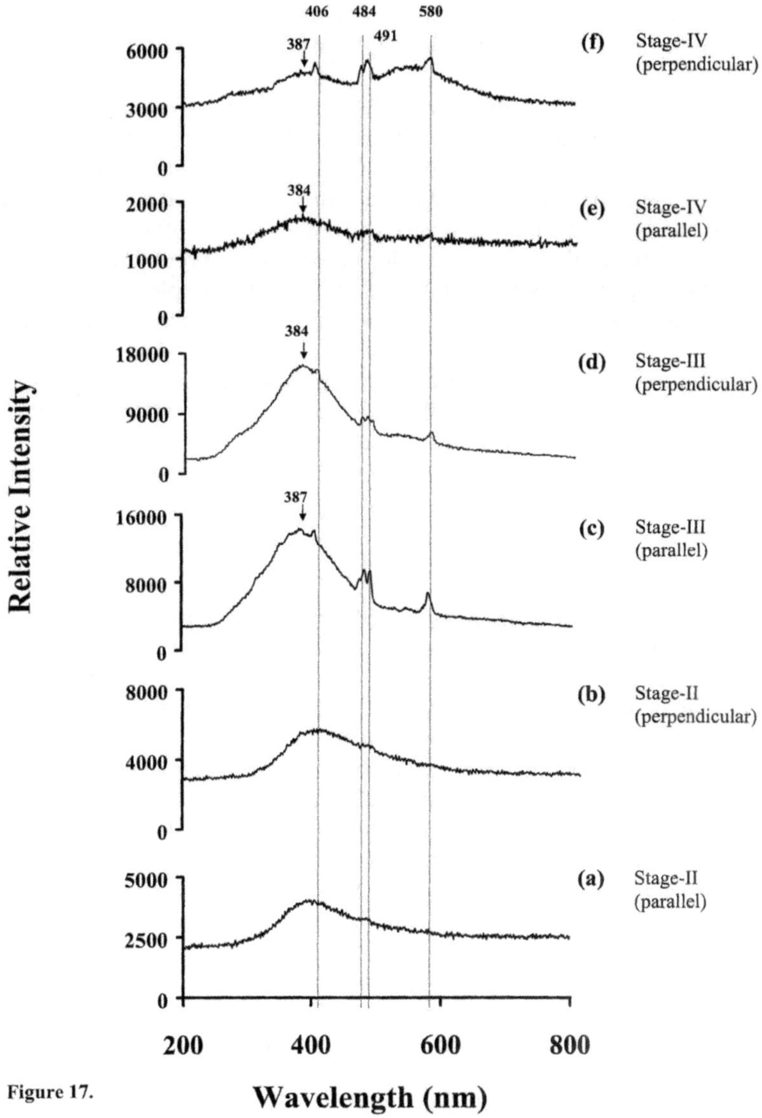

Figure 17.

Fig. 17. Cathodoluminescence spectra of naturally shock deformed zircon specimens from the Ries impact crater, which were cut parallel and perpendicular to the c-axis. Stage numbers relate to the average shock deformation degree determined on the host rock (see Table 1) These CL spectra show sharp emission lines in the visible light range and relatively high peak intensities of broad bands in the near-ultraviolet range. Reference lines of REE activators at: 406, 484, 491, and 580 nm. Numbers denote peak positions in [nm].

which is overlain by a minor peak at 287 nm (Fig. 15a). The broad bands in the shocked samples are centered at 386 nm in the 38 GPa sample (Fig. 15b), and at 319 nm in the 40 and 60 GPa samples (Figs. 15c,d). A minor peak at 353 nm is superimposed on a broad band centered at 319 nm in the 80 GPa sample (Fig. 15e).

The CL spectra of all 45°- samples are characterized by two sharp emission lines at 484 (doublet peak ranging from 478 nm to 491 nm) and 580 nm, and a weak band at 548 nm (in unshocked, 38 and 40 GPa samples) in the visible light range (Fig. 16). A broad band in the unshocked sample is centered at 323 nm (Fig. 16a), and broad bands in the shocked samples (from 38 GPa to 80 GPa pressures) are centered at 373 nm in the near ultraviolet range (Figs. 16b-e). An additional weak emission line appears at 630 nm in the 38 and 40 GPa samples (Figs. 16b,c), and a relatively weak band is centered at 406 nm in the unshocked specimen (Fig. 16a).

The CL spectra of both sample sets are characterized by relatively high peak intensities of broad bands in the near ultraviolet range, compared to the narrow emission lines in the visible light range. In contrast to the 45°-samples, the CL spectra of parallel-samples do not exhibit high peak intensities and do not contain the weak emission band at 630 nm (Figs. 15, 16).

5.3.2 Cathodoluminescence Spectroscopy of Naturally Shock-Deformed Zircon Samples from the Ries Crater

The CL spectra of zircon crystals from Stage-II, Stage-III and Stage-IV samples of naturally shock-deformed zircon crystals from the Ries crater (parallel and perpendicular samples) are dominated by relatively high peak intensities of broad bands in the near-ultraviolet range. The Stage-II samples (parallel and perpendicular) contain a broad band centered at 406 nm and a relatively weak band at 484 nm (ranging from 484 up to 491 nm) (Figs. 17a,b). The CL spectra of the Stage-III parallel and perpendicular (Figs. 17c,d) and Stage-IV perpendicular (Fig. 17f) samples are characterized by three sharp emission lines at 406, 484 (ranging from 478 nm to 491 nm), and 580 nm in the visible light range. The broad bands are centered at 384 nm. The Stage-IV parallel-sample exhibits relatively low peak intensities and show a weak broad band centered at 384 nm (Fig. 17e).

The emission lines at about 484 and 580 nm have previously been described from zircon from a Norwegian gneiss (Marshall 1988), from the leucogranulite from Taura (Kempe et al. 2000), and from synthetically doped (with REE activators: Tb^{3+}, Dy^{3+}, Tm^{3+}, etc.) zircon (Cesbron et al. 1995; Blanc et al. 2000). Position and shape of the broad brand observed in the CL spectra of all our samples are quite different from what has been described in previous CL studies of (unshocked) zircon. Cathodoluminescence spectra of the embedding material (epoxy resin) and of the glass slide used for mounting the zircon grains show that these materials do not produce, or contribute to, the broad bands observed in the zircon spectra of the present study.

6
Discussion

6.1
The Nature of SEM-CL Properties of Microdeformations of Naturally and Experimentally Shocked Zircon Crystals

The inverse relationship between BSE and CL brightness might be explained as follows. According to Remond et al. (2000), backscattered electrons (BSE) are primary electrons that leave the specimen as a result of a single large-angle scattering event or multiple small angle scattering processes. The fraction of primary electrons (η) that leave the specimen, $\eta = I_{BSE}/I_B$, increases with the mean atomic number of the specimen, where I_B is the beam current entering the sample and I_{BSE} is the BSE current leaving the specimen. This dependence of backscattered-electron yield on mean atomic number allows for the compositional imaging of flat surfaces with submicrometer resolution, which provides better spatial resolution and microcompositional information than standard optical methods.

The BSE contrast is sensitive to changes in the average atomic number, i.e., higher Z causes higher BSE yield (bright BSE contast) and lower Z causes lower BSE yield (dark BSE contrast). As noted above, in the case of natural zircon, the changes in BSE contrast might be caused by high concentrations of Hf, Y, Yb, U, and Th,(corresponding to high BSE contrast). In the case of natural zircon samples, REE as activators can cause more recombination centers for photon emission, increasing the brightness (CL-bright area) of the CL images (Blanc et al. 2000). The higher concentrations of high Z elements lead to quenching of the CL emission, producing CL-dark areas. The CL images of naturally shock-deformed zircon crystals from the Ries impact structure also exhibit this relationship between BSE and CL brightness. According to Leroux et al. (1999), the 40 GPa experimentally deformed sample consists of zircon that has been partially converted into the scheelite-structure phase. Planar deformation features (PDFs) of a glassy nature were observed by these authors in TEM images at the nm-scale in relics of zircon-structure material. One to several micrometer-wide micro-bands (twinning) of the scheelite-structure phase could be found crosscutting material with zircon structure parallel to the {100} planes of zircon. According to Leroux et al. (1999), this displacive phase transformation followed the following relationships: $\{100\}_z$ // $\{112\}_s$ and $[001]_z$ // $[110]_s$. The high density of narrow-spaced twinned features that are visible in the 38 GPa sample (especially in the CL images – Fig. 5b) could represent twinning as a result of zircon partially converted to the scheelite-structure phase. The zircon to scheelite-structure phase transition involves a significant volume decrease and changes in the c/a ratio, which is not accompanied by a change in primary coordination number of Si and Zr cations, but only by rotation of the SiO_4 tetrahedra. These changes may cause differences in the magnitude of the energy levels between the conduction and valence bands of the activator elements resulting in absorption of the excitation energy (dark lines in the CL images) and high-luminescent energy transfer as high frequency emission (bright lines in the CL images).

6.2 CL and Raman Spectral Properties of Unshocked and Shocked (20-80 GPa) Zircon Samples

6.2.1 Raman Spectra

Raman spectra of natural zircon were described by, for example, Nasdala et al. (1995) and Kolesov et al. (2001), and the spectral signature of the high-pressure scheelite-type phase was reported by Knittle and Williams (1993). In general, high-energy modes (at 700-1000 cm^{-1}) might be related to the antisymmetric or symmetric stretching modes of SiO_4 ,whereas low-energy modes (at 100-400 cm^{-1}) might be attributed to the lattice modes. The SiO_4 bending modes typically lie in the wavenumber region of 400-700 cm^{-1} (Williams and Knittle 1993; Kolesov et al. 2001). The spectra and wave numbers of our Raman modes are in excellent agreement with other studies (Williams and Knittle 1993; Kolesov et al. 2001). Following data presented by Williams and Knittle (1993) and Kolesov et al. (2001), Tables 2 and 3 give the zircon and scheelite-type vibrational modes and assignments of the observed (this work) Raman modes. The peak positions of scheelite-type modes are very similar to those characteristic of zircon. Even if these features were also observed in the spectra of the high-pressure scheelite-structure phase by Knittle and Williams (1993), there is a high probability that minor amounts of relict zircon could still be present in the 38 and 40 GPa samples. This suggestion is also supported by the TEM investigations of Leroux et al. (1999), who did not observe single-phase scheelite-type material in the 40 GPa sample, but an epitaxial intergrowth of both phases at a scale below the spatial resolution of the micro-Raman spectrometer used in the present study.

Table 2. Raman bands (this study) and their assignment to vibrational modes in the zircon structure (from Williams and Knittle 1993; Kolesov et al. 2001)

v_0 (cm^{-1})	Assignment
1008	v_3: SiO_4 antisymmetric stretch
974	v_1: SiO_4 symmetric stretch
439	v_2: SiO_4 bend
356	lattice mode
225	lattice mode
202	lattice mode

Table 3. Raman bands (this study) and their assignment to vibrational modes in the scheelite-type structure of zircon (after Williams and Knittle 1993; Kolesov et al. 2001).

ν_0 (cm^{-1})	Assignment
1001	ν_3: SiO_4 antisymmetric stretch
887	ν_3: SiO_4 antisymmetric stretch
847	strain-activated mode
610	ν_4: SiO_4 bend
558	ν_4: SiO_4 bend
464	ν_2: SiO_4 bend
406	lattice mode
353	lattice mode
327	lattice mode
297	lattice mode
238	lattice mode
223	lattice mode
204	lattice mode

Frequency shifts of the 1000 cm^{-1} band by a few cm^{-1} might be due to strained zircon caused by shock waves or high-pressure induced deformation or disorder in zircon. Similar frequency shifts were observed in radiation-damaged zircon (Zhang et al. 2002).

Raman spectra of naturally shock-deformed zircon crystals from the Ries Crater all show zircon-type phase. According to Raman spectral measurements and lack of shock-induced planar microdeformations, these samples do not exhibit high shock pressure. Only zircon from the Stage-III (parallel) sample exhibits the presence of traces of scheelite-type vibrational modes, which occur first around 30 GPa (Kusaba et al 1985).

6.2.2
Emission Peaks in the CL Spectrum

The luminescence properties are mostly the result of luminescence-activating ions, such as transition metals, rare-earth elements, or actinides. According to Remond et al. (2000), the trivalent rare-earth element (REE) radiative recombination centers are characterized by sharp visible and near-infrared (NIR) emission peaks in the spectrum. The luminescence results from electronic transitions between the partially filled 4f shells, which are well shielded by outer shell electrons. Broad intrinsic emission generally results from self-trapped excitons (STE), which are highly localised excitons trapped by their own self-induced lattice distortion. Self-trapped excitons are generally produced in crystals with a deformable lattice (e.g., SiO_2), which are characterised by strong electron-phonon coupling.

The emission energy of the STE is usually much lower than the band gap of the material due to energy lost by phonon emission during the electronic transition. In the present study, CL luminescence centers are dominated by Dy^{3+} indicated by the relatively strongest peaks at 478-491 and 573-586 nm in the visible light range. According to Blanc et al. (2000), a weak emission line at around 630 nm might be assigned to Gd^{3+}. The weak bands at 406 and 548 nm might be related to Tb^{3+} (Blanc et al. 2000).

6.2.3
Broad Bands in CL Spectra

Sippel and Spencer (1970) observed peak shifts from the green to the red part of the spectrum in CL spectra of shock-metamorphosed lunar feldspar, as well as peak broadening and decrease of luminescence intensity compared to unshocked samples of plagioclase. More recently, CL spectral measurements were performed on naturally and experimentally shocked oligoclase (An19.7, single crystal shocked between 10.5 GPa and 45 GPa) and plagioclase from equilibrated ordinary chondrites (Dar al Gani, Tenham) (Kaus and Bischoff 2000). These authors observed the disappearance of the crystal-field sensitive Mn^{2+} and Fe^{3+}-related peaks, resulting from a breakdown of the crystalline structure (i.e., occurrence of diaplectic glass) at around 35 GPa. According to Sippel and Spencer (1970), the broad bands in the near-ultraviolet range of the CL spectrum might be due to distortions or disordering in the crystal field. The excitation of electrons or holes trapped (dangling bonds in covalent crystals) at point defects, such as vacancies, interstitial, and point defect clusters, usually produce broad CL peaks at all temperatures (Remond et al. 2000). Electron defects at O in Ti-O or Zr-O systems, OH^- defects and broken atomic bonds were also related to broad bands in near-ultraviolet range of CL spectra of zircon (Kempe et al. 2000, and references therein).

7
Summary and Conclusions

The results from our BSE and CL studies can be summarized as follows:

1. The unshocked samples show crosscutting, irregular fractures in the CL images. The 38 GPa sample exhibits well-developed microdeformations (best visible in the CL image), which indicate that zircon structure was partly converted into scheelite-type phase. The CL images of the experimentally, at 40, 60, and 80 GPa, shocked samples show subplanar and nearly parallel microdeformations. For all naturally shocked samples from the Ries crater, an inverse relationship between the brightness of the backscattered electron (BSE) signal and the corresponding cathodoluminescence intensity of the zonation patterns was observed. This observation indicates that high Z elements cause high BSE contrast and these element cause quenching of the CL signal, leading to CL-dark areas in the CL images.

2. The CL spectra of unshocked and experimentally shock-deformed specimens and naturally shock-metamorphosed zircon samples are characterized by narrow emission lines and broad bands in the region of visible light and in the near-ultraviolet range. The emission lines result from rare earth element activators and the broad bands might be associated with lattice defects.

3. Raman spectra revealed that the unshocked samples, as well as naturally shock-deformed zircon crystals from the Ries, represent zircon-structure material, whereas the 38 and 40 GPa samples yielded additional peaks with relatively high peak intensities, which are indicative of the presence of the scheelite-type structure of zircon with zircon-structure relics. The 60 and 80 GPa samples display a Raman signature that is characteristic for the existence of only the scheelite-type phase. According to the Raman measurements, the naturally shock-deformed zircons might be related to the low-shock regime (<30 GPa), and do not show the same shock stages indicated by whole-rock petrography.

4. The combination of BSE and CL imaging and CL and Raman spectroscopy is a potentially useful tool that can be used to characterize the shock stage of zircons from impactites. Our results also give new insight into the structural changes that occur in zircons during shock metamorphism, and the pressures associated with these changes.

In Table 4 the distinctive BSE and CL image features, as well as the CL and Raman peaks that characterize unshocked zircon and various shock stages of zircon, are summarized. This table summarizes the structural changes observed in experiementally shock-metamorophosed zircon crystals as a function of increasing shock pressure. This indicates that SEM-CL technique and Raman spectroscopy might be a useful tool characterize the shock stages of zircons crystals from natural and experimental impact environments.

Acknowledgments

We would like to thank Ming Zhang (University of Cambridge, UK) for helpful comments, and Gero Kurat (Natural History Museum, Vienna, Austria) for access to the SEM-CL facility at the Natural History Museum, Vienna, Austria. Laboratory work was supported by the Austrian Science Foundation (grant Y58-GEO, to CK). This is University of the Witwatersrand Impact Cratering Research Group Contribution No 61. We thank U.Hornemann for the shock experiments. We thank A. Greshake and R. Schmidt, Humboldt-Universität, Berlin, for the naturally shock-metamorphosed zircon crystals from the Ries crater. AG was also supported by a Mobility Grant of the European Science Foundation IMPACT Programme. We are grateful to T. Andersen and S. Boggs for critical and helpful reviews.

Table 4. BSE/CL image and Raman and CL spectroscopy observations of unshocked and experimentally shocked (38, 40, 60 and 80 GPa) zircon samples (this study).

Sample	BSE image	CL image	Raman spectra	CL spectra
Unshocked	Open irregular fractures (mainly in the 45°-sample).	Patchy areas. No zonation patterns.	Seven peaks at 202, 215, 225, 356, 439, 974, 1008 cm^{-1}. Zircon-type structure.	Five emission lines at 406, 484, 491, 548 (weak), and 580 nm. A broad band is centered at 353 nm, which is superimposed by a minor peak at 287 nm (parallel-sample) A broad band is centered at 323 nm in the 45°-sample.
38 GPa	High density of irregular fractures.	Wide, curved and more or less parallel microdeformations (parallel-sample). A high density of closely spaced, crystallographically controlled features (45°-sample)	Peaks at: 202, 223, 238, 296, 325, 356, 404, 464, 556, 609, 846, and 883 cm^{-1}. Additional peak at 1000 cm^{-1} (parallel-sample). Scheelite-type structure.	Four emission lines at 484, 491, 548 (weak), and 580 nm. Additional peaks at 630 nm (45°-sample) and at 406 nm (parallel-sample). The broad bands are centered at 386 nm (parallel-sample) and at 373 nm (45°-sample).
40 GPa	Irregular fractures.	A high density of well-developed lamellae (parallel-sample). Narrow, closely spaced planar features (45°-sample). Patchy areas with high intensity (parallel-sample).	Peaks at: 204, 238, 296, 327, 356, 404, 464, 558, 610, 847, and 887 cm^{-1}. Additional peaks at 223, 437 and 1005 cm^{-1} (parallel-sample). Scheelite-type structure with zircon structure relics.	Four emission lines at 484, 491, 548 (weak), and 580 nm. Additional peaks at 630 nm (45°-sample) and at 406 nm (parallel-sample). The broad bands are centered at 319 nm (parallel-sample) and at 373 nm (45°-sample).
60 GPa	Long, narrow irrgular fractures.	Crystallographically controlled, subplanar, lamellar microdeformations.	Peaks at: 204, 238, 297, 327, 353, 406, 464, 558, 610, 847, and 887 cm^{-1}. Scheelite-type phase.	Three emission lines at 484, 491, and 580 nm. Additional peaks at 406 and 548 nm (parallel-sample). The broad bands are centered at 319 nm (parallel-sample) and at 373 nm (45°-sample).
80 GPa	High density of irregular fractures.	Subplanar, widely spaced, crystallographically controlled microdeformations.	Number of peaks at 205, 238, 297, 327, 353, 406 (strong), 464, 558, 610, 847and 887cm^{-1}. Scheelite-type phase.	Three emission lines at 484, 491, and 580 nm. Additional peaks at 406, 548, 353 nm (parallel-sample). The broad bands are centered at 319 nm (parallel-sample) and at 373 nm (45°-sample).

References

Åberg G, Bollmark B (1985) Retention of U and Pb in zircons from shocked granite in the Siljan impact structure, Sweden. Earth and Planetary Science Letters 74: 347-349

Blanc P, Baumer A, Cesbron F, Ohnenstetter D, Panczer G, Rémond G (2000) Systematic cathodoluminescence spectral analysis of synthetic doped minerals: anhydrite, apatite, calcite, fluorite, scheelite and zircon. In: Pagel M, Barbin V, Blanc P, Ohnenstetter D (eds) Cathodoluminescence in Geosciences, Springer-Verlag Heidelberg, pp 127-160

Boggs S, Krinsley DH, Goles GG, Seyedolali A, Dypvik H (2001) Identification of shocked quartz by scanning cathodoluminescence imaging. Meteoritics and Planetary Sciences 36: 783-793

Bohor BF, Betterton WJ, Krogh TE (1993) Impact-shocked zircons: discovery of shock-induced textures reflecting increasing degrees of shock metamorphism. Earth and Planetary Science Letters 119: 419-424

Chao ECT (1968) Pressure and temperature histories of impact metamorphosed rocks-based on petrographic observations. In: French BM, Short NM (eds) Shock Metamorphism of Natural Materials. Mono Book Corporation, Baltimore, pp 135-158

Cesbron F, Blanc P, Ohnenstetter D, Rémond G (1995) Cathodoluminescence of rare earth doped zircons. I. Their possible use as reference materials. Scanning Microscopy 9: 35-36

Deloule E, Chaussidon M, Glass BP, Koeberl C (2001) U-Pb isotopic study of relict zircon inclusions recovered from Muong Nong-type tektites. Geochimica et Cosmochimica Acta 65: 1833-1838

Deutsch A (1998) Examples for terrestrial impact structures. In: Marfunin SA (ed) Mineral Matter in Space, Mantle, Ocean Floor, Biosphere, Environmental Management, and Jewelry, Advanced Mineralogy, Springer-Verlag, Berlin, Heidelberg, pp 119-129

Deutsch A, Schärer U (1990) Isotope systematics and shock-wave metamorphism: I. U-Pb in zircon, titanite, and monazite, shocked experimentally up to 59 GPa. Geochimica et Cosmochimica Acta 54: 3427-3434

El Goresy A (1966) Metallic spherules in Bosumtwi crater glasses. Earth and Planetary Science Letters 1: 23-24

El Goresy A, Fechtig A, Ottemann T (1968) The opaque minerals in impactite glasses. In: French BM, Short NM (eds) Shock Metamorphism of Natural Materials. Mono Book Corporation, Baltimore, pp 531-554

Engelhardt Wv (1990) Distribution, petrography and shock metamorphism of the ejecta of the Ries crater in Germany - a review. Tectonophysics 171: 259-273

Engelhardt Wv, Stöffler D (1965) Spaltflächen im Quarz als Anzeichen für Einschläge grosser Meteoriten. Naturwissenschaften 17: 489-490

French BM (1998) Traces of catastrophe: A handbook of shock-metamorphic effects in terrestrial meteorite impact structures. LPI Contribution 954, Lunar and Planetary Institute, Houston, 120 pp

Geologische Karte des Rieses (1999) Bayerisches Geologisches Landesamt, München, 1 : 50 000

Gibson RL, Armstrong RA, Reimold WU (1997) The age and thermal evolution of the Vredefort impact structure: A single grain U-Pb zircon study. Geochimica et Cosmochimica Acta 61: 1531-1540

Glass BP, Liu S (2001) Discovery of high-pressure $ZrSiO_4$ polymorph in naturally occurring shock-metamorphosed zircons. Geology 29: 371-373

Glass BP, Liu S, Leavens PB (2002) Reidite: An impact-produced high-pressure polymorph of zircon found in marine sediments. American Mineralogist 87: 562-565

Grieve RAF, Langenhorst F, Stöffler D (1996) Shock metamorphism of quartz in nature and experiment: II. Significance in geosciences. Meteoritics and Planetary Science 31: 6-35

Gucsik A, Koeberl C, Brandstätter F, Reimold WU, Libowitzky E (2002) Cathodoluminescence, electron microscopy, and Raman spectroscopy of experimentally shock-metamorphosed zircon. Earth and Planetary Science Letters 202: 495-509

Hanchar JM, Miller CF (1993) Zircon zonation patterns as revealed by cathodoluminescence and backscattered images: Implications for interpretation of complex crustal histories. Chemical Geology 110: 1-13

Hartmann LA, Takehara L, Leite JAD, McNaughton NJ, Vasconcellos MAZ (1997) Fracture sealing in zircon as evaluated by electron microprobe analyses and back-scattered electron imaging. Chemical Geology 141: 67-72

Hayward CL (1998) Cathodoluminescence of ore and gangue minerals and its application in the minerals industry. In: Cabri LJ, Vaughan DJ (eds) Modern Approaches to Ore and Environmental Mineralogy, Mineralogical Association of Canada Short Course Series 27, pp 269-325

Kamo SL, Krogh TE (1995) Chicxulub crater source for shocked zircon crystals from the Cretaceous-Tertiary boundary layer, Saskatchewan: Evidence from new U-Pb data. Geology 23: 281-284

Kamo SL, Reimold WU, Krogh TE, Colliston WP (1996) A 2.023 Ga age for the Vredefort impact event and first report of shock metamorphosed zircons in pseudotachylitic breccias and Granophyre. Earth and Planetary Science Letters 144: 369-387

Kaus A, Bischoff A (2000) Cathodoluminescence (CL) properties of shocked plagioclase [abs.]. Meteoritics and Planetary Science 35: A 86

Kempe U, Gruner T, Nasdala L, Wolf D (2000) Relevance of cathodoluminescence for the interpretation of U-Pb zircon ages, with an example of an application to a study of zircons from the Saxonian Granulite Complex, Germany. In: Pagel M, Barbin V, Blanc P, Ohnenstetter D (eds) Cathodoluminescence in Geosciences, Springer-Verlag Heidelberg, pp 415-455

Kleinmann B (1969) The breakdown of zircon observed in the Libyan Desert Glass as evidence of its impact origin. Earth and Planetary Science Letters 5: 497-501

Knittle E, Williams Q (1993) High-pressure Raman spectroscopy of $ZrSiO_4$: Observation of the zircon to scheelite transition at 300 K. American Mineralogist 78: 245-252

Kolesov BA, Geiger CA, Armbruster T (2001) The dynamic properties of zircon studied by single-crystal X-ray diffaction and Raman spectroscopy. European Journal of Mineralogy 13: 939-948

Krogh TE, Kamo SL, Bohor BF (1996) Shock metamorphosed zircons with correlated U-Pb discordance and melt rocks with concordant protolith ages indicate an origin for the Sudbury Structure. In: Hart S, Basu A (eds) American Geophysical Union, Geophysical Monograph 95, pp 343-353

Kusaba K, Syono Y, Kikuchi M, Fukuoka K (1985) Shock behaviour of zircon: phase transition to scheelite structure and decomposition. Earth and Planetary Science Letters 72: 433-439

Leroux H, Reimold WU, Koeberl C, Hornemann U, Doukhan J-C (1999) Experimental shock deformation in zircon: a transmission electron microscopic study. Earth and Planetary Science Letters 169: 291-301

Marshall DJ (1988) Cathodoluminescence of geological materials. Unwin Hyman, Boston, 146 pp

McMillan PF, Hofmeister AM (1988) Infrared and Raman spectroscopy. In: Hawthorne FC (ed) Spectroscopic methods in mineralogy and geology, Reviews of Mineralogy, vol. 18, Mineralogical Society of America: 99-159

Nasdala L, Irmer G, Wolf D (1995) The degree of metamictization in zircon: a Raman spectroscopic study. European Journal of Mineralogy 7: 471-478

Poller U (2000) A combination of single zircon dating by TIMS and cathodoluminescence investigations on the same grain: The CLC-method - U-Pb geochronology for metamorphic rocks. In: Pagel M, Barbin V, Blanc P, Ohnenstetter D (eds) Cathodoluminescence in Geosciences, Springer-Verlag, Heidelberg, pp 127-160

Reid AF, Ringwood AE (1969) Newly observed high pressure transformations in Mn_3O_4, $CaAl_2O_4$, and $ZrSiO_4$. Earth and Planetary Science Letters 6: 205-208

Reimold WU, Leroux H, Gibson RL (2002) Shocked and thermally metamorphosed zircon from the Vredefort impact structure, South Africa: A transmission electron microscopic study. European Journal of Mineralogy 14: 859-868

Rémond G, Phillips MR, Roques-Carmes C (2000) Importance of instrumental and experimental factors on the interpretation of cathodoluminescence data from wide gap materials. In: Pagel M, Barbin V, Blanc P, Ohnenstetter D (eds) Cathodoluminescence in Geosciences, Springer-Verlag, Heidelberg, pp 59-126

Roberts S, Beattie I (1995) Micro-Raman spectroscopy in the Earth Sciences. In: Potts PJ, Bowles JFW, Reed SJB, Cave MR (eds) Microprobe techniques in the Earth Sciences. Chapman and Hall, London, pp 387-408

Rubatto D, Gebauer D (2000) Use of cathodoluminescence for U-Pb zircon dating by ion microprobe: some examples from the Western Alps. In: Pagel M, Barbin V, Blanc P, Ohnenstetter D (eds) Cathodoluminescence in Geosciences, Springer-Verlag, Heidelberg, pp 373-400

Scott HP, Williams Q, Knittle E (2002) Ultralow compressibility silicate without highly coordinated silicon. Physical Review Letters 88: 015506-1-015506-4

Sippel FR, Spencer BA (1970) Luminescence petrography and properties of lunar crystalline rocks and breccias. Levinson AA (ed) Proceedings of the Apollo 11 Lunar Science Conference, Pergamon Press, New York, pp 2413-2426

Staudacher T, Jessberger KE, Dominik B, Kristen T, Schaeffer AO (1982) ^{40}Ar-^{39}Ar of rocksand glasses from the Nördlinger Ries crater and the temperature history of impact brecciasJournal of Geophysics 51: 1-11

Stöffler D (1966) Zones of the impact metamorphism in the crystalline rocks at the Nördlinger Ries Crater. Contributions to Mineralogy and Petrology 12: 15-24

Stöffler D (1972) Deformation and transformation of rock-forming minerals by natural and experimental shock processes: II. Behavior of minerals under shock compression. Fortschritte der Mineralogie 49: 50-113

Stöffler D (1974) Deformation and transformation of rock-forming minerals by natural and experimental shock processes. Fortschritte der Mineralogie 49: 256-298

Stöffler D, Langenhorst F (1994) Shock metamorphism of quartz in nature and experiment: I. Basic observation and theory. Meteoritics 29: 155-181

Vavra G (1993) A guide to quantitative morphology of accessory zircon. Chemical Geology 110: 15-28

Wolfe JP (1998) Imaging phonons: Acoustic wave propagation in solids. Cambridge University Press, Cambridge, 411 pp

Zhang M, Salje EKH, Ewing RC (2002) Infrared spectra of Si-O overtones, hydrous species, and U ions in metamict zircon: radiation damage and recrystallization. Journal of Physics: Condensed Matter 14: 3333-3352

A Brief Introduction to Hydrocode Modeling of Impact Cratering

Elisabetta Pierazzo[1] and Gareth Collins[2]

[1]Planetary Science Institute, 620 N. 6th Avenue, Tucson, AZ 85705, USA (betty@psi.edu)
[2]Lunar and Planetary Laboratory, University of Arizona, Tucson, AZ 85721,

Abstract. Numerical modeling is a fundamental tool for understanding the dynamics of impact cratering, especially at planetary scales. In particular, processes like melting/vaporization and crater collapse, typical of planetary-scale impacts, are not reproduced in the laboratory, and can only be investigated by numerical modeling. The continuum dynamics of impact cratering events is fairly well understood and implemented in numerical codes; however, the response of materials to shocks is governed by specific material properties. Accurate material models are thus crucial for realistic simulation of impact cratering, and still represent one of the major problems associated with numerical modeling of impacts.

1 Introduction

Impact cratering is the only ubiquitous geologic process in the solar system. A comprehensive understanding of this process can only be reached by combining geologic and geophysical observations with experimental studies and numerical modeling. Remote observations of impact craters on planetary surfaces provide a great deal of information on crater morphology and ejecta emplacement. Moreover, geologic and geophysical investigations of terrestrial craters provide a very complementary "ground-truth" data set of an impact crater's sub-surface structure. However, these techniques are of limited use in exposing the dynamics of the impact processes. There have been no direct observations of impact crater formation in recorded history; large impact events are too infrequent. Much of our understanding of impact cratering dynamics has come from laboratory experiments, combined with studies of past nuclear testing explosions; however, these experimental studies have several limitations.

Processes like shock melting and vaporization in large impacts, which involve extreme pressures and temperatures, cannot be easily reproduced in laboratory. Furthermore, the dominance of gravity in influencing the later stages of crater formation implies that the results of small-scale laboratory collapse experiments cannot be extrapolated meaningfully to the scale of the largest craters in the solar system. Similarly, underground nuclear explosions, although extremely valuable in elucidating the principal features of crater excavation, are not really of an applicable scale.

Abstract computer simulations provide the only feasible method for simulating the formation of large ($D > 1$ km) impact structures. Over the last few decades, rapid improvement of computer capabilities has allowed impact cratering to be modeled with increasing complexity and realism. In particular, the most recent advances in computer power are paving the way for a new era of hydrocode modeling of the impact process, dominated by full, three-dimensional simulations. When validated against observation, computer models offer a powerful tool for understanding the mechanics of impact crater formation. This paper will introduce the important aspects of hydrocode modeling of impact cratering, as well as discussing the state-of-the-art progress being made in this field.

2
Fundamentals of Impact Cratering

A high-speed impact causes a sudden compression of the projectile and target materials at the impact surface, generating a shock wave that propagates through both projectile and target. As the shock passes, the material's thermodynamic state changes rapidly, in an irreversible process (i.e., increase in entropy), from its initial state to the shocked state. Indeed, the thermodynamic change is so fast that the shock is treated mathematically as a discontinuity in material characteristics. When the shock wave reaches the rear end of the projectile and the target surface, it is reflected back as a rarefaction wave that adiabatically releases the previously compressed material to low pressures (conserving entropy). The speed of the rarefaction wave is normally higher than that of the hemispherically-expanding shock wave, so that the shock wave ultimately achieves the shape of a thin shell delimited by the shock front and the rarefaction wave. Behind these waves the target maintains some particle velocity, which acts to open the crater during the excavation stage. The radial component of the material velocity is complemented by a tangential component, induced by the presence of the free surface, tending to deflect the particle trajectories towards the surface, thus pushing material into the target and expelling material from the expanding crater.

The excavation flow is retarded by any cohesive strength that the target material may retain, plus dry friction (resistance to shear of granular materials) and gravity. In large impacts, it is gravity that is most significant; excavation stops when insufficient energy remains to lift the overlying material against the force of its own weight. The subsequent and final stage of the cratering process is known as the collapse or modification stage. In essence, crater collapse is the gravity-driven modification of the unstable cavity formed during excavation. It ultimately results in a shallower crater geometry, that is more stable in a gravity field, as recognized in early observations of lunar craters (Quaide et al. 1965). In a recent review of crater collapse, Melosh and Ivanov (1999) reported that the detailed morphology of an impact crater is predominantly due to the collapse of the geometrically simple, bowl-shaped "transient crater", which is formed during excavation. For simple craters the collapse process is well understood; however, for larger, morphologically more complex impact structures, collapse is not well understood.

To model a complicated process like impact cratering requires sophisticated computer codes that can simulate not only the passage of a shock wave, but also the behavior of geologic materials over a broad range of stress states. Computer programs which handle the propagation of shock waves and compute velocities, strains, stresses, and so on, as a function of time and position are called "hydrodynamic computer codes," or *hydrocodes* (Anderson 1987).

2.1 Fundamentals of Hydrocode Modeling

Hydrocodes employ classical continuum mechanics to describe the dynamics of a continuous medium; to do so numerically, however, it is necessary to discretize the differential equations describing the continuum. This is normally done by generating a lattice (or grid) of points, spaced by a distance Δx, to represent the geometry of the object under study. When the adjacent points in the grid are connected by lines (or planes), the area (or volume) contained within these grid lines is called a "cell." Analogously, time is discretized by choosing a computational timestep, Δt, to be used during the integration. During each computational timestep, the program iterates through the grid, cell by cell, and updates the grid positions to account for the effect of external and internal forces. Alternatively, material can be represented by individual points, each with a certain mass and well-defined characteristics, and the hydrocode becomes known as a Smooth Particle Hydrocode (or SPH; e.g., Monaghan 1992).

The foundations of hydrocode modeling are three pillars, which are used to determine the forces acting on the mesh during each time step. These are: Newton's Law of motion, the equation of state, and the constitutive model. Newton's law of motion is implemented as a set of differential equations established through the principles of conservation of momentum, mass and energy from a macroscopic point of view. The equation of state relates pressure to the density and internal energy. It thereby accounts for compressibility effects, that is, changes in density and irreversible thermodynamic processes such as shock heating. The constitutive model relates the stress to a combination of strain and strain rate effects, internal energy, and damage; it describes the response of a material to deformation (change in shape or strength properties). All hydrocodes utilize some form of the conservation equations; however, the usefulness of a hydrocode, particularly for modeling impact cratering, depends on the sophistication of the equation of state and constitutive model. These areas of impact modeling will be addressed in detail later; first we discuss some other important issues regarding hydrocode modeling.

2.2 Resolution, Dimensionality, and Stability

The choice of resolution in space and time is an important issue when using hydrocodes. A simulation should be conducted with a high enough resolution to resolve all the important flow variations in space and time. However, these needs must be balanced by the specifications of available hardware and the time available to run the simulation. Impact hydrocodes are available in: one dimension (1D), in which the target and the impactor are represented by a concatenation of cells in a single column representing the direction of impact, with no lateral spreading allowed (Fig. 1a); two dimensions (2D), where the axes are represented by the direction of impact and the direction perpendicular to it (Fig. 1b); and three dimensions (3D). The total computer storage needed for a computation is proportional to the total number of cells in the simulation. For a mesh comprising N cells in each dimension, the total number of cells is N^r, where r is the number of dimensions. The minimum simulation run time depends approximately on the number of cells, N^r, and the number of timesteps, M. To avoid integration instabilities the computational timestep Δt is normally chosen to obey the *Courant-Friedrichs-Lewy stability condition* (Anderson 1987), which requires that no signal (information) can propagate across the shortest dimension of a cell in one timestep. This means that Δt is proportional to Δx ($\Delta t \leq |\Delta x|/c$, where c is the sound speed of the material); hence, the number of time steps M is proportional to N. The simulation run time, therefore, is approximately proportional to N^{r+1}.

The strong dependence of both computer storage and simulation time on the dimensionality of the hydrocode means that computer hardware can severely limit the potential complexity of a simulation. For this reason, the more widely used hydrocodes today are two-dimensional. 2D models allow lateral spreading, and by assuming axial symmetry (Fig. 1b) they are used to model vertical impacts and crater collapse, which can be justifiably considered as axisymmetric processes. Modeling the impact process in its full complexity, including obliquity of the incoming impactor, requires a fully 3D hydrocode, to abrogate the need for *a priori* symmetry assumptions. Historically, limitations in computer hardware have restricted impact modelers to occasional use of 3D modeling; however, the constant advance of computational power is gradually increasing the use of 3D hydrocodes.

A further concern for hydrocodes used in modeling impact cratering is that the 'discontinuous' nature of shock events may introduce instabilities in the discretized representation. The computational solution that stabilizes the numerical description of shocks is the introduction of a purely artificial dissipative mechanism, called *artificial viscosity*, which smoothly spreads the shock over a few cells. This is a purely numerical artifact that is applied only during material compression; although "artificial", it really represents small-scale physical processes, i.e., processes that occur at scales smaller than one cell size and cannot otherwise be resolved in the computation. By using artificial viscosity, therefore, shocks have a finite thickness and rise time and are better described numerically.

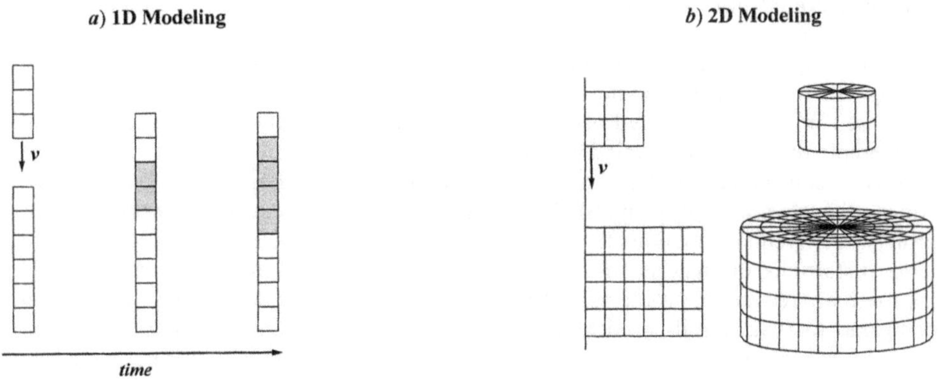

Fig. 1. Schematic of modeling impact cratering in 1D and 2D. *a)* In 1D modeling, projectile and target are represented by a row of cells. The projectile is moving toward the target with velocity ***v***. After the impact, a shock wave propagates through both target and projectile. Gray cells represent the shocked region. No lateral spread is allowed. *b)* Hydrocode modeling in 2D. The projectile is moving toward the target with velocity ***v***. The 3D view of the setup using axial symmetry shows that the simulation corresponds to a vertical impact.

2.3 Eulerian versus Lagrangian Treatment

The conservation equations at the root of a hydrocode may be formulated using a coordinate system fixed in space, known as the Eulerian approach, or one moving with the material, known as the Lagrangian approach. In the *Eulerian* approach, the mesh is fixed in space and the material flows through it (Fig. 2a). This method makes it difficult to identify material interfaces at all times during the impact computation; "mixed cells" are introduced, which include different materials, thus making the interface less sharp. As a consequence, the accuracy in the determination of material interfaces and free surfaces depends on the resolution of the mesh: the finer the mesh, the more accurately the boundary is represented, but at the price of more computational zones. In the *Lagrangian* approach the finite difference grid is fixed with the material (Fig. 2b), providing a somewhat easier treatment of the conservation equations than in the Eulerian approach. Free surfaces and contact surfaces between different materials are easily determined, and remain distinct throughout the calculation. The major limitation of the Lagrangian approach is the inaccuracy of the finite difference approximation when the cells are significantly distorted; the extreme case is when a cell folds over itself resulting in a computed negative volume (or mass). A way to overcome the problem, again at the expense of accuracy in the determination of material interfaces, is to carefully rezone the computational grid; that is, to overlay a new, undistorted grid, on the old, distorted one. This can become a tedious, time-consuming process that has to be repeated many times for simulations covering more than the very early stages of the impact event. This problem does not exist for Eulerian codes, which can easily handle flows with large distortions. Both approaches have been used in constructing hydrocodes, and both approaches can produce reliable and comparable results when handled with care.

SPH is considered an alternative to typical Eulerian and Lagrangian "cell-codes". Among the advantages of using SPH are that the conservation equations become sets of ordinary differential equations that are easy to describe, and there are no problems with large distortions or material boundaries. Among the main limitations of the SPH approach are resolution problems (SPH is intrinsically a low-resolution approach, when compared to typical cell-based hydrocodes), problems with the constitutive equation (strength models are hard to implement), and problems with the edges of the modeling region (the shock can be reflected at the edges and be transmitted back if the modeling region is not large enough).

Well-known Lagrangian hydrocodes used in impact cratering are SALE (Amsden et al. 1980) available in 2D and 3D and its 2D successor SALES-2; while among Eulerian codes are Sandia's 2D CSQ (Thompson 1979) and 2D/3D CTH (McGlaun et al. 1990), and the Russian equivalent 2D/3D hydrocode SOVA (Shuvalov 1999).

3
Material Modeling: the Equation of State and Constitutive Model

Specific material properties govern the response of materials to stress, resulting in different behaviors of different materials for nominally the same impact conditions. In a hydrocode, the stress is generally divided into a volumetric and a deviatoric component. The *equation of state* relates changes in the density and internal energy of the material to pressure, which is the volumetric component of the stress. The deviatoric stress is related to the amount of strain (distortion) required to produce that stress through the use of a *constitutive* (or rheologic) *model.* The equation of state is critical in the early stages of the impact, when the deviatoric stress is small compared to the pressure involved. The constitutive equation is fundamental for modeling the late stages of impact cratering, when material strength determines the final shape and characteristics of the crater. The difficulty of building accurate material models remains the major problem for hydrocode modeling of impacts, and one that is receiving increasing attention.

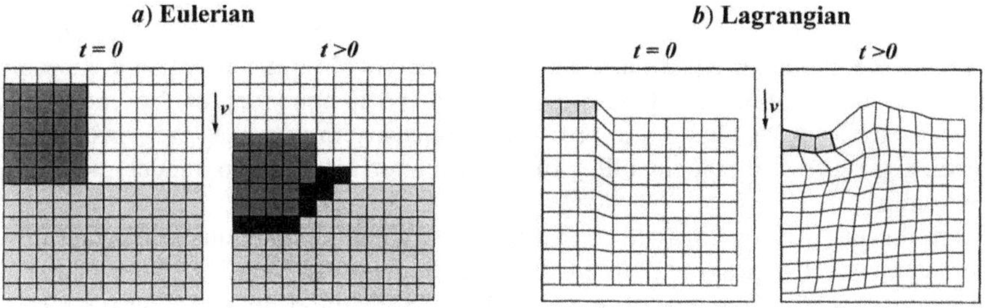

Fig. 2. *a)* Eulerian 2D representation of a cylinder (gray) impacting on a surface (light gray), at the instant of impact, $t = 0$, and at a later stage, $t > 0$. Mixed cells are shown in dark gray. *b)* Lagrangian 2D representation of a cylinder (gray) impacting on a surface at the instant of impact, $t = 0$, and at a later stage, t > 0. Cell deformation, typical of Lagrangian codes, is evident at $t > 0$. Vertical arrows represent the impact velocity.

3.1 Equation of State

The relationships between the parameters across a shock were first derived by P.H. Hugoniot in 1897 from the conservation of mass, momentum, and energy (e.g., Melosh 1989). The Hugoniot equations are entirely general, regardless of the phase of the medium through which the shock wave propagates. However, they do not completely specify the conditions on either side of the shock. A fourth equation, the equation of state, is necessary, relating pressure, specific volume, and internal energy of the medium. The final states of shocked materials are usually represented graphically in pressure-volume or in shock velocity-particle velocity plots. The combination of these individual shock events, each independent from the others, lies on a curve, called the Hugoniot.

The equation of state is unique for each material (it includes all the complexities of its atomic, molecular, and crystalline structure), and describes its thermodynamic state over a wide range of pressures, temperatures, and specific volumes (or densities). The simplest known equation of state is that of a perfect gas: $P=\rho RT$, where P is pressure, T is temperature, R is the gas constant per unit mass, and ρ is the gas density. The behavior of solids and liquids compressed by shock waves is, however, much more complex because of the strong interaction between the atoms (or molecules) of the medium. Consequently, a much more complex equation of state is required. The most widely used analytical equation of state for impact studies is the Tillotson equation of state, which was specifically derived for high-speed impact computations (Tillotson 1962). It is characterized by two complex forms, one to describe compressed regions where the internal energy, *E*, is less than the energy of incipient vaporization, and one for expanded states, where *E* exceeds the energy for complete vaporization. A series of parameters must be provided to describe a given material (e.g., see Melosh 1989). However, the Tillotson equation of state, and analytical equations of state in general, provide no information about how to compute the temperature or the entropy of a material. Furthermore, changes in pressure and density in two-phase regions cannot be described, resulting in the inability of the equation of state to model melting and vaporization.

More recent equations of state use increasingly complex computer codes that rely on different physical approximations in different domains of validity. One of the best examples of such equations of state is ANEOS (Thompson and Lauson 1972), a FORTRAN code designed for use with a number of hydrocodes. Pressures, temperatures, and densities are derived from the Helmholtz free energy and are, hence, thermodynamically consistent. Material properties are described in ANEOS by an array of 24 to 40 parameters, in addition to information on the atomic number and mass fraction of each element present in the material. A major advantage of ANEOS over other analytical equations of state is that it offers a (limited) treatment of phase changes, which is especially important when they interfere with the shock state. Although clearly superior to analytical equations of state, ANEOS has some limitations too, the most important being the treatment of gases as monoatomic species, which causes it to overestimate the critical point of most material of geologic interest. A treatment of biatomic species has recently been introduced in ANEOS by Melosh (2000) in an attempt to obtain a more realistic behavior of gaseous species in ANEOS. The upgrade represents a clear step forward toward more realistic equations of state, but it does not completely solve the problem (especially when more than bi-atomic species are involved in the vapor phase, as is the case for water). ANEOS equations of state for several materials of geologic interest have been developed in the recent years, but more work is still needed. Another equation of state code that has recently become available is PANDA (Kerley 1991), which is similar to ANEOS, but has more options for the construction of multi-phase equations of state in thermodynamic equilibrium (high-pressure polymorphs, solid-liquid, and liquid-gas coexistence lines, etc.). However, with the exception of multi-phase calcite (Kerley 1989 1991), no PANDA equations of state for rocks and minerals are published.

To avoid time-consuming calculations during hydrocode simulations, the best approach is to use tabular equations of state. One database of tabular equations of state is called SESAME, developed by the Livermore National Laboratory. Unfortunately, the usage of this database is user-restricted; furthermore, SESAME EoSs for materials of geologic interest have been developed from available EoS codes, such as ANEOS, and, thus, are characterized the same inherent limitations.

3.2 Constitutive Model: Material Rheology and Strength

The constitutive, or rheologic, model describes the response of a material to stresses that induce deformation. It is a mathematical approximation of the empirically determined behavior of geologic materials when subjected to stress. The two simplest constitutive models are for a perfectly elastic solid, where the stress is linearly proportional to the strain; and a Newtonian fluid, where the stress is linearly proportional to the strain rate. A somewhat more complicated constitutive model is that for a perfectly plastic material. Perfect plasticity is a good approximation of the stress response of many metals. A plastic material exhibits linear elastic behavior until a yield stress is reached; at stresses in excess of this threshold, a plastic material offers no further resistance to stress, permitting arbitrarily large strains to develop. When the applied stress is reduced to zero (for example, after the shock wave has gone through) the elastic strain is recovered, but the plastic strain remains. Several more complicated rheologic models can be constructed for use in a hydrocode by carefully combining the equations governing elasticity, plasticity and fluid flow. For example, rocks are often represented as plastic materials whose yield strength depends on pressure.

The typical constitutive model used in hydrocodes that simulate the early stages of an impact event is elastic-perfectly plastic; that is, the material loads elastically to the yield stress and then it begins to permanently deform plastically. This formulation, however, is far too simplistic for a correct interpretation of the late stages of the impact, those primarily responsible for the final morphology of impact craters. Simulations of the impact process in its entirety, or just crater collapse, must consider the complicated rheology of the post-shock target, which will have been heated, fractured, shaken and deformed.

The critical stress at which permanent deformation occurs in a given material is defined as the strength. The magnitude of a material's strength depends on the orientation of the applied stress. In other words, the strength of a material may be different when subjected to compression, tension, or shear (for example, Jaeger and Cook 1969). In most geologic materials the compressive strength is the greatest and the tensile strength is the smallest. The shear (or yield) strength of geologic materials, which is of most importance to the constitutive model, is controlled by many factors. Laboratory experiments reveal that, at low pressures and temperatures (well below melting), the yield strength of rock materials may be considered to have two components, a cohesive strength that is independent of overburden pressure, and a frictional component that is a function of overburden pressure and, hence, depth (Lundborg 1968). As pressure increases, the shear strength of rock materials asymptotically approaches a constant value, commonly termed the Von Mises plastic limit. A number of strength models, which approximate the strength of cold rock materials as measured by laboratory experiments, have been formulated for use in hydrocodes, most notably by Johnson and Holmquist (1994) and Ivanov et al. (1997). At higher temperatures (approaching melting) the yield strength is reduced relative to the "cold" value. Hydrocodes generally simulate this effect by multiplying the yield strength by a "thermal-softening" factor proportional to $(1 - E/E_m)^n$, where E and E_m are the local internal energy and internal energy at melting respectively, and n is usually taken to be 1 or 2.

During an impact event, much of the target material becomes comprehensively fractured (Nolan et al. 1996). In this case, any cohesive strength the material might have is destroyed, thus leaving the broken rock material with yield strength approximately linearly proportional to confining pressure (for example, Stesky et al. 1974).

Algorithms for describing dynamic fragmentation during an impact event have been implemented in several hydrocodes (Melosh et al. 1992; Ivanov et al. 1997; O'Keefe et al. 2001). In essence, the approach is to assess the amount of tensile fracture in a cell and reduce the tensile and shear strength accordingly. The amount of fracturing in the cell is parameterized by a quantity called damage D (D = 1.0 for wholesale fracturing in the cell; and D = 0.0 corresponds to totally intact material in the cell). The amount of damage is then used to modify the cohesive strength of the material Y_0, usually by multiplying it by $(1-D)$. The implementation is such that resistance to compression is unaffected by the damage accumulated; only resistance to extension and shear is affected. These strength algorithms have become quite sophisticated; however, inconsistencies remain in how shear strength and tensile strength are coupled (for example, damage is generally only accumulated during extension and not during shear failure). The most recent advances in constitutive model complexity have introduced phenomena associated with dynamic rock failure such as acoustic fluidization (Melosh and Ivanov 1999; Collins et al. 2002), strain localization (faulting) and bulking (the decrease in density associated with the fracturing of a material and the movement of broken rock debris; O'Keefe et al. 2001).

4
How Hydrocode Modeling Has Advanced the Study of Impact Cratering

4.1
Modeling Melt Production During the Early Stages of an Impact Event

Melting and vaporization in impact events are governed by the thermodynamics of shock compression and release, and depend, therefore, on accurate material equations of state. The quantity of melt and vapor produced in an impact influences various aspects of the impact cratering event and its effects, from heat deposition to the development and composition of the vapor plume, to the crater shape and impact lithology. Due to the impossibility of performing sizable high-speed laboratory impact experiments, no experimental data on substantial melt production are available. Only few observational data from terrestrial craters, in many cases still controversial, are available. The best available compilation of melt production in crystalline targets on the

Earth is given in Grieve and Cintala (1992). Many of the melt estimates are minimum values based on the observations of the glass content of interior crater deposits, and are believed to have uncertainties of about a factor of two (Grieve and Cintala, 1992). Numerical modeling of impact cratering has thus been crucial for the study of impact-produced melting/vaporization of planetary surfaces.

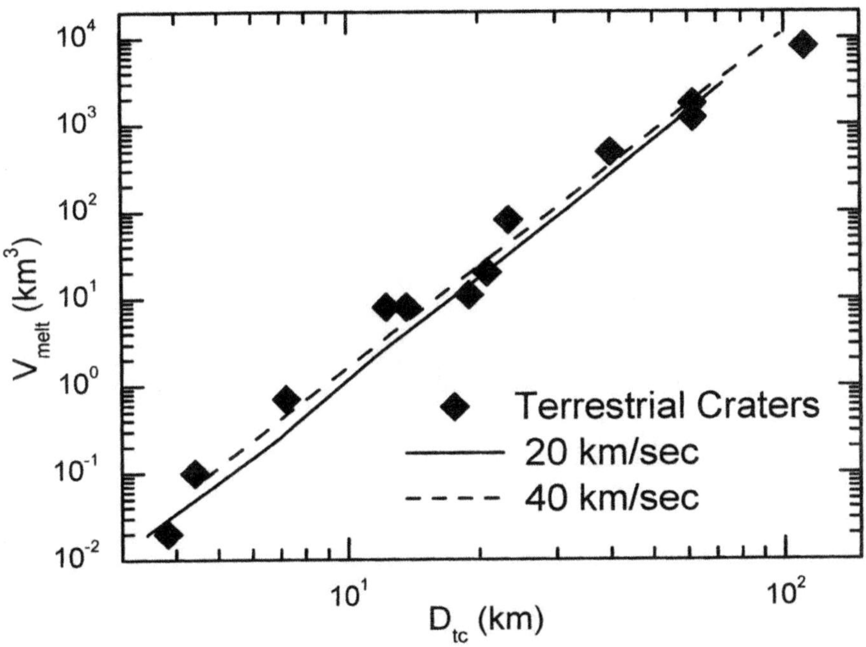

Fig. 3. Hydrocode-derived impact melt volumes versus transient crater diameter (*solid and dashed lines*), compared with terrestrial data on crystalline targets (*diamonds*), as compiled by Grieve and Cintala (1992). Simulations are for dunitic projectiles impacting dunitic targets at 20 and 40 km/s. The lines connect 2D hydrocode calculations corresponding to D_{pr}= 0.2, 0.5, 1, 2, 3, 4, 5, 10 km. From Pierazzo et al. (1997).

2D numerical modeling studies indicate that melt/vapor production scales with the energy of the impactor for vertical impacts (e.g., O'Keefe and Ahrens 1977; Pierazzo et al. 1997), and that the region of impact melt/vapor appears roughly spherical around the impact point. In particular, Fig. 3 (from Pierazzo et al. 1997) shows that numerical estimates of melt production in crystalline targets are in good agreement with available observational estimates for terrestrial structures. In oblique impacts the shape of the region of melting/vaporization is not symmetrically distributed around the impact point, as inferred from vertical impacts, but it occurs downrange of the impact point (Fig. 4, from Pierazzo and Melosh 2000b). Through a series of 3D hydrocode simulations, Pierazzo and Melosh (2000b) found that for constant impact conditions but varying impact angle, the volume of impact melt decreases by at most 20% for impacts from

90° (vertical) down to 45°. The reduction of melt volume is about 50% for impacts at 30°, and more than 90% for a 15° impact. An energy scaling law does not seem to hold for oblique impacts, even if the impact velocity is substituted by its vertical component.

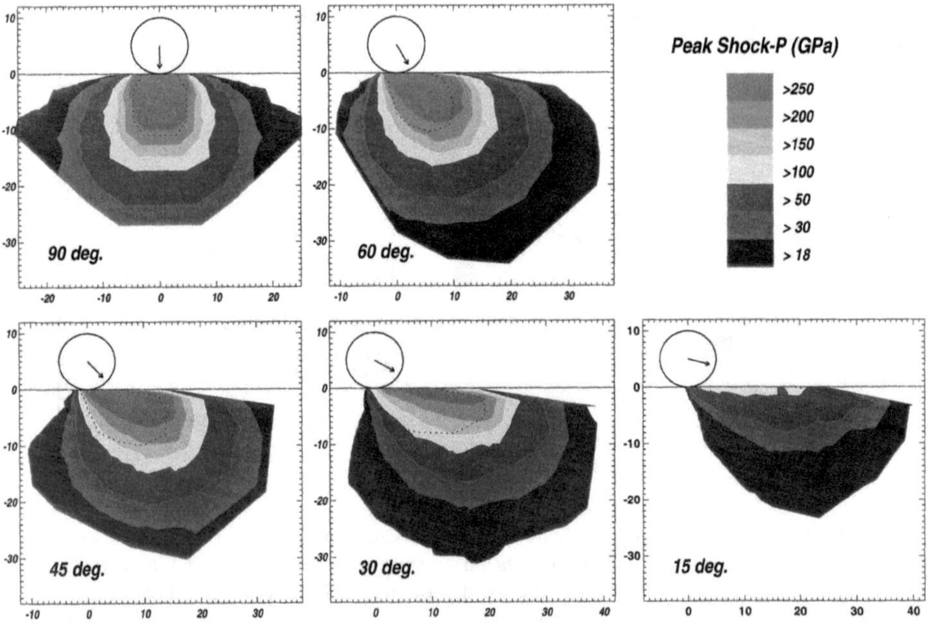

Fig. 4. Peak shock pressure contours in the plane of impact (plane perpendicular to the target surface that includes the projectile's line of flight) for various oblique impacts (angles are measured from the surface) of a projectile 10 km in diameter impacting at 20 km/s (vectors from the center of the projectile outline show the direction of impact for the various cases). From Pierazzo and Melosh (2000b).

(Pierazzo and Melosh 2000b). In an attempt to find some empirical scaling relation between melt production and some crater dimension for oblique impacts, Pierazzo and Melosh (2000b) found that the volume of melt appears to be roughly proportional to the volume of the transient crater generated by the impact for all but the lowermost impact angles (≤15°). However, more work, based on 3D modeling, is still necessary for a more complete characterization of melt/vapor production in oblique impacts.

4.2 Modeling Crater Collapse

The detailed morphology of impact craters is now believed to be mainly caused by the collapse of a geometrically simple, bowl-shaped "transient crater" during the late phase of crater formation. Impact crater collapse is controlled by the competition between the gravitational forces tending to close the excavated cavity and the inherent material strength properties of the post-shock target. Thus, accurate simulations of crater collapse require a realistic constitutive model to represent the target material, and a good understanding of the fundamentals of dynamic rock failure. This is the major difficulty in simulating complex crater collapse: numerous modeling studies using well-understood, standard strength models for rock materials (e.g., Dent 1973; Melosh 1977; McKinnon 1978; O'Keefe and Ahrens 1993; Melosh and Ivanov 1999; Collins et al. 2002; Ivanov and Artemieva 2002) have concluded that modification of the transient crater should not involve any uplift of material from beneath the crater floor or slumping of the transient crater walls. The former limitation precludes the formation of a central peak, peak ring, or external rings; the latter precludes formation of terraces and significant widening of the crater. In other words, to reproduce the observed morphologies of complex craters, collapse requires significant, but temporary, weakening of the target material beneath the crater floor.

Recently, hydrocode simulations of crater collapse have been carried out by employing temporary strength-weakening mechanisms like acoustic fluidization (Melosh and Ivanov 1999; Collins et al. 2002, Ivanov and Artemieva 2002), and strain localization and thermal softening (see O'Keefe and Ahrens 1999; 2001). Both strength weakening models seem to significantly improve the hydrocode's ability to model crater collapse; however, the relative importance of each mechanism is still poorly constrained. Clearly, more work needs to be done to better understand the strength of rock materials during impact conditions.

Hydrocode simulations of complex crater collapse that invoke acoustic fluidization as the temporary strength-weakening mechanism illustrate that collapse of the transient crater is controlled by the duration and extent of the temporary strength-weakening process (Fig. 5, from Collins 2002). For short durations of strength weakening, the collapse involves inward slumping of the transient crater rim, accompanied by the uplift of the floor of the crater (Fig. 5b). Such simulations produce final crater morphologies analogous to central peak craters. However, if the strength-weakening process persists for longer, the collapse continues with the central uplift overshooting the original target surface, and then collapsing downwards and outwards (Fig. 5d,e). Peak-ring formation appears to be a consequence of this material overriding the collapsed transient crater rim (Collins et al. 2002).

Simulations of impact crater formation have produced a consistent paradigm for how large craters might collapse to form the final complex form. However, current models do not provide a complete explanation for why large impact craters collapse in this manner. Developing a full mechanical understanding of large impact crater formation requires further testing and refining of numerical crater collapse models, based on geological observation, geophysical data, and drill cores.

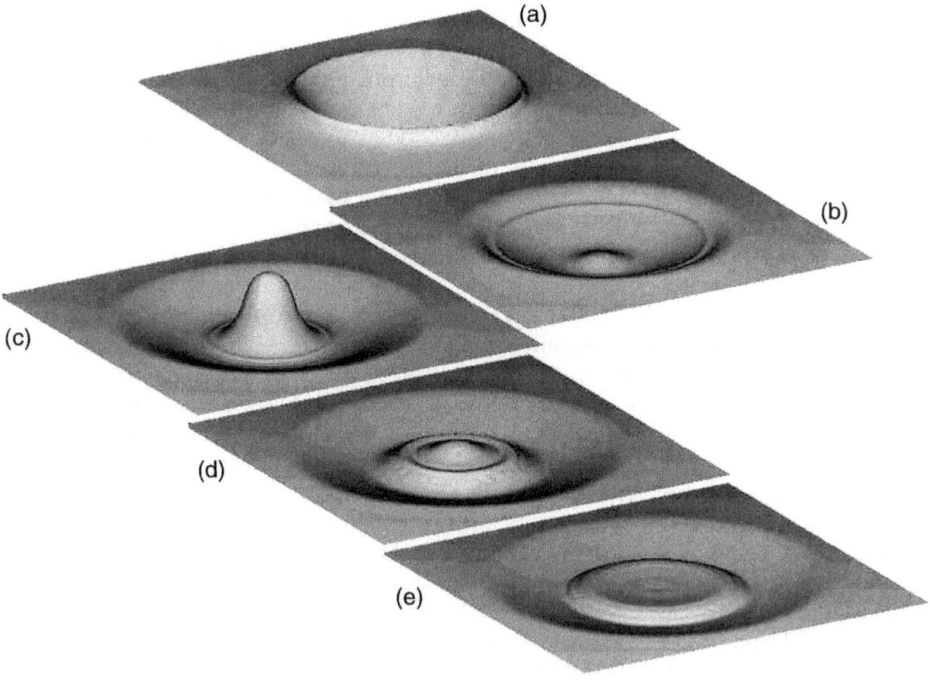

Fig. 5. The general paradigm for complex crater collapse, as described by the hydrocode simulations of Collins (2002). For short durations of strength weakening, collapse involves inward slumping of the transient crater rim (*a*), accompanied by the uplift of the floor of the crater (*b*). If the strength-weakening process persists for longer, the collapse continues with the central uplift overshooting the original target surface (*c*), and then collapsing downwards and outwards (*d* and *e*).

5
Summary and Discussion

Numerical modeling is an important facet of impact cratering research. Hydrocodes offer a means for studying various aspects of the impact process that cannot be investigated by other methods. Moreover, they provide detailed information regarding all variables of interest for the entire simulation; in a sense they may be regarded as the best instrumented experiment (Anderson 1987). Hydrocode modeling has provided much insight into the cratering process. In particular, recent hydrocode simulations have constrained the amount and distribution of melt and vapor produced during the early stages of an impact, and the evolution of the complex morphologies observed in large craters.

Hydrocodes are continually increasing in sophistication, both in the complexity of the physics that is implemented in them, and the amount of data that can be processed in a reasonable timeframe. Material models employed in hydrocode simulations are improving in realism all the time. The current state of these models has allowed the construction of a predictive, quantitative first-order model governing the dynamics of impact crater formation. Still, many questions concerning the behavior of geologic materials under the extreme conditions associated with an impact event remain unanswered, and many scientific fruits lie in wait to be discovered.

Constraints imposed by computer hardware have meant that most impact modeling work to date has concentrated on the isolated case of a vertical impact, because the axial symmetry of the process allows the simplification of the model to two dimensions. Moderate (20° to 30°) deviations from vertical impacts have small consequences on modeling outputs like melt production and crater's shape and size, allowing results from 2D simulations to be valid and useful for many applications. However, natural impacts in which the projectile strikes the target vertically are virtually nonexistent. Probability theory shows that regardless of the planet's gravitational field, the angle of impact of maximum frequency is 45°, with 50% of impacts occurring between 30° and 60° whereas the probability of near-vertical as well as grazing impacts is negligible (see Pierazzo and Melosh 2000c for a review). Furthermore, impact angle – especially low impact angles – appears to have played a major role in various problems of geologic interest, like the ejection of matter from planetary surfaces into interplanetary space, the giant impact theory for the origin of the Moon, the environmental effects of large planetary impacts, and the spin rate of asteroids. In non-vertical impacts, the axial symmetry typical of vertical impacts is broken, and more computationally intense 3D hydrocodes are required to simulate the impact. Thus, the increased use of 3D hydrocodes, facilitated by the continual advances in computer hardware, is providing an additional and important level of realism for simulations of impact events.

Systematic high-resolution modeling work on oblique impacts was first carried out by Pierazzo and Crawford (1998). The analysis of that modeling work produced a wealth of much needed information (e.g., see Pierazzo and Melosh 1999; 2000a 2000b), and inspired several further modeling studies of scientific problems involving oblique impacts (for example, Artemieva 2000; 2001; Ivanov and Artemieva 2000; 2001, 2002; Artemieva and Ivanov 2001; 2002; Pierazzo et al. 2001; Stöffler et al. 2002). This new era of hydrocode modeling of oblique impacts will doubtless significantly improve our understanding of the role of impact cratering in the evolution of the solar system's bodies.

Acknowledgments

This work was supported under grant NAG5-9112. We are grateful to Prof. H. Jay Melosh for enlightening scientific discussions. We thank two anonymous reviewers for the helpful comments and suggestions in improving the manuscript. This is PSI contribution No. 366.

References

Amsden AA, Ruppel HM, Hirt CW (1980) SALE: A simplified ALE computer program for fluid flow at all speeds. Los Alamos National Laboratories LA-8095: 101 pp

Anderson Jr CE (1987) An overview of the theory of hydrocodes. International Journal of Impact Engineering 5: 33-59

Artemieva NA (2000) Tektite origin in oblique impact: Numerical modeling [abs]. In Meteorite Impacts in Precambrian Shields. ESF-IMPACT Workshop 4 (Lappajärvi, Finland): 56

Artemieva NA (2001) Tektite production in oblique impacts [abs]. Lunar and Planetary Science XXXII: #1216 (CD-ROM)

Artemieva NA, Ivanov BA (2001) Numerical simulation of oblique impacts: Impact melt and transient cavity [abs]. Lunar and Planetary Science XXXII: #1321 (CD-ROM)

Artemieva NA, Ivanov BA (2002) Ejection of Martian meteorites – Can they fly? [abs]. Lunar and Planetary Science XXXIII: #1113 (CD-ROM)

Collins GS (2002) Numerical Modelling of Large Impact Crater Collapse. PhD Thesis. Imperial College, University of London.

Collins GS, Melosh HJ, Morgan JV, Warner MR (2002) Hydrocode simulations of Chicxulub crater collapse and peak-ring formation. Icarus 157: 24-33

Dent B (1973) Gravitationally induced stresses around a large impact crater. [abs] EOS, Transactions, American Geophyscal Union 54, 11: 1207

Grieve RAF, Cintala MJ (1992) An analysis of differential impact melt-crater scaling and implications for the terrestrial impact record. Meteoritics 27: 526-538

Ivanov BA, Artemieva NA (2000) How oblique should be impact to launch Martian Meteorites? [abs]. Lunar and Planetary Science XXXI: #1309 (CD-ROM)

Ivanov BA, Artemieva NA (2001) Transient cavity scaling for oblique impact [abs]. Lunar and Planetary Science XXXII: #1327 (CD-ROM)

Ivanov BA, Artemieva NA (2002) Numerical modeling of the formation of large impact craters. In Koeberl C, MacLeod KG (eds) Catastrophic events and mass extinctions: Impacts and beyond. Geological Society of America Special Paper 356: 619-630

Ivanov BA, DeNiem D, Neukum G (1997) Implementation of dynamic strength models into 2D hydrocodes: Applications for atmospheric breakup and impact cratering. International Journal of Impact Engineering 20: 411-430

Jaeger JC, Cook NGW (1969) Fundamentals of rock mechanics. Chapman and Hall, London 593 pp
Johnson GR, Holmquist TJ (1994) An improved computational constitutive model for brittle materials. In: Schmidt SC, Shaner JW, Samara GA, Ross M (eds) High-Pressure Science and Technology - 1993, American Institute of Physics Press, Woodbury NY, pp 981-984
Kerley GI (1989) Equation of state for calcite minerals. I. Theoretical model for dry calcium carbonate. High Pressure Research 2: 29-47
Kerley GI (1991) User's manual for PANDA II: A computer code for calculating equations of state. Sandia Report SAND88-2291, Sandia National Laboratories, Albuquerque, NM
Lundborg N (1968) Strength of rock-like materials. Int. Journal Rock Mech. Min. Sci. 5: 427-454
McGlaun JM, Thompson SL, Elrick MG (1990) CTH: A three-dimensional shock wave physics code. International Journal of Impact Engineering 10: 351-360
McKinnon WB (1978) An investigation into the role of plastic failure in crater modification. Proceedings of the Lunar and Planetary Science Conference 9: 3965-3973
Melosh HJ (1977) Crater modification by gravity: A mechanical analysis of slumping. In Roddy DJ, Pepin RO, Merrill RB (eds) Impact and explosion cratering. Pergamon Press, New York, pp 1245-1260
Melosh HJ (1989) Impact Cratering. A Geologic Process. Oxford University Press, New York 245 pp
Melosh HJ (2000) A new and improved equation of state for impact computations [abs]. Lunar and Planetary Science XXXI: #1903 (CD-ROM)
Melosh HJ, Ivanov BA (1999) Impact crater collapse. Annual Reviews of Earth and Planetary Science 27: 385-415
Melosh HJ, Ryan EV, Asphaug E (1992) Dynamic fragmentation in impacts: Hydrocode simulation of laboratory impacts. Journal of Geophysics Research 97: 14735-14759
Monaghan JJ (1992) Smoothed particle hydrodynamics. Annual Reviews of Astronomy and Astrophysics 30: 543-574
Nolan M, Asphaug E, Melosh HJ, Greenberg R (1996) Impact craters on asteroids: Does strength or gravity controls their size? Icarus 124: 359-371
O'Keefe JD, Ahrens TJ (1977) Impact-induced energy partitioning, melting, and vaporization on terrestrial planets. Proceedings of the Lunar Science Conference 8: 3357-3374
O'Keefe JD, Ahrens TJ (1993) Planetary cratering mechanics. Journal of Geophysics Research 98: 17001-17028
O'Keefe JD, Ahrens TJ (1999) Complex craters: Relationships of stratigraphy and rings to impact conditions. Journal of Geophysics Research 104: 27091-27104
O'Keefe JD, Stewart ST, Lainhart ME, Ahrens TJ (2001) Damage and rock-volatile mixture effects on impact crater formation. International Journal of Impact Engineering 26: 543-553
Pierazzo E, Vickery AM, Melosh HJ (1997) A reevaluation of impact melt production. Icarus 127: 408-423
Pierazzo E, Crawford DA (1998) Modeling Chicxulub as an oblique impact event: Results of hydrocode simulations [abs]. Lunar and Planetary Science XXIX: #1704 (CD-ROM)
Pierazzo E, Melosh HJ (1999) Hydrocode modeling of Chicxulub as an oblique impact event. Earth and Planetary Science Letters 165: 163-176
Pierazzo E, Melosh HJ (2000a) Hydrocode modeling of oblique impacts: The fate of the projectile. Meteoritics and Planetary Science 35: 117-130
Pierazzo E, Melosh HJ (2000b) Melt production in oblique impacts. Icarus 145: 252-261

Pierazzo E, Melosh HJ (2000c) Understanding oblique impacts from experiments, observations, and modeling. Annual Reviews of Earth and Planetary Science 28: 141-167

Pierazzo E, Spitale JN, Kring DA (2001) Hydrocode modeling of the Ries impact event [abs]. Lunar and Planetary Science XXXII: #2106 (CD-ROM)

Quaide WL, Gault DE, Schmidt RA (1965) Gravitative effects on lunar impact structures. Annals of the New York Academy of Science 123: 563-572

Shuvalov VV (1999) 3D hydrodynamic code SOVA for interfacial flows, application to the thermal layer effect. Shock Waves 9(6): 381-390

Stesky RM, Brace WF, Riley DK, Robin PYF (1974) Friction in faulted rock at high temperature and pressure. Tectonophysics 23: 177-203

Stöffler D, Artemieva NA, Pierazzo E (2002) Modeling the Ries-Steinheim impact event and the formation of the Moldavite strewn field. Meteoritics and Planetary Science 37: 1893-1908

Tillotson JH (1962) Metallic equations of state for hypervelocity impacts. Technical Report General Atomic Report GA-3216, San Diego, 140 pp

Thompson SL (1979) CSQII – An Eulerian finite differences program for two-dimensional material response. Part 1. Material Section. Technical Report SAND77-1339. Sandia National Laboratories, Albuquerque, NM, 87 pp

Thompson SL, Lauson HS (1972) Improvements in the chart-D radiation-hydrodynamic code III: Revised analytical equation of state. Technical Report SC-RR-61 0714. Sandia National Laboratories, Albuquerque, NM, 119 pp

MIX
Papier aus verantwortungsvollen Quellen
Paper from responsible sources
FSC® C105338

If you have any concerns about our products,
you can contact us on
ProductSafety@springernature.com

In case Publisher is established outside the EU,
the EU authorized representative is:
Springer Nature Customer Service Center GmbH
Europaplatz 3, 69115 Heidelberg, Germany

Printed by Libri Plureos GmbH
in Hamburg, Germany